Second Edition
CREATING FRESHWATER WETLANDS

Donald A. Hammer

CRC Press
Taylor & Francis Group
Boca Raton London New York

CRC Press is an imprint of the
Taylor & Francis Group, an **informa** business

Acquiring Editor:	Neil Levine
Project Editor:	Suzanne Lassandro
Marketing Manager:	Greg Daurelle
Direct Marketing Manager:	Arline Massey
Cover design:	Denise Craig
PrePress:	Kevin Luong
Manufacturing:	Sheri Schwartz

CRC Press
Taylor & Francis Group
6000 Broken Sound Parkway NW, Suite 300
Boca Raton, FL 33487-2742

First issued in paperback 2019

ISBN-13: 978-1-56670-048-1 (hbk)
ISBN-13: 978-0-367-40117-7 (pbk)

Library of Congress Card Number 96-32195

Library of Congress Cataloging-in-Publication Data

Hammer, Donald A.
    Creating freshwater wetlands / Donald A. Hammer. -- 2nd ed.
        p.   cm.
    Includes bibliographical references and index.
    ISBN 1-56670-048-5 (alk. paper)
    1. Wetlands.   2. Restoration ecology.   1. Title.
QH87.3.H36   1996
639.9'5--dc20   96-32195
                                                                                    CIP

Visit the Taylor & Francis Web site at
http://www.taylorandfrancis.com

PREFACE TO THE SECOND EDITION

Five years have seen dramatic change in U.S. attitudes towards wetlands. The "webb-foots" have arrived! In fact, membership in the Society of Wetland Scientists now rivals that of many older scientific organizations. But the widespread attention on wetlands hasn't always been positive. On occasion in social circumstances, I have even been reluctant to reveal my profession without lengthy follow-up explanations that I was not involved in wetland regulations and didn't always agree with some interpretations.

These are contentious times for wetlands and for wetland ecologists but we finally have national attention, legislative action and hopefully, a new national wetland policy will be forthcoming in the near future. As with most legislation, the present versions (as will the final) represent compromise that certainly won't please all webb-foots but it's a start. Future Congresses will doubtless make additional changes when the Clean Water Act comes up for renewal. But it seems very unlikely those changes will ever return to a national policy of wetland drainage.

Most encouraging, the past five years have seen a considerable number of wetland creation projects with much greater success than some earlier ones. Appearance of reports of less than 100% of some species, group, coverage or some other measure in the created system after 4 to 5 years as compared with a reference wetland should not be discouraging. Other reports show virtually the same conditions in the new and the reference wetland in similar intervals. Much depends on the type of wetland and the techniques used by those creating the new system or restoring a previously existing wetland. After all, 4 to 5 years is merely an Augenblick, a micro-moment, a nano-second in geological time. Of course, the new wetland with only 80% of the natural after 5 years is very likely to have close to 100% after 10 to 15 years. Though I've worked in wetlands for almost my entire career, I am still amazed at how quickly many created wetlands approach conditions in the natural systems. The exponential increase in the number of species and diversity of many created wetlands during the early stages will doubtless taper off as its diversity approaches that of the reference wetland. But isn't that the pattern of normal succession in any wetland system? And isn't that our goal?

Furthermore, the dramatic changes that have occurred in the agricultural sector in the last five years are most heartening. We appear to be close to "no net loss" and may even be approaching "net gains" in total wetland resources largely because of agricultural policy reversals.

Though some disagree, I believe that one positive aspect of pending legislation is the codification of wetland mitigation banking. Mitigation banking is a fact and a rapidly growing phenomenon but current legal uncertainties hinder broader acceptance and implementation. Successful mitigation banking will require appropriate guidelines including technical expertise, long-term financial guarantees, and judicious location of wetland banks to optimize functional values. Mitigation banking not only provides a mechanism to replace continual losses of often degraded, "postage-stamp" wetlands but carefully locating banks in watersheds will enhance the value of wetland benefits. Enhanced functional values coupled with improved understanding will lead to broader realization and acceptance of the value of wetlands. And I can't think of any better method to increase our wetland resources than to provide a profit motive for those that would do so. If someone can generate income from creating, restoring, and protecting wetlands our need for protective regulations will decline precipitously.

This edition goes beyond the general guidelines and philosophies of the first edition, providing considerably more detail on specific techniques useful in creating a wetland. Although titled *Creating Freshwater Wetlands*, the methods and approaches are just as important in any wetland restoration project. The variety of expertise and information needed for a successful project is wide and certainly beyond the scope of a single effort so the Appendices have been substantially expanded. But three aspects are crucial to developers of a new wetland or those attempting a restoration project: the absolute requirement for appropriate hydrology or hydroperiod, an understanding that wetlands are complex, dynamic systems that often need deliberate management to supply the disturbance factor and finally, a little patience. Expecting to precisely duplicate a system that has developed over the eons in 5 to 10 years is unrealistic. We have the ability to accelerate development so the new system resembles the natural one in much less than eons and the rapidly increasing database on creating and restoring techniques will increase our abilities. With the right hydrology and limited management, over time, the new system will gradually become indistinguishable from the reference wetland. Follow the guidelines and techniques herein, provide a little guidance to early succession, and sit back and enjoy the gradual development of one of the most fascinating ecosystems on the planet.

D. A. Hammer
Norris, Tennessee
10 May 1996

PREFACE

Brian Lewis conceived of this little handbook as a supplement to the 1988 conference proceedings *Constructed Wetlands for Wastewater Treatment*, and at his urging, I finally committed to begin compiling the information.

As a student I was fortunate in working on several National Wildlife Refuges, and in one case, my research benefited from the historical information on wildlife populations in old annual reports. However, in digging out information on snapping turtles from the earliest days of the refuge, I discovered the tremendous wealth of information on methods developed to restore and create the marsh that I believed was largely natural. Later I discovered similar knowledge in the earliest reports from virtually every refuge and management area that I visited.

Since many of our "natural" wetlands today were in fact restored or even created, I have been surprised when colleagues state that we know so little about wetlands that we can't restore or create one as mitigation or for any other reason. In fact, the predominant conclusion from most recent wetland meetings would be our total ignorance of basic wetland ecology. Certainly we don't know everything about wetlands — they might not be as fascinating to some of us if we did — but to conclude that we lack the information to restore or create many types of wetland is to ignore or negate a considerable body of information gathered by competent, dedicated scientists and managers over many years.

Doubtless some will respond that this information only applies to marshes. Certainly much of it does, but some is also available for other types, because bogs, swamps, sloughs and fens were included in wildlife refuges and management areas. But more importantly, marshes are precursors, early successional stages for many other wetland ecosystems. If we can create a healthy marsh and then allow it to progress along the successional gradient to another wetland type, are we not well on the way to creating the bog or swamp that is the project goal? And do we know how to create a marsh and allow it to transition? Of course we do. We simply do not supply the disturbance factor that is critical to maintaining the site in the early marsh stage.

Although today's society places considerable importance on natural and created wetlands, it was not always so. While most of past society had little use or regard for "dark and dismal swamps," and wetland drainage was national policy, a few conservationists began raising the alarm over lost wetlands as early as the turn of the century. Extensive drainage projects coupled with the severe drought of the 1930s caused calamitous declines in North American waterfowl populations. Alarmed duck hunters, many wealthy and influential in society, initiated legislation taxing sport hunters to support wetland acquisition, management and research, which provided the basis for the National Wildlife Refuge System and comparable state wildlife management areas. Faced with a dearth of water and wetland habitats but charged with restoring waterfowl populations, early managers slowly developed methods to create and restore natural wetland ecosystems to enhance waterfowl habitats throughout the country.

Creation of various public works programs (CCC, WPA, etc.) concurrent with the drive to restore waterfowl provided heretofore unheard of funding and manpower resources in the newly emerging field of wildlife management. Substantial efforts in these programs were directed towards developing methods to design and construct dikes and water control structures, plant and/or seed wetlands and terrestrial vegetation, re-stock wildlife and fish populations, manipulate water levels and manage fish and wildlife populations in the restored or created wetland ecosystems. Probably because of the bureaucratic penchant for documentation, most of the methods developed and tested were reported in the "gray literature" — the annual reports for each refuge or management area. Not surprisingly, virtually all types of wetland ecosystems in the U.S. are encompassed. Simple examination of a distribution map for the National Wildlife Refuges reveals refuge locations from coast to coast, including most types of freshwater wetlands and some brackish/saltwater systems. Restoration and creation techniques were investigated and reported in the late 1930s, 1940s and 1950s on most of these areas.

Early managers unrestrained by the policy of not reporting "negative results" prevalent in scientific journals today, included failures as well as successes. Unfortunately much of this knowledge never found its way into scientific journals and, hence, is largely unknown to the new crop of wetland ecologists and managers. Some has not even been written down though it has been passed along and is still in use by present managers. But that information is still available in the files or from the staff of the local refuge or management area office.

Unfortunately many sincere and dedicated wetland enthusiasts are unaware of this wealth of information, or choose to disregard it — "they were only interested in growing ducks." I am afraid I must disagree with the latter. Techniques developed in the late 1930s and 1940s to establish or manage *Scirpus, Potamogeton, Spartina, Quercus, Larix,* or *Taxodium* or a host of other wetland plants need not be reinvented today, nor must we use the most expensive methods. Though propagation and hand-planting were used, early wetlands managers slowly discovered the growth requirements for important species and developed methods to foster or retard vegetative types through inexpensively manipulating water levels in appropriate seasons.

Methods developed for management of duck production, migratory or wintering areas have successfully provided wetland habitats for innumerable other species of wetland wildlife, fish, and plants. Ducks may have been the priority but "they didn't just grow ducks." In fact, many of the larger "natural" wetlands extant today are located in refuges, management areas and a few parks because they were restored, enhanced and, in a few instances, created by early waterfowl managers.

Finally, my earlier experiences creating or managing wetlands with the life support function as a primary goal, and more recent work with the water purification function have convinced me that a wetland ecosystem that supports diverse and abundant vertebrate populations does so because it has the complex, requisite structure. If the structure is present to provide significant life support, then that system likely has the structure to provide much of the water purification, hydrologic buffering and other important functions that we value in wetland ecosystems. Regardless of the prime objective, designing and constructing a wetland that will attract and support representative vertebrate species will likely achieve the prime objective plus support many other important functional values.

This then is a small attempt to organize and present some of the information on methods to create or restore freshwater wetlands accumulated by wetland scientists and managers during the last 50 years. It is my hope that the reader will interpret this volume as a general guideline and employ specific, localized information from nearby offices of wetland and waterfowl research/management agencies before starting a project.

D. A. Hammer
Norris, Tennessee
2 March 1991

ACKNOWLEDGMENTS

As wetlands develop from early primitive forms through interim stages to finally mature as full-blown, complex ecosystems, so do wetland ecologists. Neither would succeed without nurturing, assistance and management. Both need and benefit from the influences of others.

Although the number of people and wetland systems that shaped and focused my development is too great to enumerate, a few had inordinate influence. I've worked in wetlands from the Arctic Circle to some 30 degrees south and on every continent except Africa and Antarctica, including extended periods in the fresh- and saltwater systems of Maine, bottomland hardwoods of the Southeast and fresh and saline wetlands of the Intermountain West, but the prairie pothole marshes of my early years in North Dakota and Lacreek National Wildlife Refuge in South Dakota had far greater impact on me than any other wetland ecosystem. Even today, I am awed by the diversity and almost explosive productivity of a prairie marsh in June.

Not surprisingly, most of the individuals I am most deeply indebted to shared their love of and experience with these same systems. They include Bob Seabloom (University of North Dakota), Bob Dahlgren and Ray Linder (South Dakota State University), Juan Spillett and Jess Low (Utah State University), Mal Coulter and Howard Mendall (University of Maine), and Merrill Hammond, Jim Monnie, and John Ellis (U. S. Fish and Wildlife Service).

Finally, neither this nor any other project during the last 31 years would have been accomplished without the interest, support, and patience of my wife, Joan. From counting snapping turtle eggs (her small hands could reach into the nest) to providing lunch for the whole crew at almost inaccessible field sites, keeping the household during my extended forays, coordinating field activities with colleagues, securing osprey or eagles so they wouldn't interfere while I landed the aircraft, and tolerating my increasing cantankerousness with days or weeks beyond the deadline, her dedication and assistance has been immeasurable.

TABLE OF CONTENTS

LIST OF FIGURES

Figure 3-3.   Even though this thunderstorm over the Pantanal of Brazil may create heavy localized downpours, supporting water supplies originate from precipitation over wide areas running off surrounding uplands.

Figure 3-4.   Periodic drying exposes and oxidizes nutrients and other substances in bottom substrates, releasing these materials into the water column following subsequent flooding.

Figure 3-5.   Duckweed and giant duckweed, though only a few millimeters across, form dense mats that contribute substantially to biomass productivity and host large populations of invertebrates.

Figure 3-6.   Emergent stands of bulrush and cattail occupy the shallow areas, with dense growths of a submergent (Sago pondweed) barely exposed in the deeper regions.

Figure 3-7.   Swollen bases (butt swellings) improve gas transfer between roots and the atmosphere for these cypress in a Louisiana bayou.

Figure 3-8.   A cross sectional view of bulrush stems reveals the tube-like aerenchyma structures that transport gases between soil regions and the atmosphere in wetland plants.

Figure 3-9.   In areas with periodic flooding, cypress "knees" increase oxygen supplies for roots growing in anaerobic environments.

Figure 4-1.   Upthrusting mountain ridges interrupt drainage patterns, forming lakes that gradually fill and are transformed into wetlands.

Figure 4-2.   A shallow caldera on the Caribbean Island of Grenada supports rooted and floating stands of cutgrass (*Cladium*) along with a few other wetland plant species.

Figure 4-3.   Mountain valley, as in the Copper River region of Alaska, and continental glaciers reshape eroded landscapes, creating broad areas with shallow relief and poor drainage that provide the hydrologic conditions suitable for wetland systems.

Figure 4-4.   Almost half of the world's wetlands occur at northern latitudes because of the combined influences of continental glaciers, low evaporation rates, and permafrost conditions.

Figure 4-5.   Ice blocks melting after glacial retreat left numerous depressions that later received clay fines from surrounding hills to form sealing liners. The myriad potholes in the Coteau du Missouri of the Dakotas and Saskatchewan comprise the principal duck factory of North America. Note the small road in the foreground.

Figure 4-6.   Deeper depressions initially filled with glacial meltwater, are slowly transformed by floating bogs growing out from shorelines.

Figure 4-7.   Encroaching vegetation gradually changes the lake to a marsh, then a bog, and finally a forest in this Newfoundland basin.

Figure 4-8.   Extensive fresh to brackish water marshes of the Mississippi Delta ebb and flow with changes in sea level and river course.

Figure 4-9.   Changing sea level and sedimentation have converted a stream into a marsh at the former river boat landing for Boone Hall Plantation near Charleston, South Carolina.

Figure 4-10. Rivers meandering across broad flood plains create and destroy wetlands with every flood or change of course.

Figure 4-11. Cutoff oxbows and old river channels form significant wetlands in river valleys with low relief. Unfortunately, many have been ditched to improve drainage for agricultural purposes.

Figure 4-12. Many of the rich marshes of the Nebraska Sandhills intercept groundwaters and are oriented in a northwest-southeast direction caused by prevailing winds moving sand and forming dunes and depressions.

Figure 4-13. Permafrost restricts downward water movement in summer, and ice heaving in winter forms ridges creating the typical polygon shapes of arctic wetlands near Churchill, Manitoba.

Figure 4-14. The beaver lodge in the background reveals the perpetrator of an Appalachian marsh that purifies acid mine drainage waters.

Figure 4-15. Beaver flowages have meandering streams in the low-gradient terrain formed behind beaver dams over hundreds of years.

Figure 5-11. Camouflage plumage and deliberately swaying with wind-blown plants improves a bittern's chances for avoiding detection. Down-looking eyes improve detection of fish as well as potential danger.

Figure 5-12. Peat (for fuel, packing material, and gardening) has long been a commercially important product of natural wetlands.

Figure 5-13. Common reed harvest in natural wetlands is an important activity in central Europe. Most of the product is shipped to the Low Countries for roof thatching. Even though the Hungarian portion represents only 23 percent of the total area (Fertos Lake in Hungary, Neusiedlersee in Austria) annual harvest of nearly 400,000 tonnes is an important income source for area residents.

Figure 5-14. Thatch roofs (reed) provide excellent insulation while needing only minor if any maintenance for 15 to 20 years.

Figure 5-15. Hunting (note blind on far right) and fishing are perhaps the most important recreational uses and consequently, economic values from many wetlands.

Figure 5-16. Note the difference in river dimensions on the east and west but not the middle of this schematic diagram of the Pantanal, Mato Gosso do Sul, Brazil.

Figure 5-17. The channel of the Rio Taquiri gradually disappears with distance into the Pantanal concurrent with initiation of hundreds of small streams that coalesce to form rivers on the western shores.

Figure 5-18. A large wetland complex furnishing recreational and educational opportunities in Coyote Hills Regional Park, as well as providing stormwater treatment for Fremont, California.

Figure 5-19. Fallen stems of bulrush produce the matrix that performs as a thin film bio-reactor in wastewater treatment.

Figure 5-20. Floodplain wetlands desynchronize and reduce peak flows and then gradually release flood waters that augment base flows, creating buffered moderate river flows during wet and dry seasons. But clearing for agriculture has dramatically reduced wetland coverage and modifying effects.

Figure 9-4.   Carefully planned grading and planting transformed the gravel pit into an early stage wetland in only one year (Photo by Photo Hawk, courtesy of Normandeau Associates).

Figure 9-5.   Small backyard wetlands can add diversity to over-developed suburbia and support a variety of plants and small animals for educational and recreational uses.

Figure 9-6.   Blue-winged teal are typically identified with Northern Plains wetlands but they use many different types of wetlands in different regions of the country for breeding, migrating, or wintering including this parrot feather bed in a South Carolina swamp.

Figure 9-7.   A viable raccoon population needs a sizable wetlands to provide adequate supplies of food and cover for all its members.

Figure 9-8.   Larger wetlands, simply because of their size, have the capability to support many more of the important functional values than do small systems. The diagonal line running from far left to the intersection at the lower right is one mile long.

Figure 9-9.   Boardwalks need not be elaborate or expensive and are essential for visitor access in most wetlands.

Figure 9-10.   Adding small benches encourages visitors to pause a minute while natural vegetation screening allows wildlife to resume their normal activities.

Figure 9-11.   This boardwalk provides visitor access as well as supporting a pipe carrying highly treated wastewater for polishing in a Carolina Bay near Myrtle Beach, South Carolina.

Figure 9-12.   Educational signs can enhance visitor understanding and enjoyment of a wetland developing support for specific projects as well as other wetlands.

Figure 10-1.   Dikes in Bear River marsh not only regulate water levels in upstream pools but also protect freshwater marshes from salt water influences when water levels rise in Great Salt Lake. Note the stoplog control structures beneath each bridge.

Figure 10-2.   Simple berms, often too narrow for foot traffic, have controlled water for centuries in Thai rice fields but frequent maintenance is needed and water level control is only through breaching the dikes.

Figure 13-4. Annual high waters eliminate most woody species on this sand-bar in the Little Missouri Badlands of North Dakota but willows and smartweeds proliferate and re-cover it each summer.

Figure 14-1. Unless the new wetland is far removed from any natural system, frogs, toads, and salamanders will quickly take up residence.

Figure 14-2. Hatchling turtles emerging from underground nests follow open skylines leading them to the nearest water and most new wet-lands will be well supplied after a year or two.

Figure 14-3. Great blue herons and other wading birds will locate a new wetland but are unlikely to establish nesting colonies until shrubs or trees have matured to provide suitable nesting sub-strates.

Figure 14-4. Extensive programs to erect and maintain nesting boxes have replaced tree cavities lost to logging and restored wood ducks to much of their former range.

Figure 14-5. Wood duck nest boxes (far left) substitute for natural cavities increasing wood duck numbers and simple board ramps (fore-ground) provide basking sites for alligators and turtles. Unless located in deep water, nest boxes should have metal predator guards.

Figure 14-6. Mallards and pintails feed largely on aquatic plants and inver-tebrates but in fall and winter its not unusual for their twice daily feeding flights to be 20-40 kilometers to find waste grain in cropfields.

Figure 14-7. Mice and other small mammals follow bands of dense riparian vegetation leading them to new habitats. Many are important foods for larger mammals and birds as well as adding further diversity and interest for recreational users.

Figure 15-1. Constructed wetlands often only have a few species planted and if the wastewaters are high strength, the plant community may be dominated by one or two species as in this *Typha* stand.

Figure 15-2. After death, marsh vegetation falls to the surface of the sub-strate, creating the litter/detritus/humus layers that provide enormous quantities of reactive surface area — attachment sites for microbial organisms.

Figure 15-3.   The two major classes of constructed wetlands are differentiated by substrate media. Soil wetlands have loam soils whereas gravel or crushed rock is used in gravel wetlands.

Figure 15-4.   A 30-ha constructed wetland designed into an area bounded by treatment lagoons on the right and an encircling railroad bed at Weyerhaueser's paper mill near Columbus, Mississippi.

Figure 15-5.   Mine drainage wetlands often have a slight slope due to the nature of the terrain in which seeps are located. Initially, treatment occurs in the upper shallow portion of each cell but as masses of iron are deposited, the shallow zone moves farther and farther down the cell.

Figure 15-6.   A typical rectangular wetlands cell with inlet and outlet distribution piping and an impermeable liner. Inlet and outlet piping should extend across the width of the cell to enhance sheet flow across the entire treatment area.

Figure 15-7.   The marsh-pond-marsh design concept to enhance ammonia removal in constructed wetland treatment systems. Alternating zones of shallow water/emergent plants and deeper water/submergent plants provides the combination of environments required for nitrification/denitrification.

Figure 15-8.   The main inlet is on the left (upwelling) and outlets with "V" notched weir plates distribute flows to three wetlands. Note that flow is equally divided between two outlets with the third (right foreground) closed off.

Figure 15-9.   Checking and if necessary, adjusting individual flows is facilitated with "T" fittings on the inlet distribution pipe to insure equal influent flows to each portion of the cell.

Figure 15-10. The flashboard or stoplog control structure in the background provides reliable water level regulation despite considerable fluctuation in inflows and discharge flow rates are measured in the Parshall flume in the foreground.

Figure 15-11. Pest outbreaks can devastate simple wetlands with only a few plant species.

Figure 15-12. Broad-leaved species such as arrowhead add to diversity and aesthetics but in the fall, they tend to drop all their leaves simultaneously adding measurably to the organic loading on a constructed wetland.

Figure 15-13. *Phragmites* is an aggressive, weedy species that dominates most wet environments in Europe and consequently, is the most commonly used species in European constructed wetlands.

Figure 15-14. The Minot constructed wetlands system includes multiple shallow water/emergent plants and deeper water/submergent plants to provide high removal rates for NH$_3$ and fecal coliform bacteria.

Figure 15-15. Ammonia removals have been excellent during normal operating periods but nitrification rates are very low in cold waters and discharge levels increased during the coldest months.

Figure 15-16. The circular history of development in the constructed wetland wastewater treatment technology progressing from systems that emulate natural wetlands through complex, artificial designs and eventually returning to designs that simulate natural wetlands.

Figure 16-1. Wetland plants are capable of surviving relatively dry conditions without undue stress although a prolonged dry spell would allow invasion of terrestrial species.

Figure 16-2. Turtle tracks highlight the drought impacting this wetland. Although many wildlife species will be temporarily displaced, most will profit from the rejuvenated productivity following re-flooding.

Figure 16-3. Most forested wetlands thrive on winter flooding but will only tolerate short periods of inundation during the growing season.

Figure 16-4. Two herbicide applicators commonly used by cranberry farmers for treating individual plants (weeds).

Figure 16-5. Excessive depths and duration of flooding during the growing season caused substantial mortality in this well-established swamp allowing buttonbush, river birch, and other weedy species to invade much of the area.

Figure 16-6. Monitoring plants is relatively simple yet provides easily interpreted indication of basic changes in the wetlands system.

Figure 16-7. Aerial photography and/or fixed point ground photography records plant community development and coverage during the season and annually.

Figure 16-8.   Birds occupy a variety of niches and are easily monitored indicators of system health and well being.

Figure 17-1.   NRCS employees that designed drainage systems only 10 to 12-years ago are now efficiently designing water control structures and plugging ditches to restore wetlands.

Figure 17-2.   Though reservoirs often drown floodplain wetlands, a few projects have merely shifted the locations. Lateral dikes restrict reservoir waters and upstream waters are pumped into the reservoir during summer but allowed to flood during winter in some 10,000 acres of valuable bottomland hardwoods along Kentucky and Wheeler lakes in Tennessee and Alabama.

Figure 17-3.   Many of our large "natural" wetlands — Lake Agassiz, Lake Mattamuskeet, Okeefenokee Swamp — have been restored following drainage attempts and are currently maintained only through deliberate management efforts. Mitigation banking provides similar opportunities today.

LIST OF TABLES

Plate 1. Undulating terrain in terminal moraines coupled with depressions left by melting ice blocks in the Coteau du Missouri of the Dakotas supports myriads of pothole marshes in wet years.

Plate 2. A "farmed" wetland in North Dakota. Since the turn of the century, sedges in this four square mile marsh have been cut for hay in dry years but returning rains transform it into a lush productive wetland.

Plate 3. In winter, linear, bottomland hardwood forests occupying low-lying regions starkly contrast with pines on slightly higher ground along the Waccamaw River near Myrtle Beach, South Carolina.

Plate 4. Multiple inter-dune depressions paralleling the shoreline near Veracruz, Mexico have become diverse productive marshes that provide important wintering habitat for many migratory birds.

Plate 5. The combination of shallow water marshes, deep water zones with submergent species, nesting islands and very high algal and invertebrate populations has proven highly attractive for wildlife and unusually effective wastewater treatment at Minot's (North Dakota) 64-hectare constructed wetland.

Plate 6. Only a few cypress surviving in the deepest water portions of this swamp allow for a profusion of floating-leaved species. Many herbaceous wetland plants require 80% or greater of full sunlight.

Plate 7. Extensive stands of bulrush in a created marsh in the Dakotas provide important nesting habitat and shelter for a variety of wetland wildlife.

Plate 8. The many different kinds of plants in this prairie marsh are evident from a low altitude overflight or an aerial photo.

Plate 9. Drought-tolerant thistles (*Cirsium arvense*) are restricted to dryer sites during wet years but will readily invade the marsh after several dry years. With the return of wet conditions, muskrats (note feeding platform) re-populate isolated wetlands from refuges in permanent marshes, lakes and streams.

Plate 10. Students and visitors often confuse the wastewater treatment system with the display pool at the Institute for Administration Development on the outskirts of Bangkok, Thailand.

Plate 11. Orlando's 480-hectare Wilderness Park plays host to several endangered species and over 150 species of plants, 141 birds, 22 reptiles, 16 mammals, 16 fish and 8 amphibians that visitors view along nature trails though its primary purpose is polishing 75,000 m^3 of wastewater daily.

Plate 12. The American Crystal Sugar Company constructed wetlands near Hillsboro, North Dakota successfully polishes sugar beet processing wastewaters accomplishing high removal rates and low discharge levels for BOD$_5$, TSS and NH$_3$.

Plate 13. A 220-hectare constructed wetland provides high level polishing treatment for Beaumont, Texas' wastewater and substantial wetland wildlife benefits.

Plate 14. Algal and *Lemna* blooms wax and wane but are screened by the final marsh component at Minot's constructed wetland. Extra fill material was used to form nesting islands in the deeper ponds in each cell and shorebird and waterfowl productivity rivals the best natural marshes.

Plate 15. A New England depression hosting a shallow lake that was invaded by mosses and sedges to form a bog is now being invaded by shrubs and trees. Over time, this low-lying area will become indistinguishable from the forest surrounding it.

Plate 16. Maintaining high water levels into early summer and then quickly dewatering resulted in this lush growth of rice cutgrass, a moist site species.

Second Edition

CREATING FRESHWATER WETLANDS

CHAPTER 1

MARSHES, BOGS, SWAMPS, SLOUGHS, FENS, TULES, AND BAYOUS

INTRODUCTION

Today, wetlands are on everyone's tongue, but just 20 years ago only a few biologists knew of the term and fewer wetland specialists partially understood these complex systems. Now, even U.S. Presidents proclaim their dedication to preserving these important resources, a fashionable restaurant on Hudson Street in downtown Manhattan is named "Wetlands" and the U.S. Senate has a Wetlands Subcommittee. In fact, the term has only recently come into common usage to provide a generic, all encompassing word that includes virtually all types of shallow water environments.

In the past, a few sportsmen, conservationists, and scientists were interested in managing and protecting marshes, swamps and bogs — wetlands. The vast majority of society viewed swamps and bogs as obstacles to progress since most wetlands were thought to be reservoirs of disease and unfit for farming or development, if not actually haunts of unimaginable monsters. In 1977, President Carter issued Executive Order 11990 — Protection of Wetlands — and 19 years later, the change in society's attitudes has been dramatic. The change reflects the accomplishments of a few dedicated conservationists that managed to bring the issue to the public's mind, but the surprising speed of the shift suggests a significant proportion of society was sympathetic. However, few were likely to have been simply supportive of wetlands — most of society had little or no concept of wetlands or their values to society. Most likely, the wetlands issue was assimilated into the overall concern for world environments during a period of elevated human consciousness.

Unfortunately, few supporters understand wetland resources, wetland ecology, or real functional values of wetlands any better than acid rain, the ozone holes, or rain forests — all current issues with fervent, vocal proponents and substantial public concern. Many wetland protectors could not define wetlands, much less describe the complex biological communities and processes within the wide range of wetland types. Many that would use wetlands today have little understanding for the complex systems they attempt to replicate or reduce for their own purposes. Even among wetland scientists, there remains much disagreement on precise definitions and limited understanding of biological communities and hydrologic, physical, and chemical processes.

1

Our ideas of wetlands are vague and ambiguous, partially because of the bewildering variety and broad range of environments encompassed within the term. Some types of wetland are commonly recognized and the names widely known; but names vary confusingly in different regions and are often encumbered with historical human perceptions. For example, a "slough" is a freshwater marsh in the Dakotas, a brackish marsh along the West Coast, and a freshwater swamp on an old river channel in the Gulf Coastal Plain. Marsh has had limited use, but slough, swamp, and bog in common usage have become burdened with fearsome, foreboding, difficult, or hindering connotations. Since widespread interest is only recent, limited information is available, even in scientific fields; and because definitions are important for regulatory actions, considerable research and discussion has occurred during the last 20 years. Because land and water can mix in many ways and biotic components modify and blur the boundaries, it can be perplexing to define wetlands or determine where wetlands begin or end strictly on wetness or dryness (Figure 1-1). Not surprisingly, definitions, especially on precise wetland boundaries, are difficult to derive and apply. That is simply the indefinite nature and complexity of the subject, not a reflection on the ability of those studying it.

Figure 1-1. Riparian zones in arid regions are critically important for wildlife habitats and water quality improvement. Although germination and early establishment is often related to flood events, essentially terrestrial, woody species are often present because of the beneficial effects of extra (often subsurface) water supplies in contrast to the detrimental effects of excess water in wetlands.

Much of the confusion derives from the indefinite character of wetland ecosystems. Wetlands are ecotones (edges), transition zones between dry land and deep water, environments that are not always wet nor obviously dry. Any sizable wetland often includes portions that are clearly dry land and clearly deep water, as would be expected in any transition region. Boundaries are imprecise and may vary with seasons and different years. Gradual changes in wetness, soil, and vegetation types occurring across the transition band confound attempts to precisely measure boundaries and subsequently process descriptions. Furthermore, wetlands need not be continuously wet, nor are they continuously dry. Many wetlands are only wet during certain years, seasons, times of day, or after heavy rains. At other times, they may be dry. However, the unique plant and animal communities in wetland ecosystems depend on environmental conditions created by alternating inundation and drying during different seasons or different years.

The rich variety of plants and animals found in most wetlands results from their transitional position in the landscape and subsequent production rates (Figure 1-2). Not only are many unique organisms restricted to wetland environments, but most wetlands receive extensive use by animals characteristic of terrestrial or purely aquatic environments. Some use wetlands seasonally — various fish spawn in shallow water wetlands, but spend most of their adult lives in deeper waters. Others visit daily — fox or coyotes on their nightly rounds; while others may reside for extended periods, depending on availability of other foraging and shelter conditions — deer or pheasants weathering winter storms or antelope and elk browsing succulent marsh vegetation. Many birds found in terrestrial and wetland habitats frequently have their highest numbers in the diverse, productive habitats of wetlands.

Diversity and abundance vary greatly between different types of wetlands and within a single wetland. Some wetlands — acidic bogs, monotypic cattail (*Typha*), or reed (*Phragmites*) marshes and many saltwater wetlands — have low diversities (i.e., large numbers of a few types of plants or animals). Others, river swamps and fresh/brackish marshes, have high diversities; that is, many types of plants or animals, but only a few individuals of each type. In either case, basic productivities measured as biomass produced per unit area per unit time commonly exceed the production rates for the most intensively managed agricultural fields.

Variation in productivity and diversity within a wetland system is readily apparent from casual observation of the "hummocks" within the Everglades. These wet, forested islands situated in large expanses of wet prairie, "the river of grass," support a more diverse assemblage of plant and animal species than the adjacent sedge marshes and mangrove swamps. Hummocks also provide critical seasonal habitats for animals normally found in the marshes when the latter become too wet or too dry for certain species. Consequently, the diversity and productivity of hummocks varies substantially during the course of a year, and their influence extends far beyond their boundaries.

Figure 1-2. The rich variety and high productivity of this prairie marsh result from its transitional location between dry upland and deep water environments.

WETLAND DEFINITIONS

Problems in defining wetland for all uses is reflected in the variety and types of earlier definitions, many of which were devised for different needs and purposes. Most avoid the how-wet-is-wet question by describing wetlands in terms of soil characteristics and the types of plants capable of growing in these wet transitional habitats. Even shallow standing water or saturated soil quickly cause the atmospheric gases that filled interstitial pore spaces in the soil to be replaced by water, and microbial metabolism rapidly consumes available oxygen. Since gaseous diffusion from the atmosphere into soil water is much slower than microbial consumption, all except a thin top layer of the soil becomes anoxic or without oxygen.

Roots of normal, terrestrial plants obtain oxygen for respiration from gases within soil pore spaces and if those spaces are filled with water lacking oxygen, their roots die and the plant dies. Hydrophytic or wetland plants have developed specialized physical structures, aerenchyma, loosely similar to bundles of drinking straws, to transport atmospheric gases including oxygen through leaves and stems down to the roots to provide oxygen for respiration. Aerenchyma also transport respiratory by-products and other gases generated in the substrate back up the roots, stem, and leaves for release to the atmosphere, reducing potentially toxic accumulations in the region of growing roots. Because of these specialized structures, wetland plants are able to survive and grow in habitats with hostile root-growing conditions that would kill other plants. Consequently, wetland plants are often the best indicator of a wetland system even though many wetland plants can grow in drier environments if competition with terrestrial plants is limited.

Inundation and anaerobic conditions also cause specific changes in chemical substances found in most soils that serve as indicators of wet soils. Anoxic substrates with reducing environments cause many elements and compounds to occur in reduced states, creating characteristic colors, textures, and compositions typical of hydric soils. Due to the prevalence of iron in many soils and its color in reduced states, wet soils often have a gray or grayish color and fine texture.

In 1979, the U.S. Fish and Wildlife Service developed a generic definition and classification system to encompass and systematically organize all types of wetland habitats for scientific purposes. It broadly recognizes wetlands as a transition between terrestrial and aquatic systems, where water is the dominant factor determining development of soils and associated biological communities and where, at least periodically, the water table is at or near the surface, or the land is covered by shallow water. Specifically, "Wetlands must have one or more of the following three attributes:

1. at least periodically, the land supports predominantly hydrophytes;
2. the substrate is predominantly undrained hydric soil; and,
3. the substrate is nonsoil and is saturated with water or covered by shallow water at some time during the growing season of each year."

This definition broadens the three essential components of wetlands in the definition contained in the wetlands protection Executive Order. It concentrates on areas containing undrained or poorly drained (hydric) soils or areas with nonsoil substrates (rock or gravel) that are covered by water during a portion of the growing season. In either instance, continued inundation (soil or rock) precludes establishment or long-term survival of plants lacking special adaptations to growing in flooded substrates. Basically, these areas are wet enough for a long enough time to produce anaerobic substrate conditions that limit the types of plants that can survive there. Only wetland plants (hydrophytes) with the ability to provide oxygen for root respiration from atmospheric sources will be present.

But importantly, two of three attributes include the qualification "predominantly" since only a rare wetland would be completely lacking in at least a few small areas of normal aerobic soil or other substrates supporting typical terrestrial plants. Since few wetlands have perfectly flat surfaces or uniformly consistent elevation changes, most also have portions that are essentially terrestrial habitats.

Conditions for wetland soils and vegetation are produced by the impact of water, and extent and duration of flooding may vary substantially in some areas with only a few centimeters difference in elevations (Figure 1-3). In addition, even a perfectly flat swamp will support some terrestrial vegetation on living tree trunks and most certainly on the remains of stumps and fallen

Figure 1-3. Depending on gradients and flows, portions of riparian zones often support wetland soils and vegetation, substantially increasing diversity.

boles that extend slightly above the typical flood line. Muskrat houses, ice heaves, and herbivore wallows form similar high spots in prairie and coastal marshes. It is important to bear in mind that most (majority, dominant, etc.) of the area must consist of hydric soils and hydrophytic vegetation, but not necessarily 100% or even 80%. Since wetlands are transition zones, a mixing or merging of environmental conditions is expected and, in fact, an important characteristic that contributes to the diversity and productivity of our wetland resources.

Also significant is the concept and implication of "periodically" and "at some time" in two attributes of this and other definitions. Both encompass alternating wet and dry periods — not necessarily continuously wet, but not continuously dry — that are critical in determining the types of vegetation that can survive there. Bottomland hardwoods (swamps) frequently sustain deep and often long-term inundation during winter, but similar conditions in spring and early summer cause physiological stress leading to death if flooding occurs over more than one growing season. Absence of winter flooding would remove the competitive advantage of wetland trees vs. upland trees and also reduce production (biomass) since the annual fertilization and watering phenomenon would be lacking. Conversely, continuous flooding in a prairie marsh (over 5 to 10 years) causes falling productivity, eventually plant stress, and finally mortality leaving deeper, open water environments of shallow lakes. Alternating periodic flooding and drying are crucial to maintaining the complexity, diversity, and productivity of natural wetlands, but the concept and its manifestations in soils, hydrology, and biological communities confound attempts to develop simple definitions and precise boundary determinations.

Similar, but not always identical, definitions have been included in the Food Security Act, the Clean Water Act, the Emergency Wetlands Resources Act and the 1987 USACE Wetland Delineation Manual, the 1989 Federal Manual for Identifying and Delineating Jurisdictional Wetlands, and the proposed Comprehensive Wetlands Conservation and Management Act of 1995.

NEW WETLAND DEFINITIONS

In 1990, a joint definition for wetlands was developed and formally adopted by the Fish and Wildlife Service, the Environmental Protection Agency, the Soil Conservation Service, and the Corps of Engineers that provided a common definition for federal agencies. Due to the pervasive nature of regulations promulgated by these agencies and their direct impact on wetland resources, this definition was expected to gain acceptance by state and local governments and other organizations. Though its language was only slightly different, it was interpreted in a manner that placed more emphasis on hydric soils and lessened the importance of wetland vegetation in wetland determinations. By extension, the historical conditions at any specific site became important since hydric soils are formed over fairly long time periods and at least some characteristics often persist for long periods even after adequate drainage is established.

Consequently, regulators were faced with explaining their description of a soybean field or other cropland as wetland even though little or no wetland vegetation was present. This was almost "once a wetland always a wetland" even though truly hydric soils and wetland vegetation may not have been present for tens of years. However, blocking drain lines or ditches would likely lead to gradual restoration of the historic wetland ecosystem, providing a basis for defining the present-day crop field as a wetland. New legislation and regulations embodying this concept penalized landowners for subsequent drainage and conversion to agriculture if a field had been abandoned for a prescribed time period. However, abandonment was defined in temporal terms and little or no consideration was given to whether or not hydric soils and wetland vegetation had returned or other components of natural wetland form and function had or were likely to be reestablished.

Despite acquisition and restoration programs (The North American Water-fowl Plan, Ducks Unlimited, etc.), the total acreage was limited and many acres of potential wetlands or cropfields would likely have become low-value hybrids in the very earliest stages of succession. Restoration of the nation's wetlands is important and valuable to our society but expecting an individual landowner to forego crop production on his land was expecting inordinate contributions from a small minority of society. Neither would the reverting croplands provide significant functional values expected of natural wetlands in any reasonable time interval.

Society must develop effective means of compensating the landowner for lost acreages while simultaneously providing him with the methods and perhaps financial means to initiate active restoration efforts. Since many areas of converted wetlands are marginally productive agricultural lands, expansion of the current efforts to prioritize locations, quantities, and qualities of national wetland resources are urgently needed. Once a general consensus for high-priority areas has been reached, leasing, easements, or acquisition and active restoration efforts should be initiated.

Rigid application of wetland definitions based largely on soil conditions and criticisms of the 1989 wetland manual as too all-inclusive resulted in a public outcry and consequent executive and legislative efforts to modify definitions and subsequent regulations. In 1991, the White House Competitiveness Committee developed a wetland manual including a definition of wetlands that would have excluded some or major portions of, large wetlands highly valued by the public. The ensuing outcry led the Congress, in 1993, to authorize the National Academy of Sciences to conduct a definitive study and the report was released in 1995. While it presents a number of excellent recommendations on coordinating agency approaches and programs, the new definition which emphasizes hydrology as the defining characteristic of wetlands may be no more practical in field situations than earlier versions. Quite sensibly, the committee recommended protecting riparian ecosystems but not by defining them as wetlands. This does not detract from the importance of riparian systems, especially in arid regions of the West, but it recognizes that many of these riparian areas lack the high water table and saturated or hydric soils typical of wetlands. But the wetland definition debate still rages and is likely to for some time.

Although a regulatory wetland definition acceptable to all parties still eludes us (see Chapter 2), efforts to protect, restore, create, or use wetlands for specific purposes have already required additional modifiers. Foremost among these efforts has been creating wetlands for mitigation purposes and close behind are projects building wetlands for water purification. In each instance, a need arose to clearly distinguish between wetlands built for specific functions, especially water treatment, those built to mitigate wetlands impacted by development, and natural wetlands for regulatory purposes. Since revisions to the Clean Water Act, specifically Section 404, have made it the principal regulatory tool for protection of natural wetlands, the Portland Regional Office of EPA developed a set of definitions and interpretations to differentiate between natural wetlands and man-made systems for application in the 404 permit review process:

> **Constructed wetland**: Those wetlands intentionally created from non-wetland sites for the sole purpose of wastewater or stormwater treatment. These are not normally considered waters of the U.S. Constructed wetlands are to be considered treatment systems (i.e., not waters of the U.S.); these systems must be managed and monitored. Upon abandonment, these systems may revert to waters of the U.S. Discharges to constructed wetlands are not regulated under the Clean Water Act. Discharges from constructed wetlands to waters of the U.S. (including natural wetlands) must meet applicable NPDES permit effluent limits and state water quality standards.

> **Created wetland**: Those wetlands intentionally created from non-wetland sites to produce or replace natural habitat (e.g., compensatory mitigation projects). These are normally considered waters of the U.S. Created wetlands must be carefully planned, designed, constructed, and monitored. Plans should be reviewed and approved by appropriate state and federal agencies with jurisdiction. Plans should include clear goal statements, proposed construction methods, standards for success, a monitoring program and a contingency plan in the event success is not achieved within the specified time frame. Created wetlands should be located where the 'return' to the environment will be maximized (not necessarily on-site) and should be protected in perpetuity, to the extent feasible, through easements, deed restrictions, or transfer of title to an appropriate conservation agency or organization. Site characteristics should be carefully studied, particularly hydrology and soils, during the design phase and created wetlands should not be designed to provide habitat and provide stormwater treatment."

Natural wetlands were not newly defined, but as in the above, guidelines and restrictions on use were emphasized.

> Discharges to a natural wetland must not degrade the functions/beneficial uses of the wetland (i.e., must meet state water quality standards applicable to the wetland and comply with EPA and state anti-degradation policies). All practicable source control best management practices must be applied to minimize pollutants entering the wetland, consistent with NPDES permit requirements. Source control BMPs would generally include erosion controls, oil/water separation, presettling basins, biofilters, etc. Inlet/outlet structures requiring fill must be permitted under Section 404 of the Clean Water Act, preferably via an individual permit. Natural wetlands may not be used for instream treatment in lieu of source controls/advanced treatment; may be used for 'tertiary' treatment or 'polishing' following appropriate source control and/or treatment in a constructed wetland, consistent with the preceding guidelines.

The emphasis on careful planning, construction, contingency plans, etc. for created wetlands, most of which are built for mitigation purposes, reflects a regulatory perspective and pessimism over the ability of designers and developers to successfully replace lost or damaged natural wetlands. Many have failed. Some because of overly ambitious plans, limited time for natural successional stages, or poor design and construction. However, evaluation is often difficult since objective, quantitative goals were rarely included in original plans. Obviously, the created wetland definition and interpretations are designed to establish a new wetland that will replace one lost through development. However, only two possible uses are discussed though many others could be incorporated and additional guidance should specify management, as necessary, to ensure that the new wetland will not only develop the form (structure in terms of water, soil, and biological communities) but also the functions performed by the replaced wetland that are valued by society.

The perspective embodied in the created wetland discussion derives from the regulatory need to clearly distinguish mitigation wetlands from wetland wastewater treatment systems and to ensure the success of wetland mitigation projects; that is, those proposed as replacements for damaged or destroyed wetlands. However, as written, it would as easily apply to wetlands built for wildlife habitats, hydrologic buffering, or recreational purposes. In most cases, the latter would be considered waters of the U.S. and subject to the provisions of Section 404. In the current regulatory climate, fear of potential regulatory complications is likely to discourage landowners interested in building a wetland for recreation or other nontreatment functions. Furthermore, most newly built wetlands will require considerable active management before becoming fully established and many will require periodic management to maintain a specific successional stage. Since maintenance might include water level manipulation, even total drying, controlled burning, or other disturbances, these activities would seem to be regulated and perhaps would be prohibited under this interpretation. Quite obviously, we wish to protect and preserve wetlands built as replacements but just as clearly, we must not discourage increasingly widespread and significant efforts to build wetlands in nonwetland sites as a means of increasing our total wetland resource base. Consequently, created wetlands should also include those wetlands built for nonmitigation purposes, but should exempt the latter from current and future inclusion in the waters of the U.S.

It is also clear that deliberate use of natural wetlands for water treatment purposes is unacceptable except for advanced or polishing treatment. Though we understand how to build and operate a constructed wetland for efficient water treatment, most are simply that, wastewater treatment plants, and the above discussion properly recognizes their status and emphasizes needs to manage and monitor constructed wetland treatment systems. At present, our limited understanding of water purification mechanisms and processes within constructed wetland treatment systems is inadequate to recommend application rates that would not impair other functional values of natural wetlands. Until this new technology has progressed considerably, natural wetlands should be protected from deliberate applications of sediments and anthropogenic pollutants. In addition, many natural wetlands are presently receiving moderate to high loadings of pollutants generated by point and nonpoint sources and the impacts of these, much less additional deliberate applications, are unknown.

A number of terms and wetland descriptors have been used synonymously and need precise definition to ensure common understanding:

Natural wetlands are those areas wherein, at least periodically, the land supports predominantly hydrophytes and the substrate is predominantly undrained hydric soil or the substrate is non-soil and is saturated with water or covered by shallow water at some time during the growing season of each year. Natural wetlands have and continue to support hydric soils and wetland flora and fauna.

Restored wetlands are areas that previously supported a natural wetland ecosystem but were modified or changed, eliminating typical flora and fauna and used for other purposes but then subsequently altered to return poorly drained soils and wetland flora and fauna to enhance life support, flood control, recreational, educational, or other functional values. Natural and restored wetlands are "waters of the U.S." and subject to regulation under the 404 permitting process.

Created wetlands formerly had well-drained soils supporting terrestrial flora and fauna but have been deliberately modified to establish the requisite hydrological conditions producing poorly drained soils and wetland flora and fauna to enhance life support, flood control, recreational, educational, or other functional values. Created wetlands may or may not be wetlands built for purposes of mitigating (in the replacement sense not in the sense of minimizing harm) detrimental impacts to natural wetlands. If they are mitigation (created) wetlands, then they are subject to 404 regulations since the purpose was to replace a natural wetland that is subject to 404. If the created wetland is not a mitigation wetland, application of 404 permit regulations is dependent upon these systems having a direct surface connection with "waters of the U.S."

Constructed wetlands consist of former terrestrial environments that have been modified to create poorly drained soils, wetland flora and fauna for the primary purpose of contaminant or pollutant removal from wastewater. Constructed wetlands are essentially wastewater treatment systems and are designed and operated as such though many systems do support other functional values. As such, they come under the purview of the NPDES program regulations and not 404 regulations.

Floating aquatics systems are a related type of natural treatment system that consist of specialized applications of floating plants in sewage treatment lagoons, i.e., the water hyacinth (*Eichhornia*) or duckweed (*Lemna*) systems. These are not constructed wetlands because they use a different conceptual design nor should they be considered wetlands. Floating aquatics systems have been properly regulated under the NPDES program.

WETLAND CLASSIFICATION

The wetland classification system is a hierarchical system similar to those used for classifying plants and animals. It starts with 5 large systems; these are progressively divided into 10 subsystems, 55 classes, and 121 subclasses, which are then characterized by examples of dominant types of plants or animals. This system provides a consistent standard of terminology for use among scientists and managers throughout the country. However, careful and consistent determination of presence or absence of wetland from definitions must be employed prior to use of the classification systems since the latter is capable of classifying almost any area that is periodically wet, even a rain puddle in a parking lot. The classification system is a tool to categorize or classify a wetland following the application of some other appropriate definition to determine whether the area in question is a wetland. It should not be used to determine whether or not a site has a wetland on it.

The classification system is commonly used by wetland specialists and it provides the framework for the National Wetlands Inventory, a comprehensive identification and mapping of wetlands led by the U.S. Fish and Wildlife Service. The standardized procedures for delineation and quantification will provide a resource database for evaluating quality and quantity of remaining wetlands and assessing negative and positive impacts of destructive as well as creation or enhancement developments.

However, nonspecialists and various legislation and agency regulations will continue to define wetlands in more general terms. With minor variations, most describe wetlands as areas flooded or saturated by surface water or groundwater often and long enough to support those types of vegetation and aquatic life that require or are specially adapted for saturated soil conditions. Such descriptions can accommodate much of the conceptual framework and detailed, specific terminology necessary for scientific classifications. Concurrently, the generic terms more closely adapt to popular conceptions of what constitute wetlands — salt- and freshwater marshes, swamps, bogs, fens, and bayous and perhaps a few subcategories of these basic types.

In popular usage, shallow-water or saturated areas dominated by water-tolerant woody plants and trees are generally considered swamps; those dominated by soft-stemmed plants such as cattail and bulrush are considered marshes, and those with mosses and evergreen shrubs are bogs.

Our principal saltwater swamps are mangrove wetlands along the southern coast of Florida. Mangroves are among the very few woody plants adapted to saltwater environments (Figure 1-4).

Coastal salt marshes (Figure 1-5) are dominated by salt-tolerant herbaceous plants, notably cordgrass (*Spartina*), blackrush (*Juncus*) or other rushes along extensive areas of the eastern and southern coasts, or cordgrass and glasswort (*Salicornia*) along the west coast. Less familiar are the inland salt marshes of the intermountain west where high evaporation rates from shallow lakes and playas concentrates salt contents favoring similar plant types.

Freshwater swamps contain a variety of woody plants and water-tolerant trees. Southern swamps typically contain bald cypress (*Taxodium*), tupelo gum (*Nyssa*), water, willow oak, swamp white oak (*Quercus*), and river birch (*Populus*). Northern swamps are more likely to include alder (*Alnus*), black ash (*Fraxinus*), black gum (*Nyssa*), northern white cedar (*Thuja*), black spruce (*Picea*), tamarack (*Larix*), red maple (*Acer*), and willow (*Salix*). Forested wetlands include bay swamps, peat swamps, white cedar swamps, red maple swamps, wet flats, muck swamps, cypress heads or strands, bottomland hardwoods, and mesic riverine forests (Figure 1-6).

Freshwater marshes are dominated by herbaceous plants. Submerged and floating plants may occur, often in abundance, but emergent plants usually distinguish a marsh from other aquatic environments. Familiar emergents include cattails (*Typha*), bulrush (*Scirpus*), reed (*Phragmites*), grasses, and sedges (*Carex*) (Figure 1-7). A wet meadow may be only intermittently satu-

Figure 1-4. Drought exposes the prop roots of newly established mangroves that will accumulate sediments and organic materials to form islands in the Everglades region of Florida.

Figure 1-5. Coastal salt marshes tend to have low species diversity because of the harsh impacts from brackish waters (and a short growing season in this Copper River delta marsh), but tidal transport of nutrients contributes to high productivity.

Figure 1-6. Cypress and a few other tolerant trees can withstand almost permanent inundation, but most hardwood swamps require drying during the growing season.

Figure 1-7. Cattails, bulrush, and sedges are typically dominant species in the freshwater marshes of the prairie pothole region.

rated or flooded with very shallow water, but it also supports marsh species, especially sedges and wet grasses. A common type of freshwater marsh, the prairie potholes of the northern Great Plains, occurs in shallow depressions formed by glaciers. Those that hold water year-round, seasonally, or following heavy rains, often support luxuriant marsh vegetation. Although most are small, there are many of them (810,000 ha in North Dakota alone), and collectively they constitute an important wetland resource, especially for waterfowl nesting.

Bogs form primarily in deeper glaciated depressions, mainly in "kettle holes" in the northeastern and northcentral regions (Figure 1-8). Bogs are dependent upon stable water levels and are characterized by acidic, low-nutrient water and acid-tolerant mosses. Other bog plants such as cranberry (*Vaccinium*), tamarack, black spruce (*Picea*), leatherleaf (*Chamaedaphne*), and pitcher plant (*Sarracenia*) may be rooted in deep, spongy accumulations of dead Sphagnum moss and other plant materials only partially decomposed under bog conditions. In the same region and also at high elevations in the Rockies, fens have water nearer to neutral and are dominated by sedges (Figure 1-9).

FUNCTIONS OF NATURAL WETLANDS

Wetlands represent a very small fraction of our total land area, but they harbor an unusually large percentage of our wildlife. For example, 900 species of wildlife in the U.S. require wetland habitats at some stage in their life cycle, with an even greater number using wetlands periodically. Representatives from

Figure 1-8. Stable water levels with low nutrients and acidic conditions favor establishment of bogs in New England depressions. Vegetation is dominated by *Sphagnum* and ericaceous shrubs with a few pioneering tamarack.

Figure 1-9. Fens often occur in the same regions as bogs but are dominated by sedges because their waters are nutrient poor and closer to neutral pH, suggesting groundwater discharges.

Figure 1-10. Western sinkholes may unexpectedly support wetland communities. White specks in the pool are ring-necks and scaup in a dense submergent bed at Montezuma Well north of Phoenix.

almost all avian groups use wetlands to some extent and one third of North American bird species rely directly on wetlands for some resource (Figure 1-10).

Due to the diversity of habitats possible in these transition environments, the nation's wetlands are estimated to contain 190 species of amphibians, 270 species of birds, and over 5000 species of plants. Many wetlands are identified as critical habitats under provisions of the Endangered Species Act, with 26% of the plants and 45% of the animals listed as threatened or endangered either directly or indirectly dependent on wetlands for survival.

Stability is neither common nor desirable in wetland systems. Unlike upland habitats, wetlands are dynamic, transitional, and dependent on natural perturbations. The most visible and significant perturbation is periodic inundation and drying. Changing water depths, either daily, seasonally, or annually, strongly influence plant species composition, structure, and distribution. Other influences, such as complex zones of water regimes, salt and temperature gradients, and tide and wave action, produce wetland vegetation that is generally stratified, much like forests. These factors combine to create a diversity and wealth of niches that make wetlands important wildlife habitat.

In addition to their vegetative productivity, wetlands team with zooplankton, worms, insects, crustaceans, reptiles, amphibians, fish, birds (Figure 1-11), and mammals, all feeding on plant materials or on one another. Other animals are drawn from nearby aquatic or terrestrial environments to feed on plants and animals at the highly productive "edge" environment of wetlands,

Figure 1-11. Waterfowl are perhaps the most commonly identified product of the life support function, and interest in the welfare of waterfowl led to early efforts in wetland protection and management.

and they in turn become prey for others from a greater distance, thus extending the productive influence of wetlands far beyond their borders (Figure 1-12).

Sport and commercial hunters and fishermen have called public attention to the economic value of wetland fish and wildlife. They were first to note the direct relationship between wetland destruction and declining populations of valuable species of fish, shellfish, birds, reptiles, and fur-bearing animals that are dependent on certain types of wetland habitats during part or all of their lives. Many studies have now linked the destruction of summer breeding wetlands and winter feeding wetlands to shifts or declines in populations of migratory waterfowl and other birds. Wetland destruction can be especially significant in regions where such habitat is least common and alternative sites may be unavailable.

Wetlands along coasts, lakeshores, and riverbanks have recently begun receiving increased attention because of their valuable role in stabilizing shorelands and protecting them from the erosive battering of tides, waves, storms, and wind. One of the greatest benefits of inland wetlands is the natural flood control or buffering certain wetlands provide for downstream areas by slowing the flow of floodwater, desynchronizing peak contributions of tributary streams, and reducing peak flows on main rivers. During dryer periods, slow releases from wetlands augment and stabilize base flows providing water to support aquatic life in streams and rivers.

Some wetlands may function as groundwater recharge areas, allowing water to seep slowly into and replenish underlying aquifers. Other wetlands

Figure 1-12. Many commercially important finfish and shellfish require wetlands for a critical portion of their life cycles.

represent discharge areas for surfacing groundwaters. Both may occur within close proximity depending upon local and regional patterns of ground water distribution. This may be one of the reasons that simply plugging the ditch from a drained wetland does not always restore a productive marsh. The previous system may have been the expression of surfacing ground water that previously was supplied by another still-drained wetland that acted as a recharge area. Some of these interdependent wetlands may lie fairly close together, but others may be many kilometers apart.

Another important but poorly understood function is water quality improvement. Wetlands provide effective, free treatment for many types of polluted waters. Wetlands can effectively remove or convert large quantities of pollutants from point sources (municipal and certain industrial wastewater effluents) and nonpoint sources (mine, agricultural, and urban runoff), including organic matter, suspended solids, metals, and excess nutrients. Natural filtration, sedimentation, and other processes help clear the water of many pollutants. Some are physically or chemically immobilized and remain there permanently unless disturbed. Chemical reactions and biological decomposition break down complex compounds into simpler substances. Through absorption and assimilation, wetland plants remove nutrients for biomass production. One abundant by-product of the plant growth process is oxygen, which increases the dissolved oxygen content of the water and also of the soil in the immediate vicinity of plant roots. This increases the capacity of the system for aerobic bacterial decomposition of pollutants as well as its capacity for supporting a wide range of oxygen-using aquatic organisms, some of which directly or indirectly utilize additional pollutants.

Many nutrients are held in the wetland system and recycled through successive seasons of plant growth, death, and decay. If water leaves the system through seepage to groundwater, filtration through soils, peat, or other substrates removes excess nutrients and other pollutants. If water leaves over the surface, nutrients trapped in substrate and plant tissues during the growing season do not contribute to noxious algae blooms and excessive aquatic weed growths in downstream rivers and lakes. Excess nutrients from decaying plant tissues released during the nongrowing season have less effect on downstream waters and biological communities.

It is no secret that natural wetlands can remove iron, manganese, and other metals from acid drainage — they have been doing it for geological ages. In fact, accumulations of limonite, or bog iron, were mined as the source of ore for this country's first ironworks and for paint pigment (Figure 1-13). Limonite deposits are most common in the bog regions of Connecticut, Massachusetts, Pennsylvania, New York, and elsewhere along the Appalachians. Wetlands were abundant in parts of the Tennessee Valley during past ages, and significant bog iron deposits were found in Virginia, Tennessee, Georgia, and Alabama. Although now of limited economic importance in the U.S., bog iron is still a significant source of iron ore in northern Europe.

Figure 1-13. This small stream in the Black Hills of South Dakota was destroyed during a bog iron mining operation in the mid-1960s.

Similarly, mixed oxides of manganese, called wad or bog manganese, are the products of less acidic wetland removal processes. Often these wad deposits also contain mixed oxides of iron, copper, and other metals.

SUMMARY

Partially because wetland is a deliberately generic term, but more importantly because of the unifying characteristic that is embodied in a fascinating but sometimes bewildering variety of forms, wetland spans the spectrum from mangrove and cypress swamps through fresh- and saltwater marshes to bogs and fens. Despite the variety, wetlands have two characteristics that are common to wetlands of most interest to us. They have soils or substrates that are saturated for long periods or much of the growing season and because of that, they have types of vegetation with specialized structures, aerenchyma, buttresses and cypress knees that transport oxygen to their roots for respiration, enabling these plants to grow in an otherwise hostile environment. Secondly, many elements and compounds occur in reduced states in saturated, anoxic soils causing characteristic colors and textures. Though a variety of wetlands definitions have been developed, these two attributes, saturated or hydric soils and hydrophytic vegetation are common to almost all.

CHAPTER 2

WETLAND REGULATION

INTRODUCTION

Between the mid-1950s and mid-1970s, approximately 700 square miles of wetlands were altered and drained nationwide each year. While losses were nationwide, most were more or less equally balanced between the upper midwest (prairie potholes) and the south (forested wetland). Nineteen states lost over 50% of their wetlands and Ohio and California lost over 90%. A second status and trends report for the mid-1970s to the mid-1980s found a significant reduction in the loss rate, but losses continued at 300 square miles per year. During this period the largest losses occurred in the south (primarily forested but also coastal wetland) (Figure 2-1). Most wetland losses were caused or induced by human activities.*

In the mid 1970s, wetland specialists were concerned over the tremendous losses of natural wetlands, roughly half of the original acreage, that was encouraged and subsidized by governmental programs. The thinking that developed into the executive order protecting wetlands was to stop federal government financing of wetland destruction on private lands and to stop federal agencies from destroying wetlands on public lands.

The last 20 years have witnessed an astonishingly rapid reversal of public attitudes and policy towards wetlands. For over 100 years, public attitudes embodied in consensus policy, considered wetlands as "wastelands" and encouraged wetland destruction and conversion with financial incentives. But in the 1960s, Massachusetts passed legislation requiring a state permit for any alteration of wetland and many other states followed. On the national scale, growing public awareness of wetland values led to incorporation in the 1972 Clean Water Act (CWA) followed by the 1977 Executive Order, Protection of Wetlands. This led to modifications in the U.S.A. Corps of Engineers (COE) regulations implementing provisions of the CWA, specifically Section 404. Later a number of states implemented similar regulations. The Coastal Barriers Resources Act of 1982 withdrew all federal subsides for development on designated coastal barrier islands and beaches, where wetlands are a critical feature of the environment. The Food Security Act of 1985 included a "swamp-buster" provision that made farmers ineligible for agricultural income-support programs if they convert wetlands and plant commodity crops on them. In

* After Hammer, D.A. et al, Mitigation Banking and Wetland Categorization: The Need for a National Policy on Wetlands, Tech. Rev. 94-1, The Wildlife Society, Wastington, D.C., 1994.

Figure 2-1. Southern wetland conversion to agriculture was accelerated during periods of high soybean prices since soybeans are a late season crop and could be grown on sites that remained wet late in spring.

1986 the Tax Reform Act eliminated most of the special tax advantages that accrued to farmers and developers for new investments, particularly in wetland areas and the Emergency Wetlands Resources Act promoted conservation of wetlands to maintain the public benefits they provide, along with fulfilling international obligations in migratory bird treaties and conventions. The intent was to protect, manage, and conserve wetlands by increasing cooperative efforts among private interests and local, state, and federal governments.

Unfortunately, none of these represented clearly defined national policy; hence, the confusion, controversy, costs, and disenchantment with those approaches. Presidential executive orders are not laws or legislation applying to all citizens; they only apply to actions by agencies of the federal government. But extension of the executive order through the 404 regulatory program and other federal programs created a public outcry leading to current legislative efforts to restrict or modify protection of various wetland types.

In contrast to wetland drainage, we had not determined and codified a wetland protection/management policy. As with the Endangered Species Act, wetland regulations had become a tool of environmental extremists — a means to block or hinder any development to which they are opposed. Hence the rigid requirements for developers to avoid any impact to 100-m^2 wet spot in a field or to replace it on site. Or similar actions for a 10,000-m^2 dumping ground

Figure 2-2. A few small waterways are the only remnants of wetland in a large tract that previously supported bottomland hardwoods.

surrounded by fill and housing or other development that had interrupted natural drainage patterns and was slowly starving the wetland. Or classifying every aspen stand in the Rocky Mountains as wetland as some are wont to do.

Natural wetlands are ephemeral components of the landscape that largely result from geological incidents and to a lesser extent, from biological and human activities. Specific location, type, and size of every wetland is dependent on a series of geophysical phenomena that created and maintain suitable hydrological and edaphic conditions at that site (Figure 2-2). Consequently, attempts to preserve every wetland or even to require on-site replacement are in fact attempts to maintain the status quo disregarding the series of unintentional events that created and maintain a wetland on that specific site. This philosophy is inherent in rigid application of in-kind, on-site restoration/creation requirements of the COE/EPA MOA on mitigation signed in January 1990. In contrast, a strategic, landscape approach might well identify more suitable locations for certain types and sizes or even different types and sizes to enhance one or more of the functional values to society.

In addition, fear of change has obscured serious consideration of arguments for strategic, landscape planning for wetland management that might increase the values of wetlands through judicious location. It also has inhibited restoration of wetlands to their original form and function especially in the coastal regions where freshwater marshes have become saltwater marshes after the intracoastal and associated canals permitted extensive salt water intrusion. For example, in Texaco's Bessy Height's field near Port Arthur, Texas, cypress stumps are still prominent in a saltwater marsh but the regulatory process discourages efforts to restore the original freshwater marshes and swamps. Required permitting contravenes the goal of restoring freshwater wetlands on sites where those wetlands previously existed despite the fact that salt-water intrusion resulted from man-induced and not natural changes. Rigid attempts to maintain the status quo totally disregard historical conditions and man-induced changes, as well as the ever changing, dynamic nature of all wetlands.

Wetland interactions and interdependencies in a watershed negate management approaches based on evaluating potential impacts to discrete wetland units. Current site specific approaches, especially regulatory measures, to wetland management are inadequate to conserve or restore wetlands. Natural wetlands are interdependent and interact with terrestrial components of the landscape and with other wetlands, especially within a watershed or biotic region, such that meaningful management must incorporate a landscape, watershed or biotic region approach. Because of these strong interactions and interdependencies, it is not possible to evaluate, assess, or categorize a wetland unit in isolation from other components in the watershed or biotic region. Wetland management must also include temporal factors since age/successional stage as well as geographical location, strongly influence both form and function of wetlands. For example, small isolated wetlands strategically located throughout a watershed may have considerably more value in terms of water quality improvement than a single, large wetland at one position, even though it may be situated at the lower end of the watershed. However, the reverse appears to be true for flood ameliorization — the larger wetland in the lower reaches seems to have greater impact on floodwaters. Relatively, narrow bands of riparian vegetation may have inordinate importance as travel lanes for some species of wildlife. Consequently, wetland management must include the context of the surroundings on a watershed, landscape, or biogeographical unit basis.

To a considerable extent, the functions and values of wetlands are related to size and location. A wet puddle or a dump provide little value to anything or anyone and a semi-dry marsh filled with car tires, appliances, and other refuse does not engender positive attitudes towards wetlands among the neighbors. Not to mention the effects of mosquitos hatching in the tires, bottles, and cans (more so than from the marsh itself!). This approach does not restore or enhance our nation's wetland resources. Nor does it provide incentives to clean up and restore the marsh, enlarge and enhance the puddle, or restore/create other wetlands. Wetland protection, as any other regulation, must be tempered with reason. Regulators must have and apply negotiating flexibility to allow development of small (insignificant) and/or poor quality wetlands in exchange for restoration/creation of larger and/or high quality wetlands on-site or in the region. This is not to say that all small wetlands are valueless — some support populations of threatened or endangered species. But in that case, the pertinent regulations are covered by the Endangered Species Act, not by wetland protection orders. There a few examples of small even isolated wetlands that are the last remnants of much larger wetland complexes. However, dogmatically applying protection to our 100-m² puddle example in a region with thousands and even tens of thousands of acres of similar, larger and valuable wetlands impugns the credibility of wetland scientists. And not unexpectedly, it has led to controversial political efforts to modify wetland definitions and categorize wetlands for regulatory purposes.

The regulatory quagmire serves neither to protect all wetlands or their functional values, accomplish no net loss or net gain, nor to accommodate economic development in an orderly, cost-effective manner (Figure 2-3). Critics point out that regulations fail to provide adequate protection while others fault the interminable, costly delays and inability to plan developments. Others cite the continued, often piecemeal, loss of thousands of acres of wetlands and our failure to implement no-net loss on a local, regional, state, or national basis much less accomplish any improvements in restoring wetlands and their functional values. Many examples of disparate implementation of regulations have also been articulated. Forceful arguments have been made for both sides of the issue.

Reversing the drainage/conversion policy probably could not have been accomplished 40–50 years ago, but an increasingly aware public supports wetland protection. The controversy regarding wetland protection is not surprising given the short time period for an almost complete reversal of a long-established drainage policy. Resolution of the controversy over wetland protection is only possible through adequate public discussion leading to consensus establishment of a

Figure 2-3. While water-related projects often are located in wetlands their overall impact is much less than imprudent housing developments that need not be.

Figure 2-4. Though many constructed wetlands are relatively large and provide many ancillary benefits, deliberate management is often required that may alter the wetland so NPDES regulations are appropriate instead of 404 regulations.

national policy on wetland protection/management embodied in national legislation and unified implementing regulations (Figure 2-4).

Current legislation in the U.S. Congress that may produce The Wetlands Act of 1997 (?) is a first attempt to establish a national policy (albeit buried in the Clean Water Act). The House version (H.R. 961) was passed on May 16, 1995 but the Senate has not (and appears unlikely to in this session) begun action on the Senate version (S. 851). Regardless, these bills embody national legislators' response to the present controversy over wetland protection and though some changes can hopefully be accomplished before passage, the final version is likely to contain many of the provisions in the House bill. This legislation is the most significant and is likely to have the most far-reaching impact on wetlands and wetland projects since the drainage acts over 100 years ago. Consequently, developers of created or restored wetlands must be cognizant (and in compliance) of these regulations and a discussion is included below. For simplicity, this legislation is referred to as the Wetlands Act or the Act even though it has yet to become law.

THE COMPREHENSIVE WETLANDS CONSERVATION AND MANAGEMENT ACT OF 1995

The wetland protection controversy became a major topic during congressional re-authorization of the Clean Water Act in 1995 culminating in inclusion of Title VIII, The Comprehensive Wetlands Conservation and Management Act of 1995, in the House-passed version. Though the Senate has yet to act, the broad purpose of the Wetlands Act is to "establish a new Federal regulatory program for certain wetlands and waters of the United States." This is a new national policy that includes regulations and incentives to reduce wetland loss, protects some wetlands, encourages restoration of drained or altered wetlands, codifies mitigation banking, and stresses wetland functions as compared to wetlands form. It emphasizes the importance of wetlands to society, reverses previous conversion policies. and encourages wetland protection, restoration, creation, and management while accommodating desirable economic development.

In fact, the Act is very specific with regard to not inhibiting economic development and to protecting the rights of private property owners — the "takings" issue. It specifies that, in general, the Secretary shall balance the objective of conserving functioning wetlands with the objective of ensuring continued economic growth, providing essential infrastructure, maintaining strong State and local tax bases, and protecting against the diminishment of the use and value of privately owned property. It includes provisions whereby property owners may request delineation and classification of wetland on their property, requires owner notification of any actions limiting the use of a property, provides for public and county court notification of wetlands, specifies that wetland delineations will be filed with property records, and provides for copies of delineation determinations to property owners and financial institutions along with provisions for landowner appeal of wetland delineations.

It goes on to provide that a property owner shall be compensated if an agency action diminishes the property value by 20% or more, stipulates that a compensated landowner will be restricted in his use of the property but states that compensation does not confer additional rights to the Federal government other than the limitation on use. In addition, the owner may receive compensation if he is unable to explore or develop oil, gas, and mineral interests if under type A or B wetlands and his use is limited; failure to provide reasonable access to mineral interests beneath or adjacent to type A or B wetlands is considered a diminution in value. However, if the decrease in value is more than 50%, at the option of the owner, the Federal government will buy the property, or that portion, at fair market value. A provision that is likely to have a chilling effect on wetland delineations stipulates that any payments made to land owners shall be made from the annual appropriation of the agency whose action occasioned the payment or judgment except that the agency may then seek reimbursement from another agency if the action resulted from a requirement imposed by the second agency.

DEFINITIONS

The Act includes a number of pertinent definitions most of which are commonly used but some are specifically tied to mitigation:

Wetlands are lands which have a predominance of hydric soils and which are innundated by surface water at a frequency and duration sufficient to support, and that under normal circumstances do support, a prevalence of vegetation typically adapted for life in saturated soil conditions. Wetlands generally include swamps, marshes, bogs, and similar areas.

Restoration is an activity undertaken to return a wetland from a disturbed or altered condition with lesser acreage or fewer functions to a previous condition with greater wetlands acreage or functions.

Enhancement of wetlands is any activity that increases the value of one or more functions in existing wetlands.

Creation of wetlands is an activity that brings a wetland into existence at a site where it did not formerly occur for the purpose of compensatory mitigation.

Mitigation banking is wetlands restoration, enhancement, preservation, or creation for the purpose of providing compensation for wetland degradation or loss.

Mitigation bank is a wetlands restoration, creation, enhancement, or preservation project undertaken by one or more parties, including private and public entities, expressly for the purpose of providing mitigation compensation credits to offset adverse impacts to wetlands or other waters of the United States authorized by the terms of permits allowing activities in such wetlands or waters.

Regulated activity:
a.   the discharge of dredged or fill material into waters of the United States, including wetlands at a specific disposal site; or
b.   the draining, channelization, or excavation of wetlands.

WETLAND DELINEATION
The Wetlands Act also requires the Secretary, in consultation with other agencies, to establish wetland delineation standards under the following restrictions such that delineated wetlands:

1.   must have clear evidence of wetlands hydrology, hydrophytic vegetation, and hydric soil during the growing season;

2. won't result in the classification of vegetation as hydrophytic vegetation if it is equally or more typically adapted to dry than wet soil conditions;
3. must have some obligate wetlands vegetation present;
4. must have water present at the land surface for 21 consecutive days in the growing season; and,
5. will not include areas created by temporary or incidental development activities.

Furthermore, delineations on agricultural lands will continue to be the responsibility of the Secretary of Agriculture and active agricultural lands shall be exempted from delineation.

Delineation standards applicable to all agencies are needed but the last attempts by various agencies and by the White House were not successful. The former was so all-inclusive and restrictive that it created the "takings" issue and the latter would have denied protection to some large, well-known wetlands. Unfortunately the Act's guidelines tend toward the latter rather than finding a middle ground between the two positions. For example, bottomland hardwoods rarely have well developed wetland hydrology or water present at the land surface for 21 consecutive days during the growing season — few species can tolerate that long an innundation period. Restricting wetlands to areas with obligate species may simplify the field delineation process by eliminating all but the core area but will omit fringe and transitional areas that provide critical connections to the surrounding upland environments.

EXEMPTED ACTIVITIES

The list of exempted activities in wetlands is very long and includes:

1. activities in those wetlands classified as Type C;
2. activities falling under a nationwide permit except that:
   a. compensatory mitigation may be imposed in some cases;
   b. and in states with substantial conserved wetlands (basically those that haven't lost more than 10% of original wetlands, i.e., Alaska, Hawaii, and New Hampshire) the Secretary shall issue a general permit without mitigation requirements;
   c. Alaska Natives lands are excluded;
3. farming, ranching, silviculture, aquaculture including haying, grazing, minor drainage and burning activities;
4. maintenance of water related structures;
5. farm, stock, aquaculture ponds, wastewater management facilities, advanced treatment municipal wastewater reuse operations, irrigation canals, maintenance of drainage ditches;
6. temporary sediment basins related to construction; construction depressions or borrow pits, and sand, gravel aggregate or minerals mining;

7. construction or maintenance of farm or forest roads, railroads up to 10 miles, mining roads, and access roads for utility transmission lines;
8. farmed wetlands;
9. activities connected with a marsh management and conservation program in a coastal parish in Louisiana;
10. activities in incidentally created wetlands unless the wetland has exhibited function and values for more than 5 years;
11. activities for preserving and enhancing aviation safety or to prevent an airport hazard;
12. activities related to aggregate or clay mining activities in wetlands;
13. activities for placement of structural members — pier, docks, bridge, transmission line towers, and pilings to elevate houses and other structures;
14. activities related to development and maintenance of transmission lines and water supply reservoirs;
15. activities undertaken in states with substantial conserved wetland areas (i.e., Alaska, Hawaii, and New Hampshire) and are:
    a. for providing critical infrastructure — water & sewer, airports, roads, communication sites, fuel storage sites, landfills, housing, hospitals, medical clinics, schools and other community infrastructure;
    b. for construction and maintenance of log handling facilities;
    c. for construction of tailings impoundments used for treatment facilities;
    d. for construction of ice pads and ice roads or snow removal and storage;
    e. related to silviculture on Alaska Native lands.
16. related to recreational hunting or shooting;
17. for cranberry production if not more than 10 acres of wetland per operator per year are modified or the activity is required by any State or Federal water quality program;
18. for an area where a State, or political subdivision thereof, has an approved land management plan.

Many of these exemptions are for relatively minor activities and unlikely to have serious detrimental impacts to our wetland resources. But others are not. Denying protection to a potentially large class of wetlands (Type C), providing blanket exclusion for Alaskan wetlands, exempting a ten-mile segment of new railroad, development and maintenance of utility transmission facilities, and placing supporting structures in wetlands is much to broad. For example, much of our forested wetland occurs as isolated segments in river floodplains, few of which are more than 10 miles wide. A new railroad or utility transmission corridor cutting a swath across these segments could cause substantial detrimental impact.

The unfortunate aspect of these exclusions is that many projects could be accomplished without causing significant impacts with judicious design and routing. However, it does not appear that requirements to avoid and minimize impacts to wetlands under the conventional approach (also included in the

Act) are applicable to any exempted activities. Exempted or activities covered under a nationwide permit, can proceed without the review and recommendations of wetland ecologists that might have avoided and/or minimized potential impacts from these projects.

WETLAND CATEGORIZATION

More significantly, the Act establishes a new classification system for wetlands, an unfortunate choice of words since classification generally means determining the biological type of wetland with the U.S. Fish and Wildlife Service classification system. Herein, the types of wetland in the Act are considered a categorization system. Three types of wetlands are defined:

Type A Wetlands have critical significance to long-term conservation of the aquatic environment of which such wetlands are a part and which meet the following requirements:

1. serve critical wetlands functions in terms of critical habitat for wetland-dependent wildlife;
2. consist of, or portion of, 10 or more contiguous acres but may be smaller in the case of prairie potholes, playa lakes, or vernal pools;
3. have identified functions served by the wetlands that are scarce within the watershed or aquatic environment; and,
4. there is unlikely to be a "higher" use for such lands than for purposes of conservation.

Type B Wetlands provide habitat for a significant population of wetland-dependent wildlife or provide other significant wetlands functions, including significant enhancement or protection of water quality or significant natural flood control.

Type C Wetlands:

1. serve limited wetland functions;
2. serve marginal wetland functions but are abundant so regulation is not necessary for conserving important wetland functions;
3. are located behind retaining walls or similar structures (fast lands);
4. lie within industrial, commercial, or residential complexes or other intensely developed areas that do not serve significant wetland functions as a result of such location.

Within one year after passage, the Secretary is directed to provide standards and procedures for:

1. the classification and delineation of wetlands;

2. administrative review of such;
3. State or local land management plans; and,
4. issuance of general permits including programmatic, State, regional, and nationwide permits.

Furthermore, the Secretary along with the Secretary of Agriculture shall complete a comprehensive program to identify and categorize (classify) wetlands in the U.S. within 10 years of passage.

Given the status of the National Wetlands Inventory, much of the identification of wetlands has been accomplished. But developing and applying a categorization (classification) system is a bit more complex especially since one category (Type C) in essence will include wetlands that have no value to society and will likely be destroyed. In addition, the Act is very specific in that categorization is to be based on functional values.

Classification and categorization are useful tools in ordering chaos whether the subjects are insects, stamps, job descriptions, or wetlands. Wetland classification is grouping wetlands based on their hydrologic, biologic, and edaphic characteristics without any attempt to include a value judgment on one group or another. Categorization, however, implies grouping wetlands based on some form of assigned value regime.

Valuation or determining/assigning values by nature must include by whom, for whom, and for what purpose. The value of something is determined by society and is not an inherent characteristic, i.e., flood alteration function of a specific wetland could have significant value to a downstream community yet lack any value to an upstream community. Value has socio-economic implications that go far beyond an assessment of presence or absence or even quantitative measurements. Valuation is also a function of time in that society's values change, and therefore the very same wetland could have a very different perceived value in the same society at a different point in time.

Evaluating natural resources is an important basis for making decisions concerning land use. Mapping and scientific assessment of soils and forest stands has long been recognized as essential to prudent management of agricultural and forest resources. Wetland resources likewise could benefit from a similar level of assessment but placing some wetlands into a category without any protection on the basis of our present knowledge of wetland functions seems imprudent.

A key element of this categorization attempt must be a means of evaluating wetlands to determine the appropriate category for each individual wetland. Widely used evaluation methods (WET, HEP, etc.) are largely technical assessment tools. None of these appears adequate to measure the true value of each function performed by the myriad of types of natural wetlands in the U.S. Consequently, any attempt to evaluate and subsequently categorize natural wetlands with existing methodologies for the purposes of determining those with lesser values, could result in irretrievable harm to our wetland resources. Unfortunately, at the present state of the art, evaluation is still largely subjective based on cursory examination or it requires detailed and costly investigations that attempt to characterize the form and function of an individual wetland system. In too many cases, cursory evaluations are highly dependent upon a few highly regarded functional values, with little avenue for encompassing the sum of the myriad functional values from even a small isolated wetland much less larger and/or multiple wetland units within geographical units. Present valuation methods are likely to underestimate the value of even the highest priority wetland and could not hope to produce a realistic value for lesser wetland systems. The latter would likely include smaller systems, isolated/disjunct systems, disturbed or degraded systems, drier-end wetlands or transitional zones of wetlands, and ephemeral wetlands. Many of these could have significant but unmeasured functional values.

With a few exceptions, we lack quantitative data on many functions in most important types of natural wetlands. The exceptions (mostly in fresh- or salt-water marshes) include components of the life support function, i.e., production of avian and mammalian fauna, finfish, shellfish, a few instances of plant products and isolated cases of water purification. For the vast majority of wetlands and even for most different types of wetlands, we lack quantitative information even on the biologic productivity, much less adequate, comparable information on other important functional values.

Furthermore evaluation of a wetland is inevitably related to time of year and age of system. Wetlands are dynamic ecosystems undergoing considerable seasonal and annual change as well as progressive change over time, as the wetland system ages. Time, techniques, and location of data collection can have substantial impact on the result of a one-time evaluation.

Historically, wetlands were grossly undervalued but later a few waterfowl hunters led efforts to protect and preserve certain types of wetlands. Recently, other life-support functions along with hydrologic buffering and water quality improvement, have been identified as significant values. Major segments of society now place high value on wetlands and government policy is to protect rather than destroy our remaining wetlands. A complete reversal in society's valuation of wetlands has occurred in less than 60 years. In fact, the most significant change took place within the last 20 years! Who can estimate the functional values of wetlands to society or society's attitudes in the 21st century? If valuation schemes are employed to categorize wetlands with one category subsequently receiving no protection, evaluators must have the ability to estimate future values as well as adequately assess present functional values of existing wetlands. Lacking an estimate or assumption of future values, evaluators could easily underrate a significant portion of our existing wetland resources, resulting in the loss of that segment before it has been evaluated (valued) under the standards of a future society. It does not appear likely that a significant new category of functional value would emerge from future investigations but it would not be surprising to discover additional functions and values or a complete reordering of priorities with further understanding of "low" value wetland systems.

MITIGATION

On the positive side, the Act codifies a philosophical approach to permit review and encourages replacement mitigation. Specifically in Type A and B wetlands, mitigation is to:

1. avoid adverse impact;
2. minimize adverse impacts on wetland functions that cannot be avoided; and,
3. compensate for any loss of wetland functions that cannot be avoided or minimized.

In making a permit determination, the Secretary is to consider the following factors:

1. means to reduce impacts through project design;
2. costs of mitigation requirements and the social, recreation, and economic benefits associated with the proposed activity, including local, regional, or national needs for improved or expanded infrastructure, minerals, energy, food production, or recreation;
3. the ability to mitigate wetland loss or degradation as measured by wetland functions; and,
4. the environmental benefit, measured by wetland functions, that may occur through mitigation.

The Secretary is directed to issue rules governing mitigation requirements including:

1. minimization of impacts through project design;
2. preservation or donation of Type A or B wetlands as mitigation;
3. enhancement or restoration of degraded wetlands;
4. creation of wetlands if conditions are imposed to insure a reasonable likelihood of the mitigation being successful;
5. compensation through contribution to a mitigation bank program;
6. off-site mitigation within the same State;
7. contribution of in-kind value; and,
8. construction of coastal protection and enhancement projects.

But the Secretary may determine not to impose mitigation if:

1. adverse impacts are limited;
2. wetland functions can be maintained without mitigation;
3. no practicable and reasonable means of mitigation are available; and,

4. similar significant wetland functions and values are abundant in or near the area.

These provisions are an attempt to require the Secretary to accomplish a difficult juggling act. Simultaneously accommodating economic development and wetland preservation may be a difficult balance in many instances. Despite our desire to protect remaining wetland resources, certain types of developments, such as water-related projects, are impossible without detrimentally impacting on-site wetlands. In Type A and B wetlands these regulations would require compensatory mitigation of wetland impacts in cases where wetlands will inevitably be impacted if the proposed development is approved. Compensatory mitigation is only to be considered after avoidance and minimization of impacts have been attempted. In its simplest form, compensatory mitigation allows the regulatory agency to say yes to development with a series of requirements. In that sense, mitigation is a tool that expands the regulator's role from a simple yes or no to one of negotiated development.

Inclusion of function within the Act is important since mitigation has been largely based on replacement of wetland form, i.e., the physical components of the impacted wetland. However, replacement of the major components may or may not replace the wetland functions depending upon the specific functions, the wetland form, and spatial and temporal locations. However, given that function is the determining factor, we must then examine the viability of the basic approach to replacement. Since mitigation assumes that the form and functions of the wetland can be replaced and includes size and location, we should examine these interrelationships.

The long list of important functional values deriving to human societies from natural wetlands may be grouped into four major categories:

1. life support;
2. hydrologic buffering;
3. water quality improvements; and,
4. recreational/historical/cultural significance.

Because wetland functions are controlled by physical, chemical, and biological processes, wetland functions are strongly related to complexity, degree of alteration (pristine, unchanged), and size and location of the wetland.

Life support is largely biological though obviously dependent on physical and chemical processes. It has moderate site dependency with moderate to high size, complexity, and pristine dependency. It includes production and maintenance of flora and fauna — forbs, grasses, shrubs, trees, fungi, invertebrates, birds, mammals, fishes, herptiles, and microbial populations that are valued for commercial products and recreation.

Hydrologic buffering is largely a physical function that is extremely site dependent and highly size related. It includes flood amelioration such as flood water storage/retention, i.e., desynchronization and reductions in magnitude of downstream flows reducing flood water damages during unusual storm events. Conversely, delayed discharges of flood waters augment base flows in rivers and streams supporting diverse aquatic life in our waterways. In some instances wetlands can have an important groundwater recharge function, supplementing other mechanisms to increase total groundwater resources. Natural wetlands protecting and supported by groundwater discharge can provide important surface water sources and, of course, some wetlands have essentially flow-through groundwater patterns.

In the water quality improvement function, chemical and physical processes tend to dominate biological processes. This function has high site dependency and lower size, complexity, and pristine dependency. It includes removal of pollutants/contaminants from inflowing waters — principally surface flows — but it can also include subsurface inflows — to purify natural water supplies.

The historical/cultural preservation function is highly site specific and strongly related to natural condition but only moderately related to size and complexity of the wetland. It includes preservation of anthropological and historical resources. Recreational functions are less strongly site specific but often closely related to natural condition. In addition to common sports, recreational aspects include open space, educational, and research activities.

Physical and chemical processes are much less dependent upon complex, diverse, and perhaps pristine wetlands. A very simple or severely degraded system may have important hydrologic buffering value and/or water quality improvement values but little or no life-support value. Generally, life-support values increase with increasing complexity and proximity to natural conditions. However, a simple wetland (low diversity/complexity) can have very high productivity for certain products. A small system (perhaps 0.1 ha) may have important water quality improvement values but little or no flood amelioration or life-support value. Exceptions include very small systems that provide habitat for unusual, threatened, or endangered species. Moderate size (>2 ha) systems may have significant hydrologic and life-support values and increasing size is related to increasing importance for these values. Obviously, location in the watershed is extremely important to the hydrologic buffering function and moderately important to water quality but may be much less important to the life-support function. Location in a state, region, flyway, country, or continent may be quite important to the life-support function, however.

COMPENSATORY MITIGATION

Numerous compensatory mitigation projects have failed, and these failures are commonly cited as reasons to deny the validity of the concept. Given the broad variety of wetland types, their geographic distribution, and diverse nature of wetland functional values, generalizations are fraught with peril. This is especially true for smaller wetlands and unique types with isolated distributions. Furthermore, the interrelationships of wetland units within a geographic area and their interdependencies on associated terrestrial environments make evaluations of replacement difficult at best. However, certain types of wetlands have been restored, enhanced, and/or created for many years. We have a considerable body of knowledge on restoration, enhancement, creation and management of marshes — especially the Prairie Potholes and other midwestern marshes. Similar though less extensive information is available for freshwater marshes in the interior valley of California, the Intermountain West, and coastal marshes along the Atlantic and Gulf coasts. Some information is available for northern bogs, less for Coastal Plain bogs, and very little for high elevation bogs. Our information on forested wetlands, especially the great river swamps of the Southeast is rudimentary at best and almost nonexistent for unique systems such as pocosins, vernal pools, riparian bands, Carolina Bays, etc.

Similarly, our information base on wetland functional values varies considerably. We have the ability to accomplish and quantify certain life-support functions — notably waterfowl, wetland mammal, fish, and timber production — but only limited information on the host of other biological products deriving from wetlands. Very few investigations have explored the hydrologic buffering functions and results have been mixed. The water quality improvement function has received considerable attention within the last few years but much of the information has derived from deliberately constructed wetlands and extrapolation to natural wetlands is largely unknown. Consequently, our ability to replace functional values, with a few exceptions, is limited because of our poor understanding of these functions.

Unfortunately, though meager, existing information has often not been used in restoration, enhancement, and creation projects. Failure of many projects lies with the lack of, or improper application of, existing knowledge rather than faulty science. Too few developers employed experienced wetland ecologists in the design, construction, and operation of wetland projects and subsequent failures were predictable.

Since compensatory mitigation projects that attempted to create new wetlands have had widely varying success rates and because opportunities for wetland restoration or enhancement are finite, the Act appropriately encourages natural wetland restoration or enhancement for mitigation rather than creation of new wetlands. Emphasis on restoring former or prior-existing wetland is pragmatic in that, in many cases, restoration of damaged or degraded wetland is much more likely to succeed than attempts to create a wetland in a formerly terrestrial environment. Quite simply, the residual hydrology, edaphic and biological components in the previous wetland make it possible to restore the wetland simply by removing or modifying the factors causing degradation.

MITIGATION BANKING

The Act specifies that within six months, the Secretary is directed to issue regulations for the establishment, use, maintenance, and oversight of mitigation banks. In fact, regulations have been issued even though the Senate has not acted (November 28, 1995 Federal Register — Federal Guidelines for the Establishment, Use and Operation of Mitigation Banks). But the Act stipulates that mitigation banks must:

1. provide for the chemical, physical, and biological functions of wetlands which are lost;
2. provide in-kind replacement of lost wetland functions and be located in, or in proximity to, the same watershed;
3. have a public or private operator with adequate financial resources including a deposit of performance bond for long-term maintenance, monitoring, protection, security;
4. employ consistent and scientifically sound methods to determine debits by evaluating wetland functions, project impacts and to determine credits based on wetland functions at the site of the mitigation bank;
5. provide for transfer of mitigation credits including posting financial bonds;
6. provide for dual use that doesn't interfere with the bank's functioning for mitigation and does not impact wetlands or other waters of the United States; and,
7. provide for public notification and comment on proposals for mitigation banks.

The complex web of regulations and the inability of some developments to avoid detrimentally impacting wetlands has resulted in this concept of replacing wetland in various forms or systems. Costs for complying with regulations and uncertainties over permit delays and/or approvals encouraged the approach of establishing banks of protected, restored, and/or created wetland that could facilitate compliance with the replacement requirements. In a further attempt to expedite regulatory reviews, this legislation authorizes establishment of specific areas where wetlands are protected/restored/created and cooperating parties may receive "credits" for wetlands in the bank that would be used to offset their liability for detrimentally impacting a wetland in a new development. Developers, often caught in a confusing, seemingly interminable web of unknowns related to potentially impacting a wetland as part of their overall development proposal demanded these changes. It is understandable that developers seek a simplified solution; many simply ask that they be told what it will cost and when a permit will be issued so they can factor the delay and cost into their project planning.

In fact, the mitigation provisions in the Act provide legal basis for active, on-going implementation of the concept of wetland mitigation banks. Mitigation banks and banking programs are increasing almost exponentially and developers are leading the efforts. A progress report on the COE mitigation banking survey showed that existing banks had increased from 13 in 1988 to 20 in 1991 and 40 in 1992 with another 60+ in planning status. The survey identified 37 existing banks with 64 planned banks that were expected to become active in 1992, and an additional 5 mitigation trusts. Of the existing banks, 38% were on the west coast, 27% in the northern plains, with 16% in each of the mid Atlantic and Gulf regions. Highway construction projects were involved in 60% of the banks and port (14%) and industrial development (11%) were the next most common. States exclusively owned 50% of the banks, 20% are privately owned, and local public bodies and federal ownership account for another 20%. The largest was 7000 acres but only 15% were >640 acres, while 51% were >40 acres and only 5% were <10 acres. Over two-thirds were located in the same hydrologic unit.

Mitigation banks provide a highly attractive alternative to minuscule, on-site attempts to replace impacted wetlands. Rigid application of in-kind, on-site replacement mitigation fosters failed mitigation projects and rarely contributes significantly to the wetland resource base. Furthermore, in-kind replacement does not always seem to benefit wetlands or society. For example, in New England, regulators discourage highway developers from routing new roads through ponds and marshes and routes tend to skirt these "high-value" wetlands through adjacent red maple swamps which are abundant and increasing. But mitigation requirements often specify replacement of "low-value" red maple swamp instead of "high-value" ponds and marshes.

Existing mitigation banks tend to be large tracts or clusters comprised mainly of altered or degraded wetlands. But the lack of methods to assess wetland functional values gravely hinders our ability to truly compensate for lost wetland functions and, in many cases, mitigation has become simply acre-for-acre replacement. Conversely, imaginative entrepreneurs are becoming involved and commercial banks could follow. Developing private mitigation credit banks and markets is most encouraging and, if properly structured, could overcome many of the pitfalls of government agency administration.

Establishing large regional wetland complexes administered by the private sector with long-term, trust-type, funding guarantees could accomplish significant protection and restoration (Figure 2-5). For example, mitigation banks could purchase private lands either containing wetlands or for wetland creation and provide management and protection in perpetuity. Not only would these wetland banks have regulatory protection but the mitigation banking concept supplies the one critical factor that has been lacking in other efforts to protect wetlands — the profit motive. Perceptions of limited profitability were the basis for most wetland destruction. Perhaps reversing that perception will provide the most durable means to protect and restore our wetland resources.

Figure 2-5. Although increased sedimentation is smothering the forest inside the levees in the Atchafalaya Basin, small intact portions outside the dike system are regrowing the original forest, much of which was heavily logged in the first half of this century.

CHAPTER 3

THREE IMPORTANT COMPONENTS —
WATER, SOIL, AND VEGETATION

INTRODUCTION

The long-term success of any wetland restoration or creation project is, to a very large extent, dependent upon restoring, establishing, or developing and managing the appropriate hydrology. Wetland hydrology (depth, period, and duration) determines abiotic factors such as water availability, nutrient availability, aerobic or anaerobic soil conditions, soil particle size and composition, and related conditions including water depth, water chemistry (pH, Eh), and water velocity. In turn, biotic components, especially plants, influence water gains through interception of precipitation, water losses through evapo-transpiration, as well as depth, velocity, and circulation patterns within the system. Plants influence water movement and even depth because vegetative resistance, "roughness", can create a slope to the water elevation. Water may mound in upstream areas to provide the necessary head to drive water through dense stands of downstream vegetation. In rare circumstances, notably beaver and alligators, wetland animals may have significant effects on system hydrology and even a large number of muskrat houses and feeding platforms in a prairie marsh or "chimneys" built by a large population of crayfish may have substantial though usually temporary effects.

Since wetlands are transitional areas between terrestrial environments and deep water aquatic systems, they are "open" systems strongly influenced by external, forcing functions such as precipitation, solar radiation, energy and nutrient inputs, and surface and groundwater flows. Wetlands are not only spatially intermediate, but they are also intermediate in terms of amounts and chemistry of water and, consequently, they are extremely sensitive to effects of the hydrologic forcing function (Figure 3-1).

HYDROLOGY

Hydrology modifies or determines the structure and functioning of wetlands by:

1. Controlling the composition of the plant community and thereby the animal community. Only a few of the many thousands of species of plants are able to grow in saturated or flooded soils. Of these, adaptations to inundation vary considerably, with fewer and fewer species able to

Figure 3-1. An aerial view perpendicular to the direction of water flow reveals the streamlined lanceolate configuration of "hummock" islands shaped by hydrologic forces in the Everglades of Florida.

survive under longer and longer periods or deeper and deeper flooding regimes. Consequently, sites with short-term and/or shallow flooding will support many different types of plants (much higher species diversity) and consequently more species of animals (Figure 3-2). The corollary, of course, is that areas with deep, prolonged flooding will have fewer kinds of plants and animals. However, this concept does not extend to productivity. Basic productivity may be as high or higher in the latter, even though they have much lower diversity; that is, the amount of biomass produced or supported in a simple system can equal or exceed that produced in a more complex system.

2. Directly influencing productivity through controlling nutrient cycling and availability, import and export of nutrients, and fixed energy supplies in the form of organic particulates and decomposition rates. Under prolonged inundation, many important nutrients are immobilized under reducing conditions in the substrate and unavailable to plants as well as separated from the water column. Periodic drying and oxidation returns these substances to active portions of the cycles within the water column and near the surface of the substrate, resulting in an explosive growth response by plants and animals. Changes in oxygen availability and concentration caused by inundation also strongly influence decomposition rates because anaerobic rates are generally only 10% of aerobic decomposition rates. Low decomposition rates in anaerobic environments

Figure 3-2. Scattered "hummock" islands add diversity and provide critical refuge for terrestrial wildlife during high water periods in the Everglades.

is the principal reason why many wetlands accumulate substantial quantities of partially decomposed organic material. Of course, in northern climates, low temperatures contribute to reduced decomposition rates.

Inflowing surface runoff contains variable but often substantial quantities of minerals, macro- and micro-nutrients, and organic material that contribute greatly to high productivities of many wetlands. Conversely, surface outflows may export significant amounts of organic material, minerals, and nutrients, reducing their contribution to wetlands productivity but enhancing productivity in downstream rivers or lakes. Similarly, groundwaters may transport minerals and some nutrients into and out of the wetland system. Hydrology in the form of circulation patterns also controls distribution of essential growth substances within the system, often enhancing spatial heterogeneity because of differential transport of nutrients into and by-products out of portions of the system. The degree of circulation also strongly influences basic productivity with stagnant water wetlands showing much lower basic productivity than flowing water or wave-influenced systems. In addition, sedimentation and erosion, depending on circulation patterns within the system, add to physical heterogeneity and consequently species richness or diversity at any one time as well as changes over time.

In summary, water moving within the system functions analogous to the bloodstream where nutrients, energy, and byproducts are physically transported throughout the system. Additional movements due to concentration gradients within free waters and in the waters near the surface of the substrate are much smaller scale. However, most wetlands are open systems much more strongly impacted by external forcing functions than the relatively closed system in the bloodstream. In wetlands with shallow impermeable substrates (clays or well-decomposed sapric peats), low hydraulic conductivities reduce the influence of exchange between the substrate and the near surface regions in the water column. In fibric or poorly decomposed peat (duff and litter layers and other highly permeable substrates), the region of active cycling may extend well below the substrate surface although exchange rates obviously decrease with depth.

The end result of all factors influencing the amount of water within the system is the water budget. The overall budget merely represents the balance between all inputs and all outputs of water, but note that this definition does not include a time-dependent variable. Generally, the water budget is determined over a 1-year period, but in some cases it may be useful to estimate the water budget for shorter periods if adequate inputs are questionable due to extremes of temperature, wind, or precipitation.

INFLOWS

Surface and subsurface inflows as well as direct inputs result from precipitation — generally the total water equivalent of all rain, ice, and snowfall in the region; in rare instances, direct condensation on surface objects may constitute a significant portion of the total. On bare soils or lake surfaces, all of the rainfall reaches the surface. In areas with low but dense vegetation — open bogs and sedge marshes — a small percentage of the total is intercepted before reaching the surface. However, in forested swamps, the intercepted proportion may reach 30 to 40%, substantially reducing the amount of precipitation reaching the surface. Since interception is directly influenced by structure and coverage of vegetation (i.e., a triple canopy forest intercepts much more than a single canopy, low-growing marsh), as well as intensity and duration of rain events and relative humidity, interception is proportionately less during a heavy downpour than during a short-term drizzle and also lower in humid climates than in arid regions.

In addition, two abiotic factors influence actual precipitation entering wetlands, either directly or from surface and subsurface inputs. In arid regions, it is not unusual to see virga — a column of rain hanging below a cloud but failing to reach the surface. Because of the very low relative humidity outside of the cloud, rainfall evaporates before reaching the ground and, hence, is not included in precipitation records. However, the water equivalent of snowfall is included in climatic data even though in dry, windy regions a considerable proportion of total moisture may be lost to the atmosphere through sublimation. This is an evaporative loss resulting from moisture changing directly from a solid to a gas form under subfreezing temperatures. Absence of the liquid state and generally frozen soils or ice-covered wetlands or lakes precludes any additions to soil moisture or to surface runoff that might become inputs to the wetlands. Under appropriate conditions, many centimeters of snow

Figure 3-3. Even though this thunderstorm over the Pantanal of Brazil may create heavy localized downpours, supporting water supplies originate from precipitation over wide areas running off surrounding uplands.

depth may simply disappear long before spring temperatures cause runoff from melting snow.

Few wetlands are supported by direct precipitation alone; most depend on water inputs from surface and/or subsurface flows (Figure 3-3). During and immediately after a storm, nonchannelized sheet flow may bring surface water overland from surrounding higher terrain. In between storms, surface flow through established channels, streamflow, often supplies the majority of the wetland's requirements for part or all of the year. Depending on the location of the wetland (i.e., bordering or adjacent to a stream or river, as in most bottomland hardwood swamps or enclosing and encompassing the stream, as in many marshes and bogs), the amount of inflow is influenced by the volume of streamflow. Obviously, in the latter case, water in the stream flows into the wetland. However, in floodplain and riparian wetlands, the amount of water entering the wetland is influenced by the duration and intensity of the rainfall, stream capacity, bank elevation, and presence or absence of oxbows or other temporary channels. Light to moderate rain may increase streamflow, but little water enters riparian wetlands until the volume exceeds the capacity of the stream and flood waters overtop river banks or channel cutoffs.

Surface runoff is often the most important source of water for natural, as well as created or restored, wetlands. Surface waters also transport quantities of fixed energy and nutrients that enhance, in some cases substantially, the productivity of wetlands. Surface flows can be measured in streams or predicted from watershed and climatic data. Although estimation often requires considerable information on the source watershed, methods have been developed and are presented in Chapter 10 to predict runoff amounts under different climatic, soil, and watershed conditions.

In contrast to surface flows, our understanding of and ability to estimate subsurface flows is poor; yet some natural wetlands are dependent on groundwater supplies. In many cases, wetland occurrence is evidence of emerging groundwaters; in others, groundwaters may flow through or transit the wetland, and still others may receive but not discharge subsurface flows and some wetlands with porous substrates add water to underground reserves. While groundwater may be critical to an individual wetland, subsurface flows generally have small amounts of minerals with little or none of the fixed energy and nutrients brought in by surface flows.

In arid regions, attempts to construct wetlands that intercept groundwaters have not been very successful due to limited understanding of locations and hydraulic gradients of underground waters. Planners attempting to use groundwater sources will need to establish a sizable network of wells and monitor elevations and flow patterns through at least one abnormally wet and dry period. Mined areas or other deep excavations with groundwater exposed are an obvious exception. Conversely, sites of emerging groundwater — seeps, springs, or artesian wells — provide excellent opportunities since only a brief review of historical flows is necessary to determine source reliability and adequacy for planned wetland sizes and configurations.

OUTFLOWS

Evapo-transpiration is the combination of water that vaporizes directly from soil or water surfaces (evaporation) and the moisture that is transported through plants to vaporize into the atmosphere (transpiration). Since few surfaces (i.e., deep lakes, bare soil, or rock) lack vegetative cover, evaporation rarely adequately estimates total losses, although the standard against which other losses are compared is based on evaporative loss from a shallow water surface known as pan evaporation. More importantly, Class A pan evaporation values are used to derive the P/E ratio for any specific region. This is the ratio of total precipitation and total evaporation, generally over a period of 1 year though in some circumstances values are calculated for monthly or shorter intervals. It is important to note that the P/E ratio compensates for sublimation losses (in the evaporative component), but not for transpiration losses since both components are developed from data collected by standard rain gauges and/or standard evaporation pans. Evaporation may also be computed from prevailing radiation, temperature, wind, and relative humidity conditions, that is, the factors that influence vapor pressures at exposed surfaces and in the surrounding air.

Evapo-transpiration increases with increases in exposed surface area, solar radiation, air and surface temperatures, and wind speed and decreases with relative humidity or, basically, the same factors that influence evaporation.

However, plants have some control over transpiration and in moisture-limiting circumstances, plants can close leaf stomata, thereby reducing exposed surface area and transpiration losses. Though rarely a factor in wetland systems, reduced soil moisture, high radiation, and other factors affecting plant physiology often activate plant water conservation mechanisms in terrestrial environments, causing substantial reductions in transpiration losses.

Estimates of evapo-transpiration from wetlands and from non-vegetated water bodies have been developed for marshes, bogs, and swamps in North Dakota, Michigan, Minnesota, New England, Florida, Utah, Nevada, and Germany. Not surprisingly, because of the difficulty of deriving accurate measurements and the variety of climates and vegetation types, the results vary. In one instance, during the growing season, a vegetated stand was believed to have lost 80% more water than nearby open water. However, most studies have shown that evapo-transpiration rates from wetlands range from 30 to 90% of losses from unvegetated or open water areas. In general, it appears that annual evapo-transpiration rates in wetlands average 80% of comparable losses from open water surfaces; that is, they are approximately 80% of Class A pan evaporation rates for that region. The North Dakota study (P/E ratio <1) in a dry but cool region estimated losses at 90% of pan evaporation, but studies in Nevada and Utah support the 80% figure.

For planning purposes, evapo-transpiration rates from wetlands can be assumed to be 80% of Class A pan evaporation from a nearby open site; hence, wetland evapo-transpiration and lake evaporation are roughly equal since Class A pan evaporation is 1.4 times lake evaporation. On an annual basis, approximately half the net incoming solar radiation is converted to water loss. Therefore, half the net solar radiation roughly equals the annual evaporation. Not surprisingly, seasonal patterns of evapo-transpiration resemble seasonal patterns of incoming radiation because wetland reflectance changes, transpiration increases and decreases, and the mulching function of litter layer increases and decreases. Combined effects result in increases during the growing season and decreases the remainder of the year. Class A pan data integrate effects of many meteorological variables and are tabulated monthly and annually in the NOAA publication Climatological Data.

Reduced rates may seem contradictory because of the large amount of surface area and the "pumping" effect of plant transpiration. For example, consider the total surface area exposed to the atmosphere by a dense stand of 2- to 3-m high cattail in a square meter compared to the water surface area in a square meter of open water. Intuitively, one would think that the cattail stand would have much higher losses, and that may be the case for limited periods during the growing season. However, even during the growing season, plant structure substantially reduces evaporation losses from exposed water surfaces by shading the surface, by occupying a substantial portion of surface space, and by obstructing air movement near the water's surface such that relative humidity is near saturation for some distance above the water's surface and the saturated air is not exchanged with drier air. In turn, limited air movement along the length of plant stems and leaves maintains high humidities near plant surfaces, thereby reducing transpiration losses compared to measurements obtained from a single exposed leaf. In addition, the litter layer can cause a mulching effect. These physical factors continue to influence evaporative losses long after the plant tops have ceased active growth and even after fall die-back. Similar effects have been noted in bogs and swamps. Consequently, it is not surprising that annual rates for evapo-transpiration from a wetland are less than regional pan evaporation rates.

WATER BALANCE

Just as surface inflows are typically the most important source for wetlands, surface outflows are likely to be the major loss component. In fact, many natural wetlands in upper portions of watersheds comprise the headwaters or source for natural streams and rivers. The significance of this is apparent from an overflight of major portions of New England, the Lake States, and eastern Canada that quickly generates the impression that almost all of the waterways originate from a bog, swamp, or lake. Farther west, beaver ponds often form the headwaters of many streams throughout the Rockies and the contributions of Okefenokee Swamp to the Sewanee River, Great Swamp and the Pamlico River in North Carolina, and numerous other examples in Louisiana, Mississippi, and Florida attest to the important role of wetlands in moderating flows in many of our streams and rivers. Without the wetland buffering effect that impedes storm flows and augments base flows, many waterways would have dramatic floodwaters after storms, followed by dry streambeds in dry intervals.

Subsurface losses are generally much less significant because most wetlands have poorly permeable or impermeable substrates; otherwise, the wetland would not be present. However, a few natural wetlands may intersect groundwaters such that subsurface waters flow horizontally through the wetland or, in a few instances, surface inflows may equal or exceed subsurface losses at least during a significant portion of the year. Though not well documented, the latter type may have an important role in recharging groundwater supplies.

Combining each of the above factors in a single term develops the water budget for the wetlands:

Imports: 1. direct precipitation
           2. surface inflows
           3. subsurface inflows

Exports: 1. surface outflows
           2. subsurface outflows
           3. evapo-transpiration

Obviously, inputs must equal or exceed exports, at least on an annual basis and, importantly, during the growing season or the site will not support a wetland system. However, if inputs exceed exports creating saturated or inundated soils that inhibit terrestrial plants for a significant portion of the growing season, the site will probably support a wetland community even though the annual balance is negative.

The hydroperiod — time of year, spatial distribution, and depth of flooding — of different types of wetlands varies substantially. Some wetland communities are adapted to almost permanent flooding, others require extended seasonal flooding while others will not tolerate more than a few days of inundation during the growing season but may endure extended flooding during the remainder of the year. Consequently, developers may need to modify their objectives, select a different type of wetland, or chose a site with a larger or multiple sources of available water. The bottom line is simply that inputs must equal or exceed outputs during the driest portion of the year if the system must be inundated during that period. Granted, managers can control surface outflows with water control structures but sub-surface and evapo-transpiration loses will continue. If the wetland type will not tolerate drying for an extended period, water inputs must equal or exceed exports during the drought period or a different wetland type or even a terrestrial community will result.

Determining values for inputs and exports and the storage volume in the wetland is useful because changes in water depths or elevations can then be estimated from:

$$\Delta V = V + I - E$$

and

$$\Delta L = \frac{L + \Delta V}{A \times D}$$

where: V = volume of storage
  I = inputs
  E = exports
  L = water level or elevation
  A = area of the wetlands
  D = depth

For some wetland projects, turnover rate or its inverse (residence time) may be important characteristics. Turnover rate (T) is simply the ratio of system volume to flow through; that is:

$$T = \frac{I}{V}$$

where I is expressed as a quantity over a time period (i.e., cubic meters per day) and T becomes a similar time delimited value. Conversely, retention or residence time (R) becomes:

$$R = \frac{1}{T}$$

or

$$R = \frac{1}{\dfrac{I}{V}}$$

Suffice it to say, hydrology is the overriding factor in presence or absence of wetland. Hydrology governs the abiotic factors which, in turn, control or influence the biotic factors that coalesce to create the form and function of complex natural systems we call wetlands. In fact, there is merit to the common belief that if you get the hydrology right, all else will follow in due time.

SOILS

Wetland soils provide support for wetland plants, are the medium for many chemical transformations, and are the principle reservoir for minerals and nutrients needed by plants as well as a variety of other substances. The principle difference with upland soils is an abundance of water that typically fills soil pores or void spaces, and the most important effect of water replacing air in soil voids is the isolation of the soil system from atmospheric oxygen. As a consequence, only a very thin (1 to 5 mm) boundary layer at the soil surface has adequate oxygen to maintain aerobic/oxidizing conditions and almost everything below is anaerobic/reducing. The exception is the rhizosphere, the thin film region around each root hair that is aerobic due to oxygen leakage from the rhizomes, roots, and rootlets. Shortly after a soil is flooded, the oxygen present is consumed by microbial organisms and chemical oxidation. Diffusion of oxygen through water is many orders of magnitude slower than diffusion through well-drained soils, and lower layers quickly become and remain anaerobic. The unique qualities of saturated soils result from the many interrelated physical and chemical changes that occur because of limited oxygen (anaerobic conditions) rather than from direct effects of excess water.

Wetland soils are generally considered hydric soils in the NRCS soil classification system because they are saturated for a long enough period in the growing season to develop anaerobic conditions that favor hydrophytic vegetation. Hydric soils are further divided into

1. mineral soils having less than 12 to 20% organic matter; and,
2. organic soils with greater than 12 to 20% organic matter. The percentage range is due to interrelated saturation and clay content factors.

In well-developed wetlands, the upper layers are often organic soils or histosols, while lower layers may consist of mineral soils though the boundary is often indistinct. Tropical wetlands with high rates of decomposition may have very thin or no organic soil layers and the substrate is almost solely mineral soil. Conversely, peat layers may be many meters thick in temperate and especially cold-climate wetlands, and the underlying mineral soil is largely isolated from the wetland systems because of the low hydraulic conductivities of well-decomposed or sapric peat.

Organic soils have a high percentage of pore spaces (>80%) and, consequently, higher water holding capacities than mineral soils (50%) and are described as having lower bulk densities, that is, the dry weight of a given volume of material is less. Organic soils generally have lower hydraulic conductivity than mineral soils (except clay) so that even though organic soils may contain large amounts of water, water movement through organic soils is inhibited. Although included within the organic soil type, boundaries between the surface litter/duff layer, fibric peat layers, and sapric peat layers are indistinct and flow can be rapid through the uppermost layers down to and including much of the fibric peat layer. Consequently, a peat soil may exhibit considerable lateral water flow through the upper fibric zone, but limited movement vertically into or horizontally within the lower sapric zone.

Organic soils have a greater cation exchange capacity (CEC) and the major cations are different than in mineral soils. CEC measures the soil's capacity to fix cations on exchange sites and commonly ranges from −300 to +500 in different soils. In addition, metal cations (Ca^{2+}, Mg^{2+}, Na^+) dominate in mineral soils, while H^+ dominates in organic soils.

Saturation and loss of oxygen generally causes wetland soils to have negative redox potentials. Redox potential (Eh) measures the soil or water's capacity to oxidize or reduce chemical substances and it often ranges from −300 to +300 millivolts (mV) in wetland soils. Oxidation is the loss of electrons and reduction is the gain of electrons. pH represents the degree of acidity or alkalinity in terms of the hydrogen ion concentration; pH of wetland soils varies from strongly acidic (3) to strongly alkaline (11), although most wetland soils are circumneutral.

Eh and pH conditions and the interactions of pH and Eh influence CEC, as well as many chemical and physical reactions in the soil. Typical wetland soils may have pH of 7 S.U. and Eh of −200 mV in which case common substances typically occur in reduced forms, that is, nitrogen as N_2O, N_2, or NH^{4+}, iron as Fe^{2+}, manganese as Mn^{2+}, carbon as CH_4, and sulfur as S^-. Phosphorus is not directly affected by pH or Eh but it is indirectly affected because of its association with metals that are affected and by changes in clay particle adsorption and CEC phenomena. Form changes in turn affect solubility and availability for plant uptake or reaction with other substances and the various transformations that occur in wetlands modify organic and inorganic substances, releasing some while trapping others.

Figure 3-4. Periodic drying exposes and oxidizes nutrients and other substances in bottom substrates, releasing these materials into the water column following subsequent flooding.

After flooding, soil oxygen is quickly consumed by microbial respiration and chemical oxidation. Subsequently, anaerobic microorganisms, able to use substances other than oxygen as the terminal electron acceptor during respiration, soon dominate the microbial community. Importantly, anaerobic decomposition rates are only 10% of aerobic decomposition rates and frequently much lower than carbon fixation or biomass production rates (Figure 3-4).

In summary, the loss of soil oxygen creates difficult environmental conditions for living organisms and unusual chemical conditions which in turn result in the unique attributes of wetland soils that contribute to one of their functional values. The wide range of redox potentials for periodically flooded soils vs. aerobic soils is important. Wetlands are often the major reducing ecosystem in the landscape and their most important function may be as chemical transformers of nutrients and other materials. These changes may transform organic inputs to inorganic outputs, inorganic inputs to organic outputs, or any combination of the foregoing. The complex of reactions may also cause retention within the wetland such that the system becomes a "sink" for a variety of substances.

VEGETATION

Many terms have been applied to plants growing in semi-wet to wet environments. Some labels differentiate the simpler forms, primarily algae, from higher, multi-cellular or vascular plants — those with physical structures to transport liquid and gaseous materials. Commonly used terms include phytoplankton, vascular aquatic plant, nonvascular aquatic plant, hydrophyte, aquatic macrophyte, vascular hydrophyte, and aquatic plant. Macro simply means larger than microscopic, whereas plankton implies small and current borne (i.e., floating or suspended in the water column without rooted attachment to the substrate). Nonvascular refers to simple plants many of which are small, even individual cells, but some such as *Chara* and *Nitella* are relatively large and possess holdfasts for attachment to the substrate but lack vascularization or internal transport structures.

A number of authors have attempted to narrowly define aquatic plant, hydrophyte, etc. to differentiate terrestrial, semi-terrestrial, shallow water, or deep water species, but little agreement has been reached. Therefore, it seems simpler to include all of these categories in a group termed "wetland plants" defined as plants capable of growing in an environment that is periodically but continuously inundated for more than 10 days during the growing season. Obviously, this would include a few species that are primarily upland types but capable of surviving 10 days of flooding or saturated soils. It would also include those occurring in the many intermediate flooding conditions, from infrequently flooded to shallowly flooded on to the furthest extent of deep water, rooted vegetation. At the extremes, rooted vascular plants may exist in water depths of 7 to 8 m in very clear waters and a few mosses and rooted algal forms have been found as deep as 27 m. However, the vast majority of wetland plants are limited to water depths of less than 1 m. At the drier, upper extreme, this definition would include species only flooded, and not necessarily every year, for 10 days during the growing season downward to those species present in areas with permanent water depths of 2 m.

Most attention has focused on herbaceous plants, those with soft, flexible stems compared to woody plants having rigid, persistent stems such as shrubs and trees. Herbaceous wetland plants are divided into free floating and rooted forms and the rooted group is then subdivided into submergent, emergent, and floating-leaved types. Although infrequently included, most woody species would be considered rooted emergents.

The free-floating category includes such types as duckweeds (Lemna) (Figure 3-5), water meal (*Wolffia*), and water ferns (*Salvinia*). Some free-floating plants (water hyacinth — *Eichhornia*) have large root systems (up to 50% of their biomass), but many salvinids and some lemnids have lost their roots and nutrients are absorbed through modified leaves. All vegetative parts of duckweeds (*Lemna*) are reduced so that they appear to be a leaf floating on the water surface, while others (*Salvinia*) have stems with sessile leaves. The productivity of floating-leaved and free-floating wetland plants is equal to or exceeds that of emergents probably because of relatively constant and favorable environmental conditions, relatively less support and respiratory tissue. They also have a considerable percentage of enclosed gas space that may enable them to trap and use CO_2 from respiration that would otherwise be lost to the atmosphere. Free-floating species have roots with numerous root hairs or modified leaves and can successfully obtain nutrients from the water column.

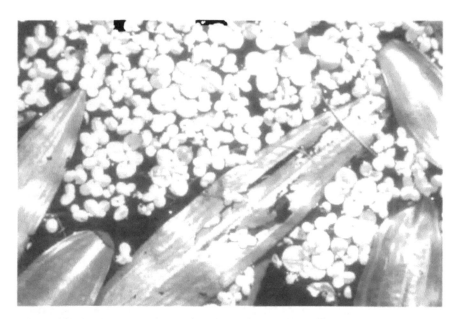

Figure 3-5. Duckweed and giant duckweed, though only a few millimeters across, form dense mats that contribute substantially to biomass productivity and host large populations of invertebrates.

Emergents are the plants most characteristic of marshes — cattail (*Typha*), bulrush (*Scirpus*), rush (*Juncus*), sedges (*Carex*) but also including bog mosses such as *Sphagnum*. Typically, emergents occur in shallow waters — 5 to 30 cm. Most emergent plants have long, erect linear leaves that reduce shading while exposing a large amount of leaf area for photosynthesis and also reduce air movement near the leaves, limiting moisture loss due to transpiration. In many plants, rates of photosynthesis increase up to approximately 50% of radiation maxima above which photosynthetic rates decline eventually to 0 at high levels. Many emergents have very high light saturation levels which probably contributes to their high biomass productivities. In addition, many wetlands plants are able to use the much more efficient C_4 pathway for carbon fixation during photosynthesis instead of, or in addition to, the normal C_3 pathway. Plants using the C_4 process can withdraw and use CO_2 at concentrations as low as 20 mg/L compared to lower limits of 30 mg/L for the C_3 pathway. Emergent plants obtain their nutrients from the substrate (Figure 3-6).

Submerged plants depend on water pressure/buoyancy for support and their stems and leaves are thin and pliable, with aerenchymal tissue and gas-filled voids providing buoyancy. Submergents typically occur in depths of 0.5 to 1.0 m. Their leaves are either long and thin, deeply dissected along the margins, or the leaf blades are separated into leaflets. These are adaptations to maximize leaf surface area to volume ratios for survival in an environment with reduced intensities of light available for photosynthesis; as little as 2 to 5% of full

Figure 3-6. Emergent stands of bulrush and cattail occupy the shallow areas, with dense growths of a submergent (Sago pondweed) barely exposed in the deeper regions.

sunlight in deep or murky waters. Production of submerged plants is generally low because of low light intensities under water and the low diffusion of CO_2 in water, although some submerged plants can use CO_2 from HCO_3^-, respired CO_2, and CO_2 in the sediments. Submerged plants use nutrients from both the water column and substrate.

A number of submergents, the floating leaved group, have a few thin, linear underwater leaves but depend primarily on broadened or rounded floating leaves. A few species have normal underwater and floating leaves but grow above the surface with leaves typical of terrestrial plants. To further complicate identification, leaf form within a single species may vary substantially depending on whether it is growing in occasionally flooded soils, shallow waters, or in deep water. In general, floating leaved species must survive with their leaves exposed to the air on the upper side and to water on the lower side, a rather extreme set of conditions. The upper surface must contend with high radiation and fluctuating temperature effects, while the lower surface has limitations on gas diffusion. Consequently, most species have rounded leaves with entire margins and a tough, leathery texture to reduce physical damage and a hydrophobic upper surface that reduces wetting. Understandably, and unlike most terrestrial plants, stomata are located on the upper surface for gas exchange with the atmosphere. Long, flexible stems and leaf petioles accommodate water level fluctuations and reduce damage from wave action while exposing maximal leaf area on the water surface.

Woody species vary from low-growing shrubs to towering cypress, spruce, and cedar. Upper portions of stems and leaves are generally similar to terrestrial forms and they may be deciduous or evergreen. Differences lie in the lower portion of the stem or trunk and in root structures. Many woody wetland plants possess specialized structures — knees, adventitious roots, prop roots, lenticels, and butt swellings — to increase gas exchange between the roots and the atmosphere (Figure 3-7).

ANAEROBIC CONDITIONS

Wetland plants often grow in substrates with inadequate concentrations of oxygen for root respiration. Most have some ability for short-term anaerobic respiration, but they grow best when oxygen is available for respiration. To overcome the oxygen limitation, wetland plants have an extensive internal lacunae system that may occupy up to 60% of the total plant volume, whereas it may be only 2 to 7% of the volume in terrestrial plants. The lacunae or aerenchyma are air spaces that allow diffusion of atmospheric gases from aerial portions of the plant into the roots (Figure 3-8). The reverse also occurs and gases formed primarily by decomposition in the substrate diffuse into the roots and subsequently into the atmosphere.

Consequently, wetland plants can satisfy the oxygen requirements of their roots by transporting oxygen from the atmosphere through the honeycomb-like lacunae down into the roots. Gas movement is believed to be primarily due to pressure gradients generated by different concentrations, but may also be influenced by temperature, relative humidity, and wind velocity differentials. In some (*Nuphar*, *Nymphaea*, *Lotus*, *Menyanthes*, or *Typha*), air flows into young green leaves during cool night hours and is then forced downward by solar heating of those leaves during the day and returned to the atmosphere via older yellow/tan leaves.

Adaptations to flooding may also include metabolic changes. In some species, anaerobic metabolism increases to support root metabolism at the onset of flooding. Later, new root systems with highly porous structure are produced to transport oxygen for aerobic metabolism (Figure 3-9). In other species, metabolism is shifted to pathways that end in nontoxic compounds (malate instead of acetaldehyde or ethanol) under flooding conditions. Since anaerobic metabolism is much less efficient than aerobic metabolism, few species depend on it for extended periods.

Oxygen transported to the roots of wetland plants can leak out of roots and oxidize the surrounding substrate. This leakage is known as radial oxygen loss (ROL). ROL creates an oxidized zone, the rhizosphere, around the rootlets, roots, and rhizomes that supports aerobic microbial populations. Aerobic microbial metabolism detoxifies potentially hazardous substances and modifies nutrients and trace organics. The juxtaposition of a thin-film aerobic region surrounded by largely anaerobic substrates is important in nitrogen, carbon, hydrogen, sulfur, and metal cycling.

Figure 3-7. Swollen bases (butt swellings) improve gas transfer between roots and the atmosphere for these cypress in a Louisiana bayou.

Figure 3-8. A cross sectional view of bulrush stems reveals the tube-like aerenchyma structures that transport gases between soil regions and the atmosphere in wetland plants.

GROWTH AND SURVIVAL LIMITATIONS

Limiting factors for wetland plants are similar to those of terrestrial species, with the obvious exception of saturated soils and anoxic root environments. Basically, wetland plants require adequate nutrients, CO_2, and sunlight to carry out photosynthesis and protection from toxic substances.

In addition to water's effects on oxygen concentrations in the substrate, water depth and clarity strongly influence light availability for submergents. Flow rate impacts oxygen and nutrient availability, substrate texture and composition, and mechanical pressure on plant structures. A number of species exhibit different leaf form if the leaves are above or below the surface and leaf form variation with increasing depth or different flow rates, as well as greater root biomass with increasing velocity is typical of others. Conversely, increased flows reduce concentrations of toxic substances and import nutrients and oxygen supporting higher production rates in slowly moving waters as compared to stagnant waters.

Most wetland plants are to some extent nutrient limited, as are terrestrial plants. In general, optimal growth is limited by concentrations and availability of nitrogen and phosphate in many wetland systems. High concentrations of various salts restrict plant establishment and growth in other systems. In addition, loose textured, highly organic soils can influence rooting depth, may provide inadequate physical support for plants, and may develop low redox potentials shifting metals to soluble toxic forms (i.e., iron and manganese).

Figure 3-9.   In areas with periodic flooding, cypress "knees" increase oxygen supplies for roots growing in anaerobic environments.

Air and water temperatures affect biochemical reactions and inhibit growth if thermal tolerances are exceeded. Many species, intolerant of low or freezing temperatures, are restricted to tropical and subtropical regions. Generally, higher temperatures promote increased production until thermal maxima are reached, after which production ceases. However, optimal temperatures for wetland plants may be as high as 28 to 32°C.

Competition with other species may be an important limiting factor for many species. For example, *Typha latifolia, Phragmites, Lythrum salicaria, Eichhornia, Myriophyllum spicatum, M. brasiliense, Pistia stratioties, Elodea densa, Hydrilla verticillata,* and *Melaleuca quinquenervia* aggressively spread, forcing out other species and forming monotypic stands. High algal populations and dense mats of *Lemna* or other floating species can interrupt critical sunlight for submergent species or create oxygen deficiencies in the waters below. In general, floating leaved plants do not oxygenate the water as well as submergent macrophytes.

Nearly 5000 species of plants may occur in U.S. wetlands, but only a small proportion of these comprise the dominant community in the different wetland types. However, these few species not only influence hydrology and soils, they are the factory that assembles organic, living materials from nonliving substances and provides the basis for all other forms of life. Establishing and maintaining an appropriate plant community (form) is essential to generating virtually all wetland benefits (functions).

CHAPTER 4

NATURE'S METHODS FOR CREATING
AND MAINTAINING WETLANDS

INTRODUCTION

Since our objective is to create or restore wetlands, understanding natural factors that formed existing wetlands is important in evaluating the feasibility of wetland construction in certain regions and comprehending how to design, construct, and manage a wetland system. Doubtless we have the ability to construct some type of wetland almost anywhere, but the continued existence of that system may be tenuous at best; it might require inordinate amounts of time and effort to maintain or it may be impossible to establish the desired type of system. Understanding the natural factors that create and maintain wetland ecosystems enhances our ability to select appropriate wetland types and to duplicate important natural processes.

LARGE SCALE: GEOLOGIC

TECTONIC

Mountain-building processes originating from plate movement and collisions result in large-scale, long-term disturbances to established drainage patterns. Where uplifting outpaces stream migration, ridges frequently interrupt rivers and streams creating basins with limited outlets that flooded become initially clear mountain lakes with wetlands fringing the margins. As material eroded from adjacent highlands is deposited in the basins, the lakes gradually become shallower and eventually water depths are suitable for various forms of wetland plants in ever-increasing portions of the basin (Figure 4-1).

Differential movement of large blocks along fault lines often creates long, linear basins or valleys with limited outlets, poor drainage, and abundant wetlands. As one block moves downward and adjacent blocks move relatively upward, former drainage patterns are cut off and the valley on the surface of the downward moving block may support extensive wetland environments. If fault blocks obstruct both ends or the downslope end of the valley, a very deep lake (such as Lake Baikal) may form and persist for thousands of years or, infrequently, the basin may support extensive wetlands such as the 11-million-ha wetland complex of the Pantanal in Mato Grosso do Sul, Brazil. More commonly, one or both ends are only partially blocked and drainage patterns are quickly re-established, although the substrate materials may be resistant

Figure 4-1. Upthrusting mountain ridges interrupt drainage patterns, forming lakes that gradually fill and are transformed into wetlands.

to erosion resulting in meandering streams and associated wetlands, as in the Sequatchie Valley of Tennessee.

If local or regional precipitation/evaporation ratios are high and inflows relatively constant, accumulation of eroded sediments and organic materials establishes suitable conditions for bog vegetation. Accumulation of organic material accelerates in the acidic, low decomposition rate environments of bogs and eventually biological succession results in reforestation with successively drier conditions until the basin is little different from surrounding terrestrial environments.

However, if the local or regional precipitation/evaporation ratios are low, as in lower elevations of the Rocky Mountains and the Intermountain Region, water levels in the basin decline over time but are subject to extreme annual and cyclic fluctuations. In addition, accumulated salt deposits from evaporative water losses severely restrict the types of wetland plants able to survive in brackish or saline and fluctuating waters. During wet years, runoff may fill and maintain adequate water levels to sustain wetland organisms throughout the growing season. Short-term runoff in drought cycles often supports a brief flurry of biological activity, followed by long periods of dry conditions. Though not resulting from tectonic forces, the playas of the western plains (Oklahoma and Texas) have similar hydrologic cycles and biological characteristics.

Gradually over geological time, normal drainage patterns are re-established and streams and rivers erode deeper into the valley or basin floors resulting in narrow V-shaped valleys with fast-flowing rivers and little wetland habitat. Mountain lakes and wetlands are ephemeral features in the landscape, and only young or growing mountains have poorly developed drainage patterns and extensive lakes and wetlands. Compare, for example, the low number of natural lakes and wetlands in the Appalachians and Ozarks relative to the many natural lakes and wetlands in the Cascades, Sierras, and Rockies despite the fact that most of the Rocky Mountain region receives substantially less precipitation and has higher evaporation rates than the much older Appalachians or Ozarks.

VOLCANOS: LAND SLIDES, LAVA FLOWS, AND CRATER LAKES

Earthquakes are often associated with tectonic activities effecting smaller scale disturbances such as landslides that relocate massive amounts of earth and rock from higher regions to block or obstruct valleys and drainage patterns, as for example Earthquake Lake in Wyoming. But earthquakes and resultant landslides blocking drainage patterns and creating lakes and wetlands are also associated with volcanic activity, which may or may not have obvious ties to recent tectonic movements. Drainage obstruction from ash, pumice, and slide materials blocked existing drainages and substantially enlarged Spirit Lake during the recent eruption of Mt. St. Helens in Washington. Less obviously, lava flows substantially reshape the topography and obstruct existing drainage patterns, but infrequently create wetlands since many lava flows have subterranean tunnels and older flows become extensively fragmented and the combination supports efficient drainage. Rarely are the basins (caldera) of quiescent volcanos flooded, creating deep lakes such as in Crater Lake, Oregon that have little wetland habitats. An exception is some small caldera on certain Caribbean islands (Figure 4-2).

Figure 4-2. A shallow caldera on the Caribbean Island of Grenada supports rooted and floating stands of cutgrass (*Cladium*) along with a few other wetland plant species.

GLACIATION: VALLEYS AND CONTINENTAL

Perhaps the most significant natural force in wetland creation in the northern hemisphere has been glaciation, both mountain glaciers and continental glaciers (Figure 4-3). Even as far south as the Sierras, mountain ranges during glacial epoches of the last million years sustained similar combinations of cool temperatures and increased precipitation, resulting in accumulations of snowfall in the higher elevations. Settling and compaction over time transformed unmelted snows into substantial thicknesses of ice that slowly began moving downward under the persistent tug of gravity. Downward migration of this large unyielding mass with crushing weight, slowly transformed well-drained V-shaped valleys into U-shaped valleys and meandering streams slowly attempting to overcome blockage by mounds of debris from lateral or terminal moraines. The slow inexorable movement of this grinding, crushing mass of ice gouged out valley walls, flattened valley floors, and deposited huge quantities of earth and rock that were bulldozed ahead of the glacial front, carried within the sheet by subterranean streams or transported on the surface of the glacier. Relocation of massive amounts of soil, rock, and debris literally dammed existing streams and simultaneous deposition of ground rock materials — "rock flour," clay particles, and other fines — produced an impermeable layer sealing the bottom of flattened valleys and basins. Extensive chains of lakes and wetlands were recreated in regions that had once known wetlands, but lost them to progressive erosion and renewed drainage patterns following original mountain-building activities. Even today, glacial meltwater and rainfall runoff flowing into poorly drained U-shaped valleys supports significant wetlands throughout the Rockies as far south as central Colorado.

On high elevation plateaus and valleys without permanent snow cover and ice sheets, the combined effects of high precipitation rates, high precipitation/ evaporation ratios, and poorly established drainage patterns created extensive lakes and wetlands throughout the immature Rockies and Cascades and some still exist in much older mountain regions such as the Poconos of eastern Pennsylvania and the Adirondacks in New York.

Similar climactic factors maintain extensive wetlands in a wide circumpolar band covering much of northern North America, Europe, and Asia. In North America, its southern borders are not coincidentally concurrent with the present locations of the Monongahela and Ohio Rivers in the east and the Missouri in the west. Projecting a line along the Ohio to the east coast and northward from the upper reaches of the Missouri along the western slopes of the Rockies to the Arctic Ocean bounds one of the largest expanses of wetlands in the world. A similar delimiting band along existing rivers in Europe and Asia to the Atlantic and Pacific Oceans outlines another vast wetland region, though the southern boundary of the Asian complex has been strongly influenced by tectonic forces shaping the Himalayas. This is not to say that

Figure 4-3. Mountain valley, as in the Copper River region of Alaska, and continental glaciers reshape eroded landscapes, creating broad areas with shallow relief and poor drainage that provide the hydrologic conditions suitable for wetland systems.

Figure 4-4. Almost half of the world's wetlands occur at northern latitudes because of the combined influences of continental glaciers, low evaporation rates, and permafrost conditions.

these regions are or were all wetlands, though a low altitude flight over major portions of Canada might generate the impression of more water than land. Though significant areas have been drained for agriculture in the Prairie provinces of Canada, most of the region to the north and east retains extensive original wetland complexes (Figure 4-4).

The vast majority of the taiga and tundra regions in northern Europe, Asia, and North America consists of wetlands that, in the aggregate, represent the largest wetland complex in the world. In fact, slightly over 45% of all natural wetlands lie between 45 and 60° North latitude. Even the combined total wetlands in the extensive deltaic and riverine swamps of the Amazon, Nile, Congo, Niger, Mekong, Ganges-Brahmaputra, Orinoco, Mississippi, and other major rivers pales in comparison to the enormous acreage of wetland lying in a wide band circling the globe in the northern hemisphere. For example, a recent compilation found almost 400 million ha of peatlands, formed by bogs, in Canada, Alaska, Finland, Norway, United Kingdom, Iceland, and the Soviet Union with only some 52 million ha combined in the rest of the world. Of course, many tropical wetlands form little peat because of high decomposition rates, but many hectares of marshes in the northern hemisphere with limited peat deposits were also not included. Furthermore, the circumpolar northern hemisphere region is interrupted only by a few mountain ranges and of course the Atlantic, Pacific, and Arctic Oceans. Wetland expanses in the rest of the world tend to occur along widely separated river systems and lack the broad, vast continuum of wetlands common in the northern hemisphere. Though a few wetland complexes occur in the southern extremities of South America, extreme cold in Antarctica and high temperatures with low precipitation in southern Africa and Australia preclude extensive wetland creation or maintenance. Furthermore, these regions have not been subjected to the most significant geological force for wetland creation.

Certainly, a much larger land mass is available in the northern hemisphere to support the shallow water conditions required for wetlands, and doubtless colder climates with high precipitation/evaporation ratios would be conducive to wetland creation and maintenance. Yet similar climactic conditions are present in portions of Chile, Argentina, Australia, and New Zealand though largely absent in Southern Africa. Much of Africa also has the gentle relief in expansive, but well-drained plains requisite for wetlands but suitable climatic regions in South America and New Zealand are occupied by rugged, high relief, mountainous terrain with limited wetlands established by tectonic forces. None of these regions has been exposed to recent influence by the major geological event in wetland creation.

Terragenic forces from continental glaciation obliterated or relocated drainage patterns, transformed mountain ranges, and leveled broad regions throughout the circumpolar wetland region in the northern hemisphere. Extensive climactic conditions similar to those creating and sustaining mountain glaciers formed huge sheets of ice over 2-km thick extending nearly from ocean to ocean in North America and Eurasia and reaching as far south as the present locations of the Missouri and Ohio in North America and the Tibetan plateau in Asia. Major rivers and other drainages were relocated — recent evidence suggests that the Missouri formerly flowed northward, emptying into Hudson's Bay and the Arctic Ocean before its course was forced southeasterly by the Wisconsin glacier. Mountains were rounded off and valleys filled in, forming the gentle relief of present-day ranges in eastern Canada and New England that supports substantial wetland habitats. Vast regions of interior Canada and the midwestern states were initially graded level by the bulldozing action and the crushing weight of massive sheets of ice. Pauses, advances, and retreats coupled with internal rivers within the ice sheet, subsequently deposited tremendous quantities of rock, sediment, and soils forming unique patterns of low-lying hills throughout the region. Lateral moraines at the edges of the ice sheet, and terminal moraines marking southern boundaries during periods when melting paced southward movement amid numerous advances and retreats, consist of materials transported hundreds and thousands of miles before, within, and on top of the ice and deposited as extensive lines of low-lying hills. Though narrow — 3- to 30-km wide — some moraines extend linearly for hundreds of kilometers, pock-marked throughout by innumerable depressions and basins of the Coteau du Missouri and other pothole regions in the Dakotas, Lakes states, and adjacent provinces in Canada. Narrower sinuous ridges (eskers) formed by materials deposited in the beds of rivers

Figure 4-5. Ice blocks melting after glacial retreat left numerous depressions that later received clay fines from surrounding hills to form sealing liners. The myriad potholes in the Coteau du Missouri of the Dakotas and Saskatchewan comprise the principal duck factory of North America. Note the small road in the foreground.

within the ice sheet have similar rocky, hummocky terrain with countless wetlands in similar basins.

In some cases, depressions formed through differential deposition of rock and soil materials, but more commonly, present-day basins represent the final resting place of ice blocks transported amid the rock, soil, and other debris ahead of, within, and beneath the glacial sheet. Although most were doubtless located in the upper regions of the 200 to 300 m of glacial till overlying much of the eastern Dakotas and western provinces, because of lower densities, melting of the few deep-lying blocks simply resulted in subsidence. Following the retreat of the ice sheet, ice blocks near the surface of the churning layer of debris melted, leaving cavities that eroded to basins (Figure 4-5). Basin bottoms were later sealed with an impermeable layer of rock flour, clay particles, and other fines eroded and wind blown from the surrounding hillsides. Hence, we find perched depressions supporting wetland communities because of a clay lens isolating surface waters from groundwaters and lacking an outlet through the surrounding hills to other surface waters and streams. Most of the myriad of pothole depressions supporting some of the most productive wetlands known are located in cavities formerly occupied by blocks of glacial ice.

In some areas, glacial meltwater flowed into large basins with limited outlets forming huge lakes — the Great Lakes and former Lake Agassiz —

that covered much of the eastern Dakotas, northwestern Minnesota, and adjacent areas of Canada. Casual observations while driving west from the Red River on the Dakota-Minnesota border reveal three distinct rises, each marking the location of an ancient beachline within about 100 km. As the glacier retreated, reducing meltwater sources, the lake waters receded leaving innumerable flooded depressions from small potholes to large lakes throughout the old lake bed. Though most depressional wetlands in this area of highly fertile soils have been drained for agriculture, many still exist. Unfortunately, some of the largest — the Black Swamp of northwestern Ohio — and others in overflow regions of the Great Lakes were drained to form some of the most productive farmlands in the world. The extensive wetlands in the Hudson Bay lowlands had a similar origin but have not been significantly altered by human development.

In eastern and northern Canada, parts of New England, and much of Scandinavia, glacial scouring removed soils exposing bedrock from which huge chunks of rock were occasionally plucked out, leaving deep irregular rock-lined cavities. Though glacial meltwater filled and subsequent precipitation and runoff maintains deep lakes in many of these depressions, lake bottoms are invariably impermeable rock strata and erosion from largely granitic rock in the surroundings provides little nutrient input for these lakes and associated wetlands. However, bog communities have adapted to and perpetuate the acidic, low-nutrient but relatively stable waters and only the deepest lakes lack extensive fringing or neighboring bogs (Figure 4-6). The progressive extension of bogs ever further onto these lakes with subsequent accumulations of deep beds of organic materials below the floating surface, gradually transforms the previous lake to the bogs and muskegs common in eastern Canada. However, these oligotrophic (and ombitrophic) lakes and wetlands are orders of magnitude less productive than the prairie potholes established on 200 to 300 m of glacial till further south and west. Lower productivities of these bedrock wetlands not only result from the very short growing season in a harsh climate, but they lack the glacial flour, clay particles, and other fines that later formed some of the world's most fertile farming soils in southern and western regions of glaciation. Acidic, low-nutrient runoff from exposed bedrock surrounding the bogs, muskegs, and taiga of eastern Canada, New England, and Scandinavia lacks the fertilizing nutrients that support the highly productive wetlands in areas of glacial deposition of the Dakotas, Lake states, Denmark, central Europe, and southern portions of the Soviet Union. Though really much more extensive — perhaps 10 to 100 times greater — the combined biomass production in these scoured out and restricted environments is considerably less than was the combined total for the original wetland resource distributed throughout the depositional zone from Alberta to Ohio (Figure 4-7).

Precipitation patterns since glacial retreat have increased the differences because the eastern bedrock regions have and continue to receive fairly high

Figure 4-6. Deeper depressions initially filled with glacial meltwater, are slowly transformed by floating bogs growing out from shorelines.

Figure 4-7. Encroaching vegetation gradually changes the lake to a marsh, then a bog, and finally a forest in this Newfoundland basin.

precipitation — at least adequate to support forest vegetation. High rainfall with subsequently high runoff rates, exacerbated by uptake of acidic products from tree leaf decomposition, rapidly erodes or leaches any nutrients that form on the uplands and, under low pH conditions, in lakes and wetlands, these nutrients fall unused to the bottom since plant uptake is hindered by acidic conditions. The deep clear lakes of this region are clear because the limited nutrients inflowing with runoff are quickly and irretrievably lost to deep sediments. In contrast, runoff draining prairie regions situated on hundreds of meters of glacial till have high nutrient loading in neutral or slightly alkaline waters flowing into shallow depressions where the biotic community rapidly exploits the bonanza.

The differences in runoff waters reflects soil and vegetation differences in the sources, which in turn result from not only soil or bedrock substrates but also substantial differences in the precipitation/evaporation ratios between the regions. Large portions of eastern Canada, New England, and northern Europe have high precipitation/evaporation ratios and forests dominate the vegetative communities. With low precipitation rates and/or low precipitation/evaporation ratios, grasses dominate the vegetative communities of the Plains states and the steppes of Eurasia. Not only did wetlands (and terrestrial communities) in prairie regions originate in much more fertile environments, decomposition of grasses did not yield large amounts of organic acids as does decomposition of wood and tree leaves. Consequently, the drier climate fostered establishment of highly productive wetlands by limiting invasion by forest types, thus influencing the amount and type of runoff that enters wetland depressions. Only very rarely is precipitation adequate to flood the prairie potholes to a high enough level to breach the surrounding landforms and initiate erosional stream formation that would eventually drain the depression.

Cooler, glacial climates with high precipitation levels also supported extensive lakes in much of the Intermountain West. During glacial periods, large amounts of runoff from nearby mountain ranges created and maintained expansive lakes. With warmer and dryer climates, only remnants remain, i.e., Great Salt Lake. But these basins, occasionally contiguous with lakes but more commonly isolated in scattered valleys, remain, and during wet years, the playas receive adequate runoff to support an explosion of wetland plants and animals. Though they may be dry for several years, Southwestern playas may actually have two wet periods in wet years — a winter flood period followed by drying and then another inundation during the summer monsoons. High evaporation rates concentrate minerals carried in by runoff waters creating saline wetlands that often have flora and fauna more characteristic of coastal brackish marshes. Interior saline wetlands in eastern Nebraska have somewhat similar plant and animal communities, though both vary with the salinity of the salt-laden, groundwater discharges that support these unique systems.

SEA LEVEL CHANGES

Sea level fluctuations over geological time have created and destroyed significant coastal and interior wetlands throughout the world. Relative emergence or subsidence of crustal plates related to plate tectonics and to glaciation are manifested as relative changes in sea level elevations. During glacial periods, large land areas are depressed by the weight of huge ice masses while water storage in this same mass removes volume from the oceans. Consequently, sea levels may rise (relatively) along coastlines forced downward from glacial weight and concurrently fall along non-glaciated coastlines. Plate movement and subduction of one crustal plate beneath another also cause changes in relative sea level elevations on each plate, with direct impacts to coastal marshes occurring on the margins of one or both plates.

The extensive wetland complex that occupies much of Louisiana today was created by a combination of sea level changes, erosion, and sedimentation. During a recent glacial epoch, ca. 18,000 years before present, sea level along the Gulf Coast was over 100 m lower than today's elevation. The Mississippi River cut rapidly downward, creating a wide but sharply V-shaped valley within previously deposited alluvial sediments, scoured channels across the continental shelf, and deposited sediments near the far edge of the shelf. As sea levels rose during the last 7000 years, erosion stopped and sediments were again deposited in the valley, gradually obliterating the harsh lines of the former V shape. With rising levels and valley filling, the river meandered across the new land surface, and sediments were deposited in a broad fan-shaped pattern with many constantly changing individual lobes at the valley mouth (Figure 4-8). Accumulated deposits tend to restrict flow to the main channel until a breakthrough occurs, after which river flow is diverted into the new region and deposition begins in that area. As those deposits built higher or after another breakthrough at some other point, flows changed again and started the same process at another point along the coast. Once major flows no longer reached the old mouth within each lobe, fine sediments accumulated and gradually filled channels and bayous until river waters were completely cut off. During the later stages of fine sediment deposition and shortly after, the shallow water environments supporting a rich wetland system existed. However, when deposition ceased, compaction and subsidence later caused the floor of these regions to sink, forming deep open-water bays with increasingly saline waters. In this case, changing sea level resulted in a broad valley containing a widely meandering river with countless oxbows and bayous in the upper reaches and a constantly shifting coastal depositional pattern that alternately created and destroyed wetlands (see Coleman in Appendix A for a thorough description).

On a broader scale, rising sea levels would have mixed impacts on North American wetlands (Figure 4-9). Coastal marshes and mangrove swamps along the Gulf Coast would likely be relocated landward with overall increases in the Florida peninsula, substantial decreases in the Louisiana delta swamps and marshes, and increases along much of the East Coast (converting the

Figure 4-8.   Extensive fresh to brackish water marshes of the Mississippi Delta ebb and flow
with changes in sea level and river course.

Figure 4-9.   Changing sea level and sedimentation have converted a stream into a marsh at the
former river boat landing for Boone Hall Plantation near Charleston, South Carolina.

Delmarva peninsula and Long Island to marsh would offset losses in the Chesapeake and Delaware Bays); but farther north, Merrymeeting Bay and Cobscook Bay in Maine would become deep-water aquatic environments. Low-lying coastal regions of Hudson's Bay and the Arctic Ocean behind the existing marshes would become roughly comparable replacements, but many delta wetlands in Alaska would be inundated and the new coast line would fringe mountain ranges only partially offset by increased wetland acreages in interior lowlands. Further south, limited marshes in Puget Sound and other areas of the Washington and Oregon coast would be drowned, as would Humboldt Bay and other California marshes, but these would be partially offset by increases in San Francisco Bay and California's Interior Valley.

Falling sea levels would renew entrenchment of the Mississippi River, destroying most of the swamp and marsh wetlands because river flows and the water table would drop well below the land surface. Deep channels would be cut through previous deposits well out into the continental shelf and sub-sequent deposition would be near the limits of the shelf. Most likely, small marshes would be created in this region but the area would be much less than existing swamps and marshes. Similar patterns would be evident along much of the Gulf Coast and Florida, but concurrently falling groundwater levels would eliminate most freshwater wetlands in peninsular Florida. Bay and coastal marshes behind barrier islands along the East Coast would be lost, as would substantial areas in the Chesapeake Bay and Long Island Sound. Much more extensive marshes would be created along Cape Cod and the Islands but virtually all Maine and Maritime Province marshes would be lost. Coastal wetlands of Hudson's Bay and the Arctic shores would become tundra wetlands with little overall net change, and similar changes would occur in much of Alaska. Along the West Coast, falling sea levels would eliminate virtually all existing wetlands and the steep gradient sea floor would preclude seaward replacement.

Overall, sea level alterations may increase or decrease wetland acreages depending upon the form, relief, and composition of surfaces on the landward side and gradients of sea floors on the seaward side. Major indirect effects on interior wetlands may also result from changes in groundwater elevations related to hydrostatic pressures, relocations of river deltas, and inundation or isolation of floodplains and riverine swamps. Though not as evident, the great river swamps of the Southeast and mid-Atlantic region were created by sedi-ment deposition along river courses whose bed elevations were not much higher than sea levels. Rising sea levels would drown out these swamps and falling levels would dry them out because erosional forces would quickly deepen river beds under the effects of increased stream/river gradients.

EROSION AND SEDIMENTATION

Land forms newly emerged from the sea are immediately attacked by the array of wind, water, heat, and cold erosional forces that act to reduce the new lands to sea level again. In some cases, erosion may almost offset uplifting

Figure 4-10. Rivers meandering across broad flood plains create and destroy wetlands with every flood or change of course.

forces and dramatically reduce elevation increases, especially in the later stages of tectonic events. In younger mountain ranges, uplifting outpaces erosion, with rapid increase in elevations. Some wetlands are created by obstructed drainages during this period, but others are caused by sediments eroded from high elevations and carried down and deposited in mountain valleys. Sediment deposition under lower gradient flows often fills the valley floor, further reducing gradients and increasing sediment deposition. Consequently, the stream or river begins to meander and undergo frequent channel alterations with cutoffs and old oxbows supporting many types of wetland systems (Figure 4-10). Cutoff oxbows and other old channels are filled with additional downward-moving sediments, but new meanders simultaneously create additional wetland habitats. Beyond the foothills and plains in the lower reaches of major rivers, sediments transported from the highest mountain peaks deposited in low stream gradient conditions similarly affect the direction and pattern of waterways. Large rivers in the lowlands frequently alter their channels, in some cases after almost every high runoff period. Extensive bottomland hardwood swamps once occupied innumerable old channels along the Mississippi, Cache, Yazoo, Tombigbee, and Appalachicola Rivers, though many have been converted to agricultural fields and levees constrain new channel changes (Figure 4-11). Sediment accumulation may cause major shifts, as is presently occurring with the Mississippi and Atchafalaya; and despite concrete structures to prevent the Great River from changing its course, in time, additional sediments obstructing the present channel coupled with an extremely high flow

Figure 4-11. Cutoff oxbows and old river channels form significant wetlands in river valleys with low relief. Unfortunately, many have been ditched to improve drainage for agricultural purposes.

event will likely reroute most of the flow down the Atchafalaya. The previous channel of the Mississippi and associated waterways could then become significant wetland habitats.

Wet meadows and marshes along the Platte and a few other western rivers often also occur in or adjacent to old river channels and are largely supported by flow-through ground waters though they often receive some overflow during flood periods.

SUBSIDENCE/EMERGENCE: KARST TOPOGRAPHY

Two types of extensive freshwater wetlands — the prairie potholes and the tundra wetlands — resulted from the effects of glaciation. A third major type represented by extensive areas of superficially isolated but often strongly interconnected wetlands resulted from erosional forces on extensive layers of limestone deposits. Throughout much of Florida, the Louisville Basin of Kentucky, and the Nashville Basin of Tennessee, differential dissolution of extensive, relatively flat layers of limestone created substantial wetland acreages. As rainfall acidified from dissolved organic materials and percolated through cracks in limestone strata, dissolution of the limestone enlarged the cracks, forming subterranean cavities and caverns. Eventually, dissolution enlarged the cavern, leaving a thin roof cap that collapsed downward forming a basin or pothole. Fallen roof material and washed-in debris may have obstructed the drainage channel in the pothole floor and subsequent runoff flooded the basin. Not surprisingly, many have circular forms and in some areas forge an irregular string of pearls. In the aggregate, the potholes or basins of karst topography originally represented substantial quantities of wetlands in the Nashville and Louisville Basins and still support significant amounts in Florida.

Though swamps with large, old hardwoods occupy many existing karst potholes in Kentucky, Tennessee, Alabama, and Florida, all of these are ephemeral systems since continued loss of limestone could reopen or create new drains at any time. In contrast, Florida potholes were eroded out during an earlier period with lower sea- and groundwater levels and the waters of many today are contiguous with present groundwater tables with limited limestone erosion. High groundwater elevations also support extensive wetlands in surface depressions of these limestone layers. Though either type is relatively secure under present sea level conditions, they would be susceptible to similar threatening erosional factors if sea and groundwater levels declined significantly.

AEOLIAN FORCES: THE NEBRASKA SANDHILLS

Wind erosion — aeolian effects — has been a significant wetland creation factor in a few instances. Light-weight sand and clay materials were carried for miles by prevailing winds and deposited in north central Nebraska, forming the Nebraska Sandhills region. Aeolian shaping and reforming causes hill and

Figure 4-12. Many of the rich marshes of the Nebraska Sandhills intercept groundwaters and are oriented in a northwest-southeast direction caused by prevailing winds moving sand and forming dunes and depressions.

depression complexes typical of sand dune regions similar to glaciated prairie potholes, but with one outstanding difference. Long axis orientation of sandhill depressions represents the direction of prevailing winds rather than round forms mirroring cavities from melted ice blocks in the glaciated Dakota potholes (Figure 4-12). Clay particles and other fines are carried by water to the bottom of sandhill depressions, eventually forming an impermeable layer or occasionally an isolated clay lens beneath each basin. Most larger, permanent marshes occupy linear, northwest-southeast oriented valleys intersecting groundwater above an extensive impermeable stratum underlying much of this 45,000-km^2 region. Smaller precipitation-dependent marshes have less pronounced long axis orientations, occur at higher elevations, and owe their existence to an isolated clay lens. Both types are highly productive wetlands reflecting the inherent fertility of the surrounding sand/soil materials.

ARCTIC TEMPERATURES AND PERMAFROST
Though precipitation rates in the high Arctic may be comparable to the southwestern deserts (only 20 to 30 cm/year), cool air temperatures reduce evaporation rates and the precipitation/evaporation ratio may be quite high. However, seemingly endless expanses of frequently more-water-than-land wetland complexes of the high Arctic are situated on infertile granitic bedrock, overall productivity is relatively low, and a large portion of the animal biomass consists of blood-sucking insects and transient birds.

Figure 4-13. Permafrost restricts downward water movement in summer, and ice heaving in winter forms ridges creating the typical polygon shapes of arctic wetlands near Churchill, Manitoba.

The cold climates of high latitudes cause another phenomenon important to creating and maintaining tundra wetlands. Permafrost, the permanently frozen zone of soil varying from less than 1 m to many meters below the surface, prevents drainage of depressional wetlands above it during the short summers. In the absence of permafrost, many tundra wetlands would quickly lose their water to deeper groundwaters and no longer support the myriads of wetland complexes. In addition, frost heaving during cold winters continually creates small ridges and depressions with unique polygon configurations, continually adding to the total wetland environment (Figure 4-13).

SMALL-SCALE FACTORS

Oddly enough, a large rodent, the beaver, may have been almost as influential in wetland creation in North America as any other factor. Beaver are one of the few animals, aside from man, that have the ability to substantially modify the existing environment to suit their own requirements. Stream and river modifications by beaver dams have profoundly influenced the nature and distribution of North American wetlands. Today, beavers are present almost everywhere south of the treeline in Canada, Alaska, and the U.S., excepting only a small area in the arid southwest and most of Florida. Their present occurrence in the deep South and Texas may be due to previously depressed alligator populations and, as alligator populations increase and reoccupy previous ranges, beaver distributions will likely retract northward.

At one time or another and in many cases repeatedly, beavers have influenced virtually every stream and small river throughout the continent, creating, expanding, and maintaining vital wetland habitats from coast to coast, from Mexico to the Arctic Circle and from desert valleys to mountain treelines. In aggregate, beaver established and maintained wetland habitats throughout eastern Canada and eastern U. S. (Figure 4-14) and, in the Rocky Mountain chain, rival combined totals for the prairie potholes or the coastal marshes. Newly formed dams may be only 3- to 10-m long depending upon the terrain, but older colonies construct and maintain multiple dams, stair-stepping a previously free-flowing stream, maintaining extensive areas of shallow productive wetlands. In relatively level terrain, the principle dams of an older beaver complex may be 2 km long or more.

Typically, beaver erect a small dam for protection and access to food supplies shortly after colonizing a new area. As proximal food supplies are depleted, the original dam is raised and/or lengthened and additional dams are constructed upstream, downstream, or both. In steep terrain, the limit of floodable area for secure access to new food supplies is reached within a few generations and, eventually, the colony abandons the site. In flat lands, beaver colonies may occupy a drainage for hundreds of years, merely shifting up- or downstream as food supplies are exhausted in one area and exploited in another (Figure 4-15). In either case, the shallow water environments created and maintained by beaver support a plethora of other wetland animals and plants, including invertebrates, fish, amphibians, reptiles, mammals, birds, and micro- and macrophytes.

Other than a narrow band of riparian vegetation and high elevation ponds, most wetlands throughout the western mountains of North America show evidence of past or present beaver influences on natural drainage patterns. Even in the older Appalachians, it is not unusual to come upon a small, unusually level sedge/grass meadow adjacent to or surrounding a fast-flowing stream (Figure 4-16). In many cases, even causal observation will discover a key indicator — a few old tree trunks (often pines) — solitary, erect, and partially buried by the deep organic sediments that underlie the meadow. Clearly, the meadow was formed by a beaver dam that flooded and killed the pines, trapped sediments, and accumulated organic materials over a number of years before the beaver left, the dam was breached, and the area was re-invaded by semi-terrestrial vegetation. Core sampling often reveals this cycle has been repeated many times as beaver colonize, deplete accessible food supplies, abandon, and then re-colonize after succession has re-established suitable vegetative food supplies.

Some prairie potholes contain a large boulder with unusually smooth, almost polished sides. The rock was deposited by the glacier and may have been transported within a larger ice block that later melted, forming most of the depression. However, the polished boulder invariably lies in the deepest

Figure 4-14. The beaver lodge in the background reveals the perpetrator of an Appalachian marsh that purifies acid mine drainage waters.

Figure 4-15. Beaver flowages have meandering streams in the low-gradient terrain formed behind beaver dams over hundreds of years.

Figure 4-16. Beaver impoundments kill terrestrial trees, but persistent snags remain long after the beaver have left and the pond has become a wet meadow. (Photo by Joan Hammer.)

portion of the pothole and the most likely explanation relies on the behavior of a large herbivore. The deepest portion of the depression was formed by buffalo rubbing the sides of the boulder during spring shedding. As each of many thousands of animals pushed against the rock to scrape off the accumulation of winter hair, their hooves churned the adjacent soil to fine powder and ever-present winds carried the powdered soil out of the basin. Lichens and subsequent erosion have modified the rock surface, but close examination reveals a band of smoothed surface around each of these boulders corresponding to buffalo shoulder height. Some boulders lie in small deep depressions caused by buffalo activity, while others are found in the deepest portion of a larger basin. Because of their depth and location, the latter are important refuges for wetland species during cyclic drought periods in the pothole region.

Critical refuges for wetland species during drought periods in the Florida Everglades, the Carolina Bays, and the Pantanal of Brazil are also created by activities of an animal. Alligators in North America and caimans in South America excavate deep and sizable depressions — "gator holes" — during drought in many areas of the Southeast, but most noticeably in the unique bays of the Carolinas and the Everglades. These oases are doubtless formed to provide water, security, and perhaps a handy foraging area for alligators or caimans, but they also provide sanctuary for myriads of other wetland animals and plants that repopulate desiccated surroundings when the rains finally return.

MAINTENANCE

To paraphrase an old adage, the three most important factors in maintaining a wetland are: disturbance, disturbance, and disturbance. Wetlands result from disturbance of normal drainage patterns and are quickly lost with re-establishment of normal drainage ways. Whether large scale (glaciation) or small scale (beaver dams), whether long term (tectonic forces) or short term (seasonal flooding), wetlands are absolutely dependent upon some disturbance factor for initial formation and for continued existence. Without disturbance — hydrologic, geologic, cyclic precipitation, fire, animal activity, etc. — wetlands are relentlessly eliminated by ecological and geological forces that gradually replace wetland environments and denizens with successively drier, terrestrial habitats and inhabitants. Consequently, preservation of wetlands requires a flexible philosophy and active management in contrast to preservation methods suitable for forests, grasslands, or deserts. Wetlands require flooding and adequate water depths, yet continuous inundation and stable water levels may hobble the normal productivity of many wetlands.

Requisite disturbances may be large scale and long term, such as changes in sea levels or lake levels, or they might be large scale but short term, such as periodic droughts in the prairie pothole region. Continuous flooding over

Figure 4-17. Disturbance is critical to maintaining many types of wetlands; fire often removes accumulated organic materials, restoring water depths, and reversing successional changes.

a period of years soon immobilizes essential nutrients in the substrates and concurrently washed-in sediments accumulate. Periodic drought oxidizes and recirculates nutrients and wind erosion, during severe droughts, removes accumulated sediments, resetting the basin wetlands to earlier stages of succession.

Fire is a very important disturbance factor for many systems. It can be small or large scale but it usually has long term effects. Fires consume peat opening up the depressions in northern bogs and similarly remove accumulated organic material as well as retard woody species in the Everglades. Marshes may be havens of refuge during prairie fires in wet years but in drought periods, the fires sweep through the marshes, burning off accumulated organic materials (Figure 4-17). In each case, fire removes the organic matter that was gradually filling the depression and resets the system to an early stage of succession, delaying and retarding the inevitable progression to drier and drier terrestrial habitats.

On a shorter time frame, extensive bottomland hardwood swamps occur in floodplains because of the annual cycle of flood disturbance. Without annual flooding, these unique systems would soon become indistinguishable from the surrounding terrestrial forest. Animal activity — alligators, caimans, ungulates — disturbing wetlands often has a similar seasonality, though it is typically on a much smaller scale. Seasonal droughts enhance conditions for fires that burn accumulated organic layers, opening up and reversing succession in bogs and other peatlands. Even daily changes can be important in maintaining some wetlands. Without the twice daily flood disturbance of tides, most of our coastal marshes would cease to exist.

Wetlands are ephemeral components of the landscape formed by drainage interruptions and maintained by geological, hydrological, and biological factors that arrest or retard the impacts of other biological and geological factors that tend to transform the wetlands into a copy of its neighboring ecosystems — upland or deep water habitats. In contrast to the latter, unique, complex, and productive wetlands thrive on disturbance and change, and soon cease to exist under stable conditions. Unfortunately, the short human lifespan causes many of us to perceive the present status and conditions of some ecosystems as desirable and therefore, worthy of protecting and maintaining. And because of our short time perspective, we assume that maintaining existing conditions will preserve the system. Though rarely appropriate for forests, grasslands, or deserts, eliminating disturbance rapidly sounds the death knell for wetlands. "Disastrous" fires, floods, droughts, etc. are often beneficial to terrestrial systems but are absolute requirements for maintaining wetland systems.

CHAPTER 5

WETLANDS: FUNCTIONS AND VALUES

INTRODUCTION

Wetlands have both value and function and, though often used interchange-ably, these terms are not synonymous. Function describes what a wetland does, irrespective of any beneficial worth assigned by man. For example, a wetland may function by producing 100 mosquitos/m^2 or 10 muskrats/ha, purifying 5000 m^3/day of contaminated water, or by temporarily storing 100,000 m^3 of flood waters. A function is an objective process or product. A value is a subjective interpretation of the relative worth of some wetland process or product, that is, the market or recreational value of 10 muskrats or the cost for eradicating 100 mosquitos. Values can be positive or negative and they can be high or low. For example, flood storage capacity upstream from a city has high value to the residents, yet the same wetland downstream might have low value to the city because it provides them limited flood protection. Of course, the downstream wetlands could provide water purification, wildlife, recre-ational, or other functional values and be of high value to city residents.

Some values, the sale of muskrat pelts or veneer quality timber, have well-defined tangible parameters but others — recreational, research, and educa-tional benefits — are intangibles that can often only be quantified by estimating replacement costs or amount and willingness to pay by users. Flood protection is intermediate, in that potential flood damages can be estimated under various flooding regimes or the cost of constructing and operating a flood storage reservoir may be taken as the value of this function of a wetland. Inconvenience or inability to use a residential yard because of mosquitos may be estimated by comparing property values and accepting the difference as the value (neg-ative) of the nearby wetland.

Although 10 to 15 functional values are often described for wetlands, most can be grouped into 6 major categories — life support, hydrologic modifica-tions, water purification, erosion protection, open space and aesthetics, and geo-chemical storage as described below:

1. Life support: includes all types of microbial, invertebrate and vertebrate animals, and microscopic and macroscopic plants;
2. Hydrologic modification: includes flood storage and conversely base flow augmentation, ground water recharge and discharge, altered pre-cipitation and evaporation, and other physical influences on waters;

3. Water quality changes: includes addition and/or removal of biological, chemical, and sedimentary substances, changes in dissolved oxygen, pH, and Eh, and other biological or chemical influences on waters;
4. Erosion protection: includes bank and shoreline stabilization, dissipation of wave energy, alterations in flow patterns, and velocity;
5. Open space and aesthetics: includes outdoor recreation, environmental education, research, scenic influences, and heritage preservation; and,
6. Geo-chemical storage: includes carbon, sulfur, iron, manganese, and other sedimentary minerals.

Of these, our information is greatest for the life support functions and values, and rapidly increasing for removal of pollutants from surface waters. Understanding and quantifying the remaining functional values remains to be accomplished and, in fact, these are most frequently estimated, rather grossly at times.

LIFE SUPPORT FUNCTIONAL VALUES

Wetlands produce many and diverse forms of life and provide habitat for many others. Life forms that must have wetlands to complete their life cycle may be termed "obligate" forms and wetlands can be said to produce these forms. Examples include fish that must have wetland for spawning or nurseries, even though adults spend most of their lives in deeper waters; snails or clams found only in wetlands, toads that lay eggs and the larval stages live in wetlands, but adults are primarily terrestrial; and of course ducks, geese, and swans. Other "facultative" life forms use wetlands when available. Though some attribute of wetlands may substantially increase survival, growth, or population sizes, these facultative forms can survive in the absence of wetlands. For example, deer and pheasant exploit food and shelter available in wetlands, but both have thriving populations in solely terrestrial environments. However, winter cover in dry or frozen marshes strongly influences individual survival in many prairie regions; northern deer herds would be much smaller without cedar swamp wintering areas. The combination of food, water, and cover needed by a species to survive and reproduce is its required habitat. Facultative forms find additional but not essential components of their habitat in wetlands, whereas obligate forms find all or critically needed components.

Wetlands are the most threatened wildlife and fishery habitats of all our natural resource base. Some 87 million ha of wetlands existed in the U.S. before colonization, but only half (approximately 44 million ha) remain. Generations of fear and antagonism toward wetlands, without understanding and appreciating the important ecological and economic benefits they provide, caused wholesale destruction during the last 300 years, most dramatically during the last 100 years. Wetlands were wastelands, obstacles to orderly development and progress, that needed to be drained, ditched, and filled to tame the wilderness. And even though misguided attitudes have changed somewhat, we are still losing nearly 40,000 ha of wetlands each year in the U.S.

Historical and current wetland losses and the importance of wetlands to fish and wildlife are reflected in the large number of threatened and endangered species that require wetland habitats. At present in the U.S., 5 of 33 endangered mammals, 22 of 70 threatened and endangered bird species, 22 of 70 endangered and threatened reptiles, and 22 of 41 threatened and endangered fishes are dependent on wetlands or found in freshwater wetland habitats during part of their life cycle. Only a few of these have potential commercial value, but without wetland additions to the biological diversity of our natural resources, education, recreation, and research activities would be significantly impoverished.

Wetlands are dynamic, transitional, and dependent on disturbances — the most obvious of which is fluctuating water levels. Changing water depths, either daily, seasonally, or annually, strongly influences species composition, structure, and distribution of plant communities. Varying from deep inundation to complete drying with all possible intermediates creates myriads of different zones at any one time and other combinations of zones during other times of the year. Salt and temperature gradients, tide and wave action, and pH and Eh variations structure plant communities along stratified patterns similar to well-developed forests and other complex systems.

The combined interactions of abiotic and biotic factors create a diversity and abundance of habitats that make wetlands our most important wildlife habitat. Even though wetlands only occupy a small fraction of our total land area, they support a disproportionate share of our wildlife and fish. For example, over 900 species of wildlife in the U.S. require habitat components only found in wetlands at some stage in their life cycle, with many more periodically using wetlands. Members of almost all bird groups use wetlands to some extent and one third of the species of North American birds rely on wetland for some critical habitat component.

MICROSCOPIC FORMS

Valued products generated by the life support function typically consist of larger animals and plants — crayfish and clams, sport and commercial fish, ducks and muskrats, fruits and berries, or thatching and timber. But these systems, as all other ecosystems, would cease to exist without the critical roles of innumerable forms and types of microscopic life that process and transform organic and inorganic substances supporting important bio-geo-chemical cycles and making energy and nutrients available to all higher life forms. We are just beginning to recognize this role in another functional value — water purification. Furthermore, microbially mediated reactions in wetland systems form long-term deposits of substances containing high percentages of carbon, oxygen, and hydrogen (hydrocarbons), sulfur and metal compounds that affect global atmospheric, climatic, and hydrographic parameters in active portions of the bio-geo-chemical cycles. In addition, many of the less obvious forms — freshwater sponges and jellyfish, orchids and fungi, algae, and insects — provide significant benefits for recreation, education, and research purposes. Finally, even the smallest virus or bacterium, the larger algae, protozoa and fungi, and on up the food chain through insects and small plants and animals provide food (nutrients and energy) for valued products, the larger animals and plants.

INVERTEBRATES

Invertebrates also process organic materials contributing to energy and nutrient cycling but, more importantly for this function, they are the foundation of most higher food chains culminating in valued vertebrates. Wetland macro-invertebrates consist largely of annelid worms, mollusks, arthropods, and insects. Most annelids, flatworms, leeches, earthworms, nematodes, and other worms burrow into substrates or adhere to submersed aquatic vegetation, but a few are free swimming. Mollusks include snails, clams, and mussels, most of which are bottom-living or associated with vegetation and almost all are important foods for fish, salamanders, turtles, mink, otter, muskrat, raccoons, and birds. Arthropods are represented by crayfish and freshwater shrimp — the most abundant and widespread of all wetland invertebrates. Both are important foods of wetland mammals, fish, and birds. Aquatic larvae of insects provide abundant foods for fish, frogs, mammals, birds, and other invertebrates. Larval forms of aquatic beetles, water striders, stoneflies, damselflies, drag-onflies, springtails, mayflies, midges, mosquitoes, and other insects are also important in nutrient and energy transformations.

VERTEBRATES: FISH, AMPHIBIANS, REPTILES, BIRDS AND MAMMALS

Small forage fishes dominate freshwater wetlands, but the larger sport and commercial species are heavily dependent on wetland minnows for food and many also breed and spawn in wetland systems. Typical minnows include such species as fathead minnows, killifishes, top minnows, shiners, mosquito fishes, and sunfishes. Larger fish species that use wetlands diurnally or seasonally are represented by gar, bowfin, pickerel, northern pike, walleye, suckers, bull-heads, carp, yellow perch, catfish, crappie, and bluegill. Larval and adult stages are important foods for birds, amphibians, reptiles, and aquatic mammals, as well as to sport and commercial fishermen (Figure 5-1).

Amphibians are represented by frogs, toads, and salamanders. Many forms may be largely terrestrial, but few of the 190 species of North American amphibians do not require wetland for at least a part of their life cycle (Figure 5-2). For example, adult toads are terrestrial, but larval stages are restricted to wetland as are many tree frogs. Bull frogs and green frogs are more closely tied to aquatic environments in all life stages; but even the terrestrial forms find critical refuge in wetlands during drought periods. Not only are they important food items for other vertebrates, they generate the night sounds that contribute to unforgettable experiences in many wetland systems and, hence, have important recreational values. Of course, frog legs contribute to local economies and culinary delight in many areas of the country.

Figure 5-1. Spring peepers calling from shallow wetlands are the first harbingers of spring in much of eastern North America.

Figure 5-2. Adult toads only return to water for egg-laying, but shallow wetlands are critical to survival of their tadpole stages.

Many types of salamanders occur in wetlands but the best known, most numerous, and widespread species is the tiger salamander. During spring in some areas of the U.S., thousands cross highways enroute to shallow wetland breeding sites. Most lay eggs in wetlands, but the adult stage is largely terrestrial. However, some salamanders, mudpuppies, and water sirens, never leave wetlands because their partial metamorphosis permits them to become reproductively mature in a morphologically immature and aquatic body form.

Egg masses and tadpoles as well as adult frogs, toads, and salamanders are important foods for fish, snakes, birds, mammals, and other amphibians. The larger frogs and toads are especially non-selective and, in some areas, smaller frogs including their own species are the most common food items.

In contrast to amphibians, reptiles lay their eggs in terrestrial environments even though many species use wetlands for food, cover, and water. However, members of the three major groups of reptiles — snakes, turtles, and alligators — often typify wetlands to many members of our society. The dark and dismal swamp is often perceived as crawling with snakes, alligators, and other dreadful creatures.

The important role of beaver in creating wetland habitat was discussed in Chapter 4. What is not as commonly known is that alligators also modify wetland habitats in ways that may be crucial to other forms of wetland wildlife. Gator "holes" dug for shelter and refuge during droughts often provide the only remaining pockets of wetland habitats for other wildlife and fish during extended droughts. Without these critical refuges, repopulation of surrounding areas at the onset of rains would be dramatically slowed due to the large distances involved. Mounds of vegetation from old alligator nests also provide resting and nesting sites for birds, loafing areas for mammals, and slightly drier points for establishment of many plants. Alligators feed on a variety of fish, birds, and mammals, in some cases so heavily that prey populations may be reduced.

Turtles are the most common, most diverse, and most visible reptiles to wetland visitors because many species bask on logs, mounds, or other elevated structures (Figure 5-3). Common turtles include painted turtles, snapping turtles, mud and musk turtles, softshell turtles, sliders, cooters, box turtles, and pond turtles. Use of wetlands varies considerably, with snapping, softshell, and mud turtles being truly aquatic, emerging only to lay eggs; box and wood turtles, being largely terrestrial, enter the water only to hibernate. Although turtle eggs are important foods for many mammals, adults are largely consumers. Softshells are carnivorous, feeding largely on fish; box turtles are principally herbivorous feeding on leaves, fruit, berries, and occasionally worms and insects, and painted and snapping turtles are omnivores, dining on algae, higher plants, insects, fish, and other vertebrates (Figure 5-4).

Although a common misconception of wetlands, snakes are much more important and abundant in terrestrial systems. In fact, only garter snakes are

Figure 5-3. Owing to their sunbathing behavior, painted turtles are often the most obvious forms of life in many wetlands.

Figure 5-4. Common snapping turtles occur in almost every wet environment east of the Rockies, but the largest populations occur in productive marshes and swamps.

Figure 5-5. Though poisonous snakes are almost synonymous with swamps, the innocuous garter snake is the most widespread snake in North American wetlands.

commonly found in far northern wetlands (Figure 5-5). To the south, water snakes vastly out number water moccasins, though the frequency with which they are misidentified would suggest that water moccasins are very abundant. Mud snakes and queen snakes are other common wetland snakes. Snakes feed on insects, crayfish, snails, worms, amphibians, fish, and birds and are, in turn, food for turtles and birds.

To many, the abundance and diversity of birds more correctly embodies the essence of wetlands. Freshwater wetland birds include many species of heron, egret, ibis and bitterns, ducks, geese and swans, rails, coot and gall-inules, cormorants and pelicans, loons and grebes, shorebirds, cranes, ospreys, eagles, falcons and owls, and many types of songbirds. In fact, representatives from almost all avian groups use wetlands, and one third of all North American birds depend directly on wetlands for some critical resource. Wetlands support many types and high numbers of birds not only because of abundant food supplies, but also because wetlands provide excellent nesting and loafing sites protected from predators. With the diversity of birds represented, virtually all other types of fish and wildlife, as well as many plants, are used by one species of bird or another. Birds, in turn, are preyed upon by snakes, turtles, other birds, mammals, and, infrequently, amphibians and fish.

Though ducks, geese and swans often typify wetlands, other types such as grebes and loons, that build floating nests and are unable to walk on land, are more dependent on wetlands. Some birds only use wetlands occasionally; for example, peregrine falcons preying on shorebirds or ducks though their nest sites may be remote dry cliffs. Of the songbirds, a few species of wrens, blackbirds, and sparrows rarely leave wetlands, but most flycatchers, warblers, and sparrows are mainly terrestrial even though many find essential food or cover during some portion of the year in wetlands. Populations of some species may be much higher (i.e., red-headed woodpeckers and chickadees) in wooded wetlands even though terrestrial forests provide adequate habitat for their populations.

Most birds are diurnal and many congregate in large numbers forming spectacular displays. Consequently, wetland birds have developed a broad constituency and are often thought of as the most important life support functional value for wetlands. In turn, the concern for wetland birds has done much to alter public attitudes towards wetlands.

Few mammals are as closely tied to wetlands as are many birds, but many mammals exploit abundant food supplies or shelter during certain periods of the year. Wetland-dependent mammals include muskrat, nutria, beaver, marsh rice rats, water shrews, and swamp rabbits. Muskrats inhabit almost all wetland types throughout North America, feeding primarily on emergent vegetation (notably cattail and bulrush), but also using shellfish and crayfish in many areas. Emergent vegetation supplies the raw materials for muskrat houses that are also important as nesting and loafing areas for birds, living quarters for other animals, and growing sites for semi-aquatic plants.

In large rivers, muskrat and beaver dig bank dens, feed largely on deep-water shellfish and a few plants in shallow backwaters, and both feed on bark and foliage (beaver) or fruits, hard mast and invertebrates (muskrat) of terrestrial forests, especially during times of stress. But these are the exceptions. Muskrat and beaver are, for the most part, dependent on and very influential in wetland creation and management.

Other mammals that use wetlands extensively but thrive in terrestrial habitats include lemmings, mice and voles (Figure 5-6), shrews, mink, weasel, otter, raccoon, wood rats, and swamp rabbits. Wolves, coyotes, bobcats, deer, elk, and moose find food or shelter but are not restricted to wetlands. However, populations of these mammals are generally much higher in areas containing wetlands because of the additional quantities and diversity of food and shelter in wetlands (Figures 5-7, 5-8, and 5-9).

WETLAND PLANTS

Wetland plants range from microscopic forms — bacteria and minute algae — to towering cypress trees hundreds of years old. Algae include microscopic, planktonic forms (blue-green and green algae), many groups that grow attached to vegetation or other substrates (mostly green algae) as part of the periphyton community, filamentous mass forms such as *Oedogonium*, and a few macroscopic green algae such as *Chara, Cladophora,* and *Nitella.* The smallest higher plants are the floating forms of duck meal (*Wolffia* and *Wolffiella*), duckweeds (*Lemna* and *Spirodela*), and the water ferns (*Azolla*). Floating forms may cover open water areas of still waters, but in areas with wind and wave action are generally restricted to protected regions along

Figure 5-6. Herbivorous meadow voles and other mice form the mainstay for fox, weasel, hawks, owls, and other predators that use wetlands.

shorelines or in stands of emergent vegetation. Many floating types have high invertebrate populations associated with them and consequently provide an important food source for many wetland animals and fish. Submergent plants, typically the most important food producers for waterfowl and some fish, include pondweeds, milfoils, coontails, bladderwort, tapegrass, and widgeon grass.

Rooted floating forms include water lotus, watershield, spatterdock, and waterlily; common emergents found almost worldwide include common reed, cattail, bulrush, arrowhead, arrow arum, sweet flag, sawgrass, rushes, and spikerushes (Figures 5-10 and 5-11). A few of the emergents and rooted floaters produce large quantities of seed used by many animals and some (cattail, bulrush, and spatterdock), have large energy stores in tubers or roots that are heavily used by muskrat, nutria, beaver, and moose. In addition, leaves and stems of many emergent and rooted floating species are grazed by various waterfowl species, the same mammals as above, as well as deer, elk, and other large herbivores. A few emergents (wild rice, blueberries, and cranberries) produce large crops of seeds or berries used by many birds and a few mammals. Emergents also provide nesting habitat and shelter for virtually all forms of birds and mammals found in wetlands. Fish and amphibians benefit from shelter and cover provided by all plants with portions in the water column and, of course, the free-floating and rooted floating species. Many of the latter also provide fishing and hunting perches for small herons, rails, and gallinules whose elongated toes are adaptations for distributing weight, permitting these birds to walk across floating leaves.

Figure 5-7. Raccoons were largely restricted to riparian wetlands and extensive swamps, but they have invaded the prairie marshes of the Dakotas and adjacent provinces during the last 40 years because they now find winter shelter in abandoned farm buildings.

Figure 5-8. Though red fox are primarily terrestrial, tracks across this frozen marsh reveal a nocturnal search for mice or rabbits in a nearby pothole marsh.

Figure 5-9. Mule and white-tailed deer find lush browsing and shelter from inclement weather in western marshes.

Figure 5-10. Moose browse riparian shrubs and are often seen shoulder deep in bogs and lakes, feeding on lush wetland vegetation.

Figure 5-11. Camouflage plumage and deliberately swaying with wind-blown plants improves a bittern's chances for avoiding detection. Down-looking eyes improve detection of fish as well as potential danger.

The transitional forms — smartweeds, grasses, etc. — provide shelter, nesting cover, forage, and seed crops that supply habitat for a variety of amphibians, birds, and mammals. Some smartweeds (*P. lapathifolium*) regularly produce abundant seed crops whereas others (*P. persicaria*) have much lower, intermittent seed production.

Shrubs and trees provide denning and nesting sites, loafing areas, hunting perches, shelter from inclement weather, and a variety of soft and hard mast to support endemic and migratory birds and mammals. Where accessible, foliage is used by deer and other herbivores but more importantly, canopy level foliage supports insect populations that in turn are food for warblers, tanagers, vireos, and other songbirds. Wading birds establish mixed nesting colonies in relatively low-growing shrubs in island situations, but prefer multi-layered, near-canopy heights over land. On land, concentrated nutrient loads from bird droppings may stress trees, but colonies in flooded cypress often have high fish populations, taking advantage of the increased algae and invertebrate production beneath the colony.

Basic productivity of many wetlands far exceeds that of the most fertile farm fields (which in many cases are former wetlands). Wetlands receive, hold, and recycle nutrients continually washed from upland regions. These nutrients support an abundance of macro- and microscopic vegetation, which convert inorganic chemicals into the organic materials required — directly or indirectly — as food for animals, including man.

Although plant species diversity is often lower in some wetlands, especially marshes and bogs, plant biomass production is greater in many wetland ecosystems than in most terrestrial habitats. For example, cattail marshes produce 20 to 34 metric tons/ha/year, reed marshes 15 to 27, bogs 4 to 14, wooded swamps 7 to 14, and duckweed stands 13 to 15 metric tons/ha/year. A number of pilot projects and a few operating systems employ aquatic plant production to generate methane gas or as fuel for direct heating. Peat, of course, has long been mined from wetlands for fuel (Figure 5-12), and recent experiments are exploring cattail use. Reeds (*Phragmites*) and bulrushes are used in various parts of the world for fuel or more commonly for thatching (Figure 5-13). Reeds are still used extensively throughout Europe and the Middle East, and bulrush and sawgrass serve a similar function in much of tropic America (Figure 5-14). Another reed (*Arundo donax*) is the preferred material for instrument reeds, critical vibrating elements in the woodwinds — clarinets, saxophones, oboes, and basoons — that grace our symphonic evenings. One wonders if the name reed, as used with the instrument, originated from the plant.

Although lowland rice is certainly more important in the Far East and rice and cranberries are important, probably the single greatest value generated by wetland plants in the U.S. is timber and nut production. In 1979, there were 13 million hectares of bottomland hardwood and cypress swamps in the southeastern U.S. containing 112 m³/ha of merchantable timber worth over $617/ha

Figure 5-12. Peat (for fuel, packing material, and gardening) has long been a commercially important product of natural wetlands.

Figure 5-13. Common reed harvest in natural wetlands is an important activity in central Europe. Most of the product is shipped to the Low Countries for roof thatching. Even though the Hungarian portion represents only 23 percent of the total area (Fertos Lake in Hungary, Neusiedlersee in Austria) annual harvest of nearly 400,000 tonnes is an important income source for area residents.

Figure 5-14. Thatch roofs (reed) provide excellent insulation while needing only minor if any maintenance for 15 to 20 years.

(or $8 billion in aggregate). Other commercially valuable products derived from plants include wild rice, blueberries, cranberries, and honey. An unusual product, medical dressings of Sphagnum moss, was commonly used prior to modern synthetics because of the acidic, antiseptic qualities and dried moss has been used for packing materials and gardening for many years.

Each of these wetland life forms has intrinsic values and some have tangible, quantifiable values; but the value of many is intangible and has been estimated through various means including market value, expressed willingness to pay for an activity, fixed unit day values, travel costs, or dollar amounts spent to engage in some activity. National surveys periodically conducted by the U.S. Fish and Wildlife Service estimate number of individuals engaged in outdoor activities and their expenditures. In 1990, 3 million hunters spent $689 million hunting migratory birds (primarily waterfowl) and 76 million wildlife watchers spent $18.1 billion observing and photographing waterfowl (Figure 5-15). During that same year, fur industry sales were almost $1 billion, principally from the sale of muskrat, nutria, mink, and beaver pelts. In 1989, the alligator harvest in Louisiana was worth $2.7 million and the crayfish harvest contributed $21 million to the state's economy.

Only a few attempts have been made to develop estimates of commercial value for all the life support functional products from wetland systems. A combined value assessment for a hypothetical 405-ha bottomland hardwood site in eastern Oklahoma in 1985 derived the following estimates:

Figure 5-15. Hunting (note blind on far right) and fishing are perhaps the most important recreational uses and consequently, economic values from many wetlands.

	$/ha/year
Retail meat value	3.45
(Game species-rabbit, squirrel, deer, etc.)	
Hunting recreational value	25.34
Other recreational activities	31.51
Furbearer harvest	0.74
Timber harvest	87.00
Pecan harvest	116.43
Total	$264.47

In some areas, lease fees for waterfowl hunting rights or membership fees in duck clubs represent the marketable value for a tract of wetlands, but these do not reflect life support products generated from that tract. Typically, duck hunting areas in California, the lower Mississippi River Valley, along the Lake states, and in the Chesapeake Bay have lease fees ranging from $100 to $2000 per hectare and inherited memberships in prestigious hunting clubs may cost $5000 to $10,000 per year. Although waterfowl hunters place a high value on these wetlands, the desirable entity represents the functional value produced by many different wetlands, ranging from Arctic or prairie nesting grounds and mid-latitude migratory areas as well as the highly sought after wintering habitats.

Determining accurate, representative estimates for life support functional values has historically been difficult, largely due to the inability to employ accepted valuation methods developed for marketing. The intangible values generated by recreational activities may prove to be the most important in preserving, restoring, or creating wetlands. Although other interest groups have begun to appreciate wetland values, the historical pattern of interest and activism by hunters, wildlife watchers, photographers, and researchers created the heightened awareness in present society that led to a variety of legal and regulatory measures to protect and restore our wetland resources. Placing a market value on an evening of frog calls from a Louisiana bayou, morning clouds of geese lifting from a marsh, or afternoon dragonflies patrolling a sedge meadow may be no more likely than determining the value of a letter or visit to a congressman. However, the dedicated appreciation of those interested in the intangible life support values has made wetlands a household word and wrought surprisingly rapid changes in public attitudes towards wetlands. Some other functional values may be more easily quantified and more persuasive to decision makers but the life support functions underlie much of our historical and current interest and they deserve the credit if we do manage to reverse present trends and expand rather than reduce our nation's wetland resources.

WATER PURIFICATION

Wetland ecosystems have intrinsic abilities to modify or trap a wide spectrum of water-borne substances commonly considered pollutants or contaminants. Doubtless our ancestors perceived and exploited these abilities; but in more recent times, casual observations fostered renewed interest, leading to investigations that documented changes in concentrations of various materials after processing by natural wetland systems. Much of the early work on constructed wetlands for wastewater treatment was stimulated by observing this purification phenomenon in natural wetland systems.

Early ecological studies of the processes, reservoirs, cycling, and substance dynamics in wetlands revealed that natural wetlands receive and process many forms of nutrients and energy transported from adjacent upland regions by inflowing waters. Ironically, many of these investigators set out to evaluate detrimental impacts to wetlands from receiving various forms of wastewater. Most were surprised to discover that little impact was evident but instead, the wetland removed many of the contaminants from the inflowing waters and discharged relatively clean waters. Both saltwater and freshwater wetlands were shown to have very important roles in the natural cycling of organic and inorganic materials. Concurrently, these studies showed that the immense productivity of wetlands was related to the inputs of nutrients and energy from terrestrial environments. In addition, the important role of wetlands as short-term and long-term sinks (reservoirs) became apparent because wetlands are often the major reducing component in the environment. Transforming nutrient and energy inputs to increased biomass and innocuous products within the wetland system while concurrently discharging cleansed waters even led to some workers labeling wetlands the "kidneys" of the environment.

Many observers have noticed accelerated soil erosion after heavy rains wash across un-vegetated soils and some were fortunate to encounter situations where silt-laden waters transiting natural wetland systems were readily compared with unprocessed waters. The striking visual differences were easily verified by sampling and analysis, and the information became an important component in a communal body of knowledge on natural wetland values. Most ecologists believed this phenomenon was widespread and a few even suggested that it might occur on a large scale, though little documentation was available. I recently had the opportunity to observe an example of water quality improvements in river waters by a natural wetland system on a very large scale.

The Pantanal of western Brazil and adjacent portions of Paraguay and Bolivia is a large basin bordered by high plateaus on the east (savannah) and north (semi-deciduous forest) and a moderate mountain range (semi-deciduous forest) on the west. Runoff from these regions causes much of the 11,000,000-ha area to be flooded from December to June, and a significant but unmeasured proportion is permanently wet. Although Pantanal means marsh or swampland in Portuguese, the region is more correctly termed a large plain with a considerable amount of permanently inundated area in old river channels, meanders, lakes, smaller depressions, and potholes. In fact, an aerial overview accentuates the striking physiographic similarities with the prairie pothole region and the high Arctic in North America (absent the woody vegetation of course). An important difference is the presence of many rivers entering the Pantanal from the eastern highlands, gradually disappearing, and then reforming on the western and southern boundaries and draining off to the south. Over geological time, alluvial deposits of highlands silt has gradually transformed a flat or concave basin floor into a convex dome-like surface with higher elevations in the center and lower on the margins.

Doubtless this region provided important water improvement functions as tectonic forces created the original basin. The accumulated deposits that formed the present cross-sectional profile are dramatic evidence of previous beneficial modification of inflowing river waters; but accelerated erosion and pollution from clearing and agricultural activities and other anthropogenic sources has tremendously increased the contaminant loading of rivers draining the plateaus on the east and north. The Rio Taquiri alone carries over 30,000 metric tons of silt per day plus a variety of agrichemicals from soybean fields on the eastern plateau. Other rivers transport lower silt loads but most receive untreated sewage, industrial and mining pollution before reaching the Pantanal. For example, one iron ore mill used 4.8 kg of detergent per day for washing ore stacks along the Rio Correntes, gold miners use (and lose) 36,000 kg/year of mercury along the Rio Couros and the Rio Aqua Branca, and eight alcohol distilleries (fuel) discharge 3,600,000 liters/day of organic waste ("vinhoto")

Figure 5-16. Note the difference in river dimensions on the east and west but not the middle of this schematic diagram of the Pantanal, Mato Gosso do Sul, Brazil.

into rivers draining the northern plateau. The combined impact of increased pollutant loadings has caused recent hydrologic and biological changes in the upper reaches of the Pantanal that are of concern to Brazilians and conservationists worldwide.

The amazing fact is that alarmingly high concentrations of silt and pollutants in inflowing river waters are reduced to innocuous levels in waters of rivers draining the region. Examination of a topographic map (abstracted in (Figure 5-16) provides insight into the overall processes if not the complex of mechanisms. Notice the size, especially width, of the Rio Taquiri as it drops off the plateau and enters the Pantanal on the east. A fairly wide, deep, and fast-flowing river courses out into the Pantanal and rather quickly its width, depth, and velocity are reduced. A third of the way into the Pantanal, numerous small braided streams arise, flowing perpendicularly out of the Rio Taquiri into the adjacent regions. Progressively increasing water loss with penetration into the Pantanal drastically reduces the Rio Taquiri until it almost disappears. A similar pattern is evident in the course of the Rio Aquidauana. Many of the rivers flowing into the Pantanal virtually disappear because of sheet flow and dissemination through very small waterways in this vast wetlands region (Figure 5-17).

In fact, the Pantanal functions as an 11,000,000-ha sponge that absorbs inflowing waters, cleanses them of impurities, and slowly releases clean water through minor streams that aggregate into larger rivers along the southern and western boundaries. This large natural wetlands complex transforms heavily

Figure 5-17. The channel of the Rio Taquiri gradually disappears with distance into the Pantanal concurrent with initiation of hundreds of small streams that coalesce to form rivers on the western shores.

polluted influent waters into clean waters collected by the Paraguay River and used throughout much of southern South America. Not only does it provide clean water, the slow release of waters collected during the rainy season augments base flow in the Rio Paraguay during the dry half of the year, supporting adequate year-round supplies for communities, navigation, and other human and natural uses.

On a large scale as well as in local areas, natural wetlands can perform substantial improvements in water quality and quantity despite abnormal conditions. Even though the system is grossly overloaded and significant changes have occurred in wetland of the upper Pantanal, the natural wetland complex of the total system still provides valuable water improvements for downstream rivers. However, major land use changes in the surrounding plateaus causing accelerated pollution have only occurred within the last 10 to 15 years and alterations in the Pantanal are already evident. Undoubtedly, continued overloading will soon destroy the water improvement function, as well as important other functional values of the Pantanal in the near future. More urgently, the Hydrovia Project, a 3,000-km waterway would dig a navigation channel through the Pantanal and likely drain much of the area. Since the complex has trapped a variety of substances, including heavy metals, over many years, drainage would likely remobilize these substances to the detriment of downstream water users. Protecting the water improvement function of the Pantanal will require extensive changes in land use, government policy, and economics of world markets. Unfortunately, other important functional values of this world-class wetland are likely to be severely depressed before water purification is significantly damaged.

Natural wetlands receive inputs of nutrients and energy from waters flowing off surrounding upland regions. The complex of plants and animals in depressional wetlands have adapted to and, in fact, thrive on these imported substances. Over time, we have discovered how natural wetlands process, recycle, and trap these nutrients and energy. This knowledge has enabled us to deliberately design constructed wetlands that emulate this function of natural wetlands to provide low-cost, low-maintenance, but highly effective treatment for a variety of wastewaters.

Over many years our knowledge of water purification in natural wetlands has slowly increased leading to fairly rapid developments in constructed wetland technology. Ironically, current and future research on wastewater treatment in constructed wetlands systems is likely to substantially improve our understanding of this important functional value in natural wetlands.

Natural wetlands provide effective, free treatment for many types of water pollution. Wetlands can effectively remove or convert large quantities of pollutants from point sources (municipal and certain industrial wastewater effluents) and non-point sources (mine, agricultural, and urban runoff) including organic matter, suspended solids, metals and excess nutrients. Natural filtration, sedimentation, adsorption, microbial decomposition and other processes help clear the water of many pollutants (Figure 5-18). Some are physically or chemically immobilized and remain there permanently unless disturbed. Chemical reactions and biological decomposition break down complex compounds into simpler substances. Through absorption and assimilation, wetland plants remove some nutrients for biomass production. One important by-product of plant physiology is the release of oxygen, which increases the dissolved oxygen content of the water and also of the soil in the immediate vicinity of plant roots. This increases the capacity of the system for aerobic bacterial decomposition of pollutants as well as its capacity for supporting a wide range of oxygen-using aquatic organisms, some of which directly or indirectly utilize additional pollutants.

Some nutrients are held in the wetland system and recycled through successive seasons of plant growth, death and decay. If water leaves the system through seepage to groundwater, filtration through soils, peat, or other substrates removes remaining nutrients and other pollutants. If water leaves over the surface in winter, excess nutrients released from decaying plant tissues have less effect on downstream waters during the non-growing season. In summer, nutrients trapped in substrate and plant tissues during the growing season do not contribute to noxious algae blooms and excessive aquatic weed growths in downstream rivers and lakes.

It is well known that natural wetlands can remove iron, manganese, and other metals from acid drainage — they have been doing it over geological time periods. In fact, accumulations of limonite, or bog iron, were mined as the source of ore for this country's first ironworks. Limonite deposits are most

Figure 5-18. A large wetland complex furnishing recreational and educational opportunities in Coyote Hills Regional Park, as well as providing stormwater treatment for Fremont, California.

common in the bog regions of Connecticut, Massachusetts, Pennsylvania, New York, and elsewhere along the Appalachians. Although now of limited economic importance in the United States, bog iron is still a significant source of iron ore in northern Europe.

Since wetlands transform or remove pollutants from inflowing waters, the ultimate fate of certain substances within the wetland ecosystem is of more than academic interest. Depending upon the source, influents may contain various natural and anthropogenic organic compounds, metals including heavy metals, pathogenic organisms, salts, etc. A few materials (i.e. selenium, arsenic) are selectively taken up by plants but most are precipitated or complexed within the substrate. Generally, only 4 to 5% of the nutrient loading on a wetland system is incorporated into plant or animal tissue. However, some metals may occur in plants at relatively high concentrations. For example, iron levels as high as 5000 mg/kg and manganese levels up to 4100 mg/kg were present in cattail leaves and stems grown in experimental cells heavily loaded with acid mine drainage. However, similar concentrations occurred in cattail parts from a natural wetland not receiving acid drainage. In either case, only traces of other metals were present. Copper was below detection limits in cattail from a natural marsh but averaged 6.1 mg/kg in cattail from two municipal wastewater treatment systems. But higher concentrations of lead were found in a natural cattail stand (1.7 mg/kg) than in cattail from the municipal systems (0.3 mg/kg).

Though only low levels of potential toxic metals occurred in these samples, long-term effects of relatively high levels of iron and manganese are not known. Short term, iron and manganese did not appear to have detrimental effects on cattail growth and vitality in the experimental cells. In fact, plants in the upstream portion of each cell were more robust than plants in the lower sections. Upper portions of each cell received raw inflowing acid mine drainage that probably contained small concentrations of micro-nutrients in addition to substantially higher concentrations of iron and manganese. Differential plant robustness within each cell was likely due to micro-nutrient uptake in the upstream portion and limited micro-nutrient availability to plants in the lower sections.

PURIFICATION PROCESSES

Recent work with constructed wetlands has shown that wetlands accomplish water improvement through a variety of physical, chemical, and biological processes operating independently in some circumstances and interacting in others. Vegetation obstructing the flow and reducing the velocity enhances sedimentation and many substances of concern are associated with the sediment because of clay particle adsorption phenomena. Increased water surface area for gas exchange improves dissolved oxygen content for decomposition of organic compounds and oxidation of metallic ions. But the most important process is similar to the decomposition occurring in most conventional treatment plants — only the scale of the treatment area and composition of the microbial populations is likely to be different.

In both cases, an optimal environment is created and maintained for microorganisms that conduct desirable transformations of water pollutants. Maintaining that environment in the small treatment area of a package plant requires substantial inputs of energy and labor. Wetland systems use larger treatment areas to establish self-maintaining systems providing environments for similar microbes, but also supporting additional types of microorganisms because of the diversity of micro-environments. The latter, along with a larger treatment area, frequently provide more complete reduction and lower discharge concentrations of water-borne contaminants. Regardless, most removal or transformation of organic substances in municipal wastewaters or metallic ions in acid mine drainage is accomplished by microbes — algae, fungi, protozoa, and bacteria. Wetlands, as do conventional treatment systems, simply provide suitable environments for abundant populations of these microbial populations.

Components

Water purification functions of wetlands are dependent upon four principle components — vegetation, water column, substrates, and microbial populations. The principle function of vegetation in wetland systems is to create additional environments for microbial populations. Not only do the stems and leaves in the water column obstruct flow and facilitate sedimentation, they also provide substantial quantities of surface area for attachment of microbes (reactive surface). In addition to the microbial environments in the water column of lagoons, wetlands have much additional surface area on portions of plants within the water column. Plants also increase the amount of aerobic microbial environment in the substrate, incidental to the unique adaptation that allows wetland plants to thrive in saturated soils. Most plants are unable to survive in water-logged soils because their roots cannot obtain oxygen in the anaerobic conditions rapidly created after inundation. However, hydrophytic or wet-growing plants have specialized structures in their leaves, stems, and roots somewhat analogous to a mass of breathing tubes that conduct atmospheric gases (including oxygen) down into the roots. Since the outer covering on the root hairs is not a perfect seal, oxygen leaks out, creating a thin-film aerobic region (the rhizosphere) around each and every root hair. The larger region outside the rhizosphere remains anaerobic, but the juxtaposition of a large, in aggregate, thin-film aerobic region surrounded by an anaerobic region is crucial to transformations of nitrogenous compounds and other substances. Wetland vegetation substantially increases the amount of aerobic environment available for microbial populations, both above and below the surface. Wetland plants generally take up only very small quantities (<5%) of the nutrients or other substances removed from the influent waters, although some systems incorporating periodic plant harvesting have slightly increased direct plant removals at considerable expense.

Recent experiments have shown that plant architecture (using temperature and relative humidity differentials between various portions of the plant) increases gas exchange beyond levels expected from passive (air tube) transport. However, attempts to compute oxygen mass introduced into the substrates by radial oxygen loss have been confounded by a number of variables that lack precise definition. Although earlier experiments suggested that plants with deep root structures had higher removal efficiencies, recent results suggest that plant species with dense, fibrous though shallow roots have lower radial oxygen loss per unit of plant biomass, but may input larger quantities of oxygen into the substrate because they tend to grow in denser stands.

More importantly, plants create and maintain the litter/humus layer that functions as a thin film bio-reactor. As plants grow and die, leaves and stems falling to the surface of the substrate create multiple layers of organic debris — the litter/detritus/humus/peat component of wetlands (Figure 5-19). This accumulation of differentially decomposed biomass creates highly porous substrate layers that provide a substantial amount of attachment surface for microbial organisms. Decomposing biomass also provides a durable, readily available carbon source for microbial populations. The water quality improvement function in wetlands is principally dependent upon the high conductivity of this litter/humus layer and the large surface area for microbial attachment. Wetland vegetation substantially increases the amount of environment (aerobic and anaerobic) available for microbial populations within the water column and below the water — substrate interface. Consequently, the most important role of the plants is to simply grow and die, which explains why removal performance efficiencies of different plant species tend to be similar within broad ecological categories, i.e., emergents, submergents, and wet-meadow plant

Figure 5-19.   Fallen stems of bulrush produce the matrix that performs as a thin film bio-reactor in wastewater treatment.

species. This also explains why vegetation harvesting is not needed and could in fact, be detrimental to pollutant removal performance of a wetland.

Plants generally take up only very small quantities (<5%) of the nutrients or other substances removed from the influent waters. Organic pollutants are often reduced to elemental forms with some (nitrogen, carbon, hydrogen) being released to the long-term bio-geo-chemical reservoir in the atmosphere. Inorganic materials are transformed to innocuous substances and/or precipitated and trapped in the substrates. By contrast, floating aquatic systems (water hyacinth and *Lemna*) must incorporate periodic plant harvesting to increase nutrient removal at considerable operating expense. Unfortunately, floating aquatics systems enjoyed wide use and the concept that the plants extract nutrients in floating aquatic systems has erroneously carried over into common understanding of processes in natural and constructed wetlands.

Microbes (bacteria, fungi, algae, and protozoa) alter contaminant substances to obtain nutrients or energy to carry out their life cycles. In addition, many naturally occurring microbial groups are predatory and will forage on pathogenic organisms. The effectiveness of wetlands in water purification is dependent on developing and maintaining optimal environments for desirable microbial populations. Fortunately, these microbes are ubiquitous, naturally occurring in most waters and likely to have large populations in wetlands and contaminated waters with nutrient or energy sources.

Invertebrate and vertebrate animals harvest nutrients and energy by feeding on microbes and macrophytic vegetation, recycling and in some cases transporting substances outside the wetland system. Functionally, these components have limited roles in pollutant transformations but they often provide substantial ancillary benefits (recreation/education) in successful systems. In addition, vertebrate and invertebrate animals serve as highly visible indicators of the health and well-being of a wetland ecosystem, providing the first signs of system malfunction to a trained observer. Some invertebrates and many vertebrates occupy upper trophic levels within the system that are dependent upon robust, healthy populations of micro and macroscopic organisms in the critical lower levels. Declines in lower level populations (including those involved in pollutant transformations) are reflected in changes in more visible animals in the higher levels. However, observations on types and numbers of indicator species must be carefully interpreted by an experienced wetland ecologist since certain species thrive in overloaded, poorly operating systems.

PURIFICATION VALUES

Economic value for water purification has only recently been appreciated but is much less difficult to estimate than the life support functions because replacement cost is an appropriate and accepted method of estimating value. However, two approaches are possible that lead to quite different estimates. For example, a wetland system that provides treatment of municipal wastewaters to permitted discharge standards is worth the cost of building and the annual operating cost for a conventional package plant producing similar quality effluent. The wetland treatment system may cost $250,000 for a community of 1000 people and a comparable conventional system may cost $2.5 million. Operating costs for the wetland system may average $10,000 per year and the conventional system may cost $100,000 per year to operate. Hence, the wetland treating this wastewater is worth $2.0 million initially and $90,000 per year thereafter.

In cases where constructed wetlands have been compared to conventional package treatment plants, the cost difference favors the wetland system and provides valid estimates of the water purification functional value of natural wetlands. However, even if the community builds a constructed wetland treatment system, the residents of the community will not equally value a comparable natural wetland system downstream on the receiving waters. The natural system may have been performing the same function, but the community is not credited for the purification function of the natural system; hence, it is of little value to them. Our regulations require that the community treat its wastewater to specified levels for various parameters before discharging to a receiving stream or a natural wetland. Consequently, the natural wetland has little value to the community. In a few cases, partially treated wastewater is discharged into natural wetlands for polishing and the community manages and receives credit for water purification performed by the natural system. Doubtless the residents of these communities appreciate the water purification functional value of wetlands each time they consider higher rates for city services.

The rest of us do not receive a direct, quantifiable benefit from natural wetlands, but we do benefit in a less measurable manner. If natural wetlands remove various pollutants from surface waters that would otherwise increase the cost of treating those waters to drinking water standards, we do benefit; but determining the value of that benefit is more difficult in all but a few cases. Where acid mine drainage with high iron and manganese concentrations flows into drinking water supplies and those metals must be removed to meet drinking water standards, the cost of that removal in a conventional treatment system may be taken as the value of the wetland systems performing similar treatment.

Unfortunately, comparisons with organic contaminants are much less clear because substantial modification often occurs in intervening water bodies even though the aquatic life in those waters may be heavily impacted by the organic loading. Poorly treated wastewater may decimate aquatic life for a considerable stretch of the receiving stream or river, but gradual improvement with travel distance often produces fairly clean water that needs little, low-cost treatment to be adequate for drinking water. In this case, a wetland between the low-quality wastewater discharge and the receiving stream would have little direct monetary value to either community though it would have considerable value to fish and other life forms and related commercial or recreational activities on the river. However, the latter value is less easily quantified and less obvious to regional residents.

Natural wetlands, especially river swamps and wetland areas in Western riparian zones, provide substantial benefits in treating sediment, nutrients, and agri-chemicals in runoff from row crop fields and pastures protecting aquatic life in streams, rivers, and estuaries. The value of that treatment can be estimated given measurements of the annual contaminant load carried into the streams and rivers, again in terms of replacement cost (i.e., the cost of building and operating conventional facilities to accomplish the same treatment). However, the direct value is again less clear because the real worth is the value for sport and commercial fisheries or other activities based on the finfish, shellfish, bird, or mammal resources in the rivers, lakes, or estuaries. Obviously, these are valued by our society or we would not have statutory requirements on municipal or industrial discharges designed to protect these resources; but we do not have similar requirements on non-point source pollution, principally agricultural wastewaters, though recent regulations will require management of urban stormwater runoff. Consequently, the value for treatment of dispersed but significant agricultural and urban pollution performed by millions of hectares of wetlands throughout the nation is imprecise and unappreciated.

Fortunately wetlands have, for thousands of years, and continue to remove contaminants from surface waters, preserving life forms in downstream ecosystems and protecting drinking water supplies regardless of whether accurate valuations are possible or whether due appreciation develops. Wetlands can continue to perform water treatment even though so heavily impacted by other activities that most of the life support and other functional values have been lost.

HYDROLOGIC MODIFICATION

Although marshes and bogs also perform this function, floodwater modification is most often identified with bottomland hardwood swamps. But the value of marshes has recently gained greater acceptance. More than one observer has suggested that flooding intensity in the summer of 1993 was greatly increased by wetland drainage in the Midwest. In fact, given the extent of wetland drainage in the region and simply reviewing a map of previous wetland distribution and riverways would suggest that the loss of wetland water retention was quite likely a significant factor in flood severity. For example, wetland drainage has eliminated almost 1,000,000 ha in North Dakota (49%), 300,000 ha in South Dakota (35%), 2,500,000 ha in Minnesota (32%), 1,450,000 ha in Iowa (89%), 1800 ha in Wisconsin (46%), 2,800,000 ha in Illinois (85%), and 1700 ha in Missouri (87%). In the north, many of these wetlands were prairie pothole marshes randomly, but in many areas, rather evenly distributed across the landscape. Many hectares were floodplain marsh or forested wetlands and in Illinois and Missouri, most were forested wetlands. It's not difficult to imagine what effect the loss of millions of hectares of storage area may have had on the severity of those floods. Nor to suspect that rapid conveyance of runoff (ditches, drains, canals, levees) accentuated downstream impacts.

Forested wetlands in river floodplains have dramatic effects on peak flows of flood waters and also on base flows during dry periods. By directly obstructing and slowing flow velocity and by acting as natural reservoirs, wetlands substantially reduce the height of downstream floodwater peaks and the frequency and duration of flooding. Conversely, by retaining flood waters and slowly releasing them over extended periods, wetlands augment base flows, thereby protecting aquatic life in rivers and streams. The combination of attenuated peaks and augmented base flows results in continuous, moderate water levels in rivers influenced by wetlands as compared to elevated flood peaks, high velocities and, soon thereafter, virtually dry streambeds in rivers without the moderating influence of wetlands. Rapidly growing appreciation of this functional value has stimulated consideration of "non-structural" flood control proposals that employ wetlands and other natural vegetation and landforms rather than earthen or concrete dams and reservoirs to protect our farmlands and cities (Figure 5-20).

Only a few attempts have been made to develop quantitative estimates of the value of this important function and most use replacement cost, either the cost for flood damages or the cost for constructing and operating a flood storage dam and reservoir. A classic evaluation done by the U.S. Army Corps of Engineers in the St. Charles River basin in Massachusetts concluded that drainage of 3400 ha of forested wetland would increase downstream flood damages by $17 million per year. A 40-ha swamp in Illinois was found to store over 8% of the total flood water runoff, and inflow-outflow measurements

Figure 5-20. Floodplain wetlands desynchronize and reduce peak flows and then gradually release flood waters that augment base flows, creating buffered moderate river flows during wet and dry seasons. But clearing for agriculture has dramatically reduced wetland coverage and modifying effects.

of a marsh-bog-forested wetland complex in northern Minnesota yielded estimates of flood peak reductions of 0.2 to 0.5 m at downstream communities. Other studies have shown that watersheds with 40% lake and wetland area have flood peaks only 20% as large as watersheds with little or no wetland area. For large floods, modeling has suggested that the flood reduction value of wetlands seems to increase with the size of the flood, the farther down the watershed the wetland is located, and increased wetland area. A large wetland in the lower reaches of the watershed during a high flood event has more effect on reducing flooding but smaller wetlands in the upper reaches of the watershed may have greater impact during low flood events.

Groundwater recharge is a related function and directing a proportion of surface waters underground during a flood event is another means by which wetlands can reduce downstream flooding while replenishing groundwater supplies. However, the significance and magnitude of this function is poorly understood, with some wetlands undoubtedly contributing waters to underground sources but others quite obviously not. The latter is not surprising since most wetlands are underlain by impermeable materials and hydrologically separated from groundwater. Groundwater recharge has been shown to occur in isolated wetlands such as prairie marshes, cypress domes, and floodplain forests. A few studied wetlands in Wisconsin, North Dakota, and Florida had direct connections and contributed significantly to groundwaters, in one case affecting groundwater over an area of over 400 km^2. But other wetlands have been shown to have little influence on groundwater or, in some cases, the wetland is present because of surfacing groundwaters. Since wetlands with substantial percolation losses may occur in the same vicinity as wetlands supported by emerging groundwaters, relationships are not well understood and rarely quantified.

Determining the value for groundwater recharge is similar to that for water purification; that is, replacement cost. If a wetlands contributes to groundwater supplies, the value of that function is the cost of supplying similar quantities of water from some alternative source or perhaps the cost of pumping water to the surface from deep wells compared to pumping costs for shallow wells. The latter could be important in irrigation areas such as the farmlands underlain by the gradually falling Ogalla Aquifer and other regions with declining water tables in the western U.S.

EROSION PROTECTION

Shoreline erosion is caused by tidal currents along coasts, river currents during flooding, and wind- or boat-generated waves on lakes and reservoirs. Wetlands reduce shoreline erosion by absorbing and dissipating wave energy, by binding and stabilizing shoreline substrates, and by enhancing deposition of suspended sediments. This function is perhaps most valuable along coastlines and barrier islands, but can occasionally be important along rivers, the Great Lakes, and on reservoir shorelines. Though few studies have documented shoreline stabilization for inland waters, a number of investigations have shown that unvegetated shorelines retreat at up to four times the rate of shorelines protected by saltwater marshes. Wave-caused erosion is a serious problem on reservoir shorelines in the Upper Missouri system due to frequent high winds and long fetches across wide water bodies. The U.S. Army Corps of Engineers has conducted a number of experiments and pilot projects to vegetate these shorelines to reduce property loss and downstream dredging and channel maintenance costs.

Shoreline protection can also be valued using replacement costs; in this case, the value of property losses, storm surge flooding damages, and costs for channel maintenance. Other benefits, including reducing turbidity of waters, reducing siltation, and smothering of fishery and wildlife habitat and aesthetics, are no less important but much more difficult to quantify.

OPEN SPACE AND AESTHETICS

Many people are attracted to natural environments — those that seem untouched by man — and wetlands offer an abundance of opportunities for a "primitive" experience. Wetlands also often rank high in aesthetic value probably because of the multitude of and complex intermingling of the land-water interface that has broad appeal (Figure 5-21). Owing to the variety of waters,

Figure 5-21. Wetlands in urban environments can increase property values, provide open space, aesthetics, and unique opportunities for education and recreational as well as stormwater treatment as at Lake Greenwood in central Orlando.

land forms, plants, and animals, wetlands are full of different shapes and textures, stimulating visual senses and smells and sounds to satisfy other sensual experiences. The natural appeal is understandable since wetlands are often some of the last areas in the landscape to undergo development.

CULTURAL RESOURCES

Delayed or limited development in combination with the anaerobic, reducing environments in wetlands protects and preserves historical and anthropological resources. Ancient and historical cultures exploited the rich natural resources in and adjacent to wetlands, leaving behind evidence of themselves and their lifestyles that may be surprisingly well preserved in the anaerobic, acidic environments of some wetlands. Since the remains are often well preserved and the sites undisturbed by development, wetland archaeological sites are often extremely valuable study sites.

RECREATION AND EDUCATION

Wetlands are optimal areas for environmental education because of the facility with which important scientific principles can be demonstrated and observed. Basic principles of ecology — succession, trophic levels, food webs, and nutrient and energy cycling — are more easily shown in a small beaver pond than almost any other type of ecosystem. Of course, research on wetlands has contributed in many ways to our overall understanding of our environment.

Wetlands support many types of direct recreation including hunting, trapping, fishing, wildlife watching, nature photography, berry picking, picnicking, hiking, walking, and boating, some of which have been discussed in the life support section. In a few instances, the values for these activities have been estimated with unit day, travel cost, or willingness to pay methods. Unfortunately, none of these are widely accepted in economic circles and these important functions are generally not well appreciated.

BIO-GEO-CHEMICAL CYCLING

Most of the functions described above are short-term, though some may extend over tens or even hundreds of years. Wetlands have another very significant but long-term functional value that is commonly overlooked and poorly understood. Wetlands function as "sinks," as traps for a variety of substances that may be immobilized in enduring wetlands or the deposits created by wetlands for long time periods from a human perspective but relatively short in a geological sense. Peat that has been mined from wetlands for ages for low quality fuel is a concentrated organic mass representing the initial stages in the formation of coal and petroleum. Most of our coal and petroleum deposits resulted from carbon fixation, or immobilization, in a wetland system and subsequent transformations under high pressure and temperatures. In fact, the glitter on my lady's finger likely originated in a dark and dismal swamp! On a worldwide scale, wetlands continue trapping carbon in gradually deepening deposits that will someday become coal or petroleum. Current thinking suggests that the carbon dioxide concentration in the atmosphere is increasing primarily because of the burning of fossil fuels — coal and petroleum. Since wetlands can immobilize carbon for thousands and millions of years instead of the tens or hundreds of years expected by growing more trees, would it not be less costly and more efficient to remove carbon from the atmosphere for long-term storage in restored or created wetlands than by encouraging tree planting?

Perhaps more importantly, trees and other planted vegetation do not immobilize significant quantities of sulfur, a major constituent in acidic precipitation. Sulfates that fall on or are washed into marshes, bogs, and swamps are reduced to sulfides which react with metallic ions to form insoluble substances that gradually accumulate in the organic mass that becomes peat, then coal or oil. Depending on inflow concentrations, much of the sulfur may be complexed with iron, manganese, or other metals. Again, much of the increase in atmospheric concentrations of sulfur are believed due to fossil fuel burning. As with carbon, the most effective and least costly method to remove sulfur from short-term biochemical cycling in the atmosphere-water-soil compartments is to restore and create wetlands. The processes in wetlands that formed sulfur-containing fossil fuel deposits are just as active today as they were millions of years ago. Should we not consider restoring and creating wetlands to enhance and improve the value of this wetland function as a long-term solution to reduce atmospheric imbalances?

Wetlands also created many of the sedimentary mineral deposits that are important sources for our metal industries. Bog iron and wad manganese deposits were some of the earliest sources of raw materials for the iron industry in Europe and North America, though both are of limited importance today. These concentrated deposits were removed from inflowing waters over thousands of years and laid down as metallic ions by natural wetland systems. The process is ongoing today and, in fact, increasing under deliberate efforts to build wetlands for treatment of acid mine drainage. Treatment wetlands, along with innumerable natural systems, remove dissolved metals contaminating inflowing waters, oxidize, and later reduce them to insoluble sulfide compounds. Though each new ore deposit is small, hundreds of constructed and thousands of natural systems are operating in the eastern U.S. and Canada alone and the aggregate over 50 to 100 years will be a considerable body of highly concentrated, easily processed iron ore. These shallow, easily mined deposits of rich ore will be less expensive and less environmentally damaging to use than present sources. The complex bio-geo-chemical processes creating iron and manganese ores also bind and remove substantial quantities of sulfur from short-term, near-surface cycles reducing one of the atmospheric components of acid rain (Figure 5-22).

Figure 5-22. Wetlands are often the major reducing environment in the landscape; consequently, they have a major role in transforming or trapping many organic and inorganic substances.

Finally, many geo-chemists have noticed that mineral deposits in sedimentary rocks often have convoluted, serpentine patterns suggestive of ancient river courses. Not only ferrous materials but other metals, heavy metals, and even uranium deposits occasionally exhibit this unusual spatial distribution. Quite likely, these deposits were laid down along ancient rivers; of course, the determining factor was not the river per se, but the wetlands in the river's flood plain where similar biochemical processes were operating as we find in bog iron and wad manganese forming wetlands today.

Amid the worldwide concern over elevated concentrations of carbon and sulfur causing atmospheric changes that exacerbate acid precipitation and global warming, the most widely accepted solutions seem to be reducing sulfur and carbon releases and planting trees for carbon fixation. Discharge reductions may slow the trends but are unlikely to bring about significant reversals and growth rates, for even tropical forests pale in comparison with the biomass production rates of a vigorous cattail marsh. Furthermore, most tree planting campaigns employ fast-growing but relatively short-lived species that are unlikely to immobilize fixed carbon for more than 50 to 100 years.

Since measured carbon fixation rates in marshes are more than double the rates for forests, marshes immobilize sulfur and myriad other metals as well as carbon, and the accumulated materials are subtracted from short-term soil-water-atmosphere cycling, should we not give serious consideration to actively promoting marsh and bog restoration and creation? Is it merely coincidental that acid precipitation and global warming trends accelerated concurrently with increased use of fossil fuels and destruction of extensive areas of natural wetlands? Perhaps we have not only increased the rate of carbon and sulfur withdrawal from long-term reservoirs (peat, coal, and oil), but simultaneously reduced the rate of deposits into those reservoirs by eliminating the carbon and sulfur storage processes of extensive natural wetland systems.

In the final analysis, trapping carbon, sulfur, and metallic ions in long-term storage reservoirs and eliminating their immediate impacts on the world's atmosphere and climate may be one of the most important functional values that wetlands provide to our society.

CHAPTER 6

DEFINING YOUR OBJECTIVES

CREATION OR RESTORATION?

Restoring a natural wetland has substantial advantages over designing and building a functional wetland from scratch. If your objective is to establish a wetland in some general area, first determine whether impacted or damaged wetlands occur nearby. If you are fortunate in finding one, then evaluate the existing wetland and try to determine previous conditions. What factor(s) has changed that degraded, reduced, or eliminated the previous wetland? Evaluate the status of the current hydrology, soils, and vegetation community in the area of interest. Is it possible or feasible to re-establish the requisite hydrology or soils? Or would it be less expensive to construct a new wetland at some other site? Examine existing land use patterns in and adjacent to the old wetland site. Would a restored wetland on this site complement or detract from adjacent land uses or would nearby activities negate your restoration efforts or detrimentally impact functional values of a restored wetland?

In the worst case, you may discover that a previous wetland has been completely filled and paved over for a parking lot. Restoration of this system may be almost as difficult as creating an entirely new wetland although past ground water levels and other hydrological factors may still be advantageous. At the other extreme, thousands of wetlands have only been ditched and drained, and installing water control structures or plugs at appropriate locations in the ditch system may be all that is needed. Simply plugging the ditches is not recommended, even though it may be initially successful, since long-term maintenance of the restored wetlands will likely require deliberate water level management to insure optimal productivities. However, in the upper Midwest, periodic drought provides the disturbance factor that enhances nutrient recycling and retards succession so that a ditch plug will likely restore the original hydrology. But carefully examine nearby wetlands in your evaluations. The selected wetland might have subterranean hydrologic connections with other wetlands in the vicinity and waters impounded in one wetland may drain underground to another ditched wetland. Restoration might require plugging ditches throughout the whole complex of wetlands.

Even if the site has been farmed for many years, the soils may contain much of the original seed bank and an appropriate schedule for re-flooding will quickly restore much of the vegetative complex. Seeds of many wetland plants exemplify maximal ability to survive under favorable conditions — in some cases for hundreds of years — and then germinate and sprout at the onset of suitable conditions. In addition, wildlife using the re-flooded area transport seeds and propagules of many wetland plants, occasionally over long distances. Inundating a previously farmed prairie pothole will restore a diverse and productive prairie marsh within 5 to 10 years, but restoring the form and function of a bottomland hardwood swamp from a soybean field may take 50 to 75 years or longer after the ditches are plugged. In both cases, much of the seed bank from the previous wetland exists within the soils of the agricultural fields, but growth and maturation times are quite different for cattail, bulrush, and pondweeds compared to oaks, cypress, and gums.

But in one sense, starting with a clean slate (previous soybean field) may be easier than attempting to restore a degraded forest even though the hydrology may be present. Much of our remaining forest wetlands have been seriously degraded by past harvesting practices. Restoring the original bottomland hardwood forest must often go beyond simply restoring the appropriate hydrology. Similar to conditions in many upland forests, over the years cutting practices have removed the most valuable species and the most valuable individuals within species, frequently leaving a mixture of culls, edge, and early successional species. In many cases these rejects will inhibit regeneration of the original dominants for many years. Consequently, in addition to planting and/or seeding methods described in Chapter 13, developers may need to employ prescribed burning, felling, or injecting culls, herbicide applications, clear cutting, selective harvesting, and other forest management practices to restore the forested wetland.

In the Mississippi Delta, simply breaching the overbank allowing river waters to re-enter cut-off meanders and oxbows is a very low-cost restoration method. Breaching levees also would allow flood waters to sustain wetlands on previously farmed fields throughout much of the floodplain of many of our major rivers. In either case, early successional stages (marshes) are likely to become established and if the objective is forest restoration within any reasonable time frame, tree planting will likely be required. In the latter case, developers may need extensive tracts of land and/or support and mutual agreement on objectives from many other landowners and even towns people. Potential flood impacts will need careful evaluation as well though many observers have suggested that existing levees have increased the severity of recent flood events.

In some cases, you may wish to expedite re-establishment of the natural vegetation or animal communities through deliberate planting or stocking. Bear in mind that the best source for planting or stocking materials is a nearby natural wetland since that stock will be adapted to the climatic and edaphic conditions of your area. Secondly, use stock from nurseries or suppliers located at the same latitude (remember that altitude translates to latitude) as your site. Do not attempt to introduce species that do not (or did not) occur in natural wetlands near your restoration site, since doing so could at best result in total planting failure and in the worst case may introduce an exotic pest species.

Figure 6-1. Studying a nearby natural wetland will provide information on distribution and density to develop quantitative objectives for plant communities in the new wetland.

Carefully follow the procedures in Chapter 13 for planting times, methods, and subsequent management. Remember that much of your planting or stocking effort could be accomplished much more easily through simple water level manipulation and a little patience.

Since a description of the unimpacted wetlands is unlikely to be available and projections of its current successional stage are dubious, the only practical yardstick is an undisturbed wetland in the vicinity (Figure 6-1). Lacking that option, the next best alternative is to use a description of a nearby natural system in reports or journal articles. The U.S. Fish and Wildlife Service has published an excellent series of community profiles and ecological characterizations of most wetland types in the U.S. (see references in Appendix A). These sources are also useful in determining exactly which successional stage of a regional wetland is most desirable and/or most easily restored.

Cost, complexity, and duration of restoration activities vary with the degree of alteration of hydrology, soils, and biotic communities from the original wetland conditions. Generally, repair, enhancement, or restoration of an existing wetland site will be much less expensive and much more likely to succeed than creating a wetland on a terrestrial site. In either restoring or creating wetlands, the original objectives should be clearly defined and recorded, and amendments should also be recorded, since on-site modifications during construction and planting tend to be the norm rather than the exception. Many wetland mitigation projects in recent years have been severely criticized because the original and amended plans or objectives were not adequately documented (or were unquantifiable) and the apparent failures are difficult to explain or understand. This is also true for management plans, modifications, and unusual events during the early years of operation. Failure to document a drought, deluge, storm, vandalism, etc. that may cause partial or complete failure in subsequent years results in future evaluators concluding the project managers were unable to create a wetland. In fact, deliberate amendments or circumstances beyond their control caused the failure but were not documented. Of course, goals should also be quantified to facilitate measurement, comparison, and progressive evaluations during the project.

OBJECTIVES

Regardless of what type of site is selected and which restoration or creation techniques are employed, it is imperative that planners clearly and precisely determine and record the objectives of the effort. If your project will restore a natural wetland, clearly specify goals and objectives, and comparative, quantitative measures that will be used to evaluate project success. Do not forget to include a time frame in your goals, but be realistic in specifying when, for example, you anticipate having 90% coverage by a selected species or a combination of species.

Irrespective of specific project objectives, the overall goal should be to develop a largely self-maintaining system, at least after the initial start-up period. Complexity and cost of operation are substantially reduced in self-maintaining systems and long-term viability is significantly enhanced. Consequently, designing for the desired function(s) should take precedence over designing specific forms. Designers must recognize that a wetland function may be achieved by a variety of forms and be cognizant that the specific form, for example type and coverage of vegetation, is likely to change over time regardless of what and how many plants were initially installed. So long as the desired function(s) is supported, project objectives are achieved and the project is successful. To do otherwise ignores the dynamic nature of wetland systems.

Be careful of specifying extent of coverage or dominance by a few species or only planted species. Describe the function to be attained and include ample latitude in describing the form needed to achieve that function. For example, a constructed wetland objective might be to produce a discharge with less than 30-mg/l suspended solids rather than to have 100% coverage by species A and B at the end of one or more years. Species A and B might very well be replaced by other, unplanted species and coverage might only be 70% but the discharge is below the project goals. If the project objective is to establish 100% coverage by species A and B (form), then the developers failed but if the objective is to accomplish a discharge with less than 30-mg/l suspended solids, the developers succeeded. Similar approaches are applicable to other wetland functions including commonly used criteria for mitigation wetlands. Specifying X hectares with Y% coverage of species A, B, and C after one year (form with a time element) totally disregards the required start-up period (rarely less than 3 to 5 years) and the dynamic nature of wetland systems (who will prevent or weed out all the other species that will invade the system?).

Objectives of the wetland creation project should include quantitative descriptions of parameters representing the function and form of the created wetland. Obviously, you know what you would like to see in the finished product, but will others view the new wetland similarly? Simply stating that you will create a bog or a marsh does not provide you or others with the benchmarks to measure your progress or success. Restoring a wetland to its former, perhaps pristine state may seem desirable, but will you have comparative measurements to assess the new wetland? It would be unusual if someone had carefully measured the structure or function of the previous system providing quantitative data for comparison with the new system. And given this unlikely happenstance, would the information be truly useful? Remember, wetlands are a transitional stage in the ecological succession of a site. Hence, careful, detailed quantitative data collected at one point in time and compared with similar information collected at another point will provide instantaneous glimpses, snapshots, of a system undergoing gradual, progressive, and, in many cases, irreversible changes. To restore the wetland to its previous condition requires arresting or reversing the successional train of events by modifying hydrological and biological determining factors. To restore the wetland from the state that succession naturally would have established may require much less deliberate management effort if that condition can be accurately described and replicated.

Obviously, developing objectives for wetland creation must have a slightly different basis since a wetland did not previously exist on the site. In this case, review reported descriptions of regional wetlands and visit, observe, and study nearby systems. Quantitative measurements of important elements of wetland structure (form) at a nearby system will not only provide a yardstick for comparative evaluation of establishment techniques, but will provide considerable insight into composition of wetland structure.

In either case, it is imperative that you develop and record quantitative parameters describing your objectives. A dense stand of cattail or some emergent plants and some open water with turtles and ducks may seem desirable but, unfortunately, none of these have the same meaning to all observers. Specifying 30% coverage by bulrush at 30 stems/m^2, 70% open water with pondweeds at 20 stems/m^2, 20 occupied nesting territories of marsh wrens, and/or 80% removal of sediment from upstream watershed runoff after 5 years are measurable parameters for readily understood and accepted comparisons. Similarly, an objective of establishing a mixed stand of willow oak (20%), cherrybark oak (20%), swamp white oak (10%), red maple (10%), tupelo gum (10%), cypress (10%), black gum (10%), willow (10%), and river birch (10%) with 1 white-tailed deer per hectare and 5 prothonatary warblers per hectare after 20 years is easily evaluated by common field methods (Figure 6-2).

These examples include objectives mixing form and function, as humans all too often tend to do. Typical objectives will include a certain number of larger or more visible plants and animals and perhaps units of water recharge or purification since those obvious characteristics provide desirable benefits

Figure 6-2. Objectives may be defined in terms of producing a certain number of a desired species of wildlife but care is needed since the highly productive system supporting this large brood of blue-winged teal is an older, well established system.

derived from natural wetland systems. However, mixed objectives may be difficult to achieve and/or quantify once attained. Since wetland functional values derive from wetland form or structure, quantifiable objectives describing the desired form or structure are commonly used to define and measure for evaluation. But including latitude in form objectives is important. Rarely is the true objective a certain form (structure); most commonly a function (s) is the true goal. Form objectives may define the desired structure in terms of the hydrology (flooding duration, timing, and water depths), soil or substrate composition (saturation periods, depths, organic content, and mineral composition), and plant community composition (species, growth form, size, and areal coverage). Animal community composition would rarely be included as a form objective largely due to animal mobility but it is commonly a functional objective. Selection of one form (structure) vs. another form will determine the type and magnitude of functions supported by the new wetland, and form selection must be guided by careful evaluation of desired functions and requisite structure to support those functions.

Conversely, certain functions may be more easily quantified than some forms. Water purification is readily evaluated by simple methods — collection, analysis, and comparison of water quality parameters in influent and effluent waters. Flood crests and duration can be visually measured or a continuous record can be obtained with various automatic recorders.

Restoration objectives may include re-establishing mid or late stages of wetland successions since much of the soil/substrate, perhaps some of the vegetation and hydrology, and likely a seed bank exist on site. However, creation objectives should describe a very early successional stage if the evaluation period is short (less than 10 years for a marsh and less than 60 years for a bog or swamp) since the complex of soils, vegetation, and hydrologic patterns necessary for many functions is unlikely to fully develop in shorter intervals. Conversely, some important functions are supported by early successional stages — young systems — and objectives defining the appropriate forms should also include descriptions of methods to periodically retard or reverse natural succession to maintain the system in an early stage.

Regardless of the functions chosen or form selected, detailed quantitative descriptions of the objectives and time interval are absolute requirements to provide guidance during development and to support subsequent evaluations. WETER, a knowledge-based system under development at the U.S. Army Waterways Experiment Station may be useful in bringing a rigorous methodology to the process of defining objectives and developing project plans.

WETLANDS FUNCTIONS AND OBJECTIVES

Development objectives typically include one or more of the commonly accepted functional values:

1. recreation
2. education
3. flood reduction
4. research
5. aesthetics
6. water purification
7. bank stabilization/protection
8. commercial products
9. base flow augmentation
10. ground water recharge
11. wetland replacement (mitigation)

Initially, most developers will identify with one functional value and select that as their objective; that is, create wetlands that provide wildlife habitat to support nature appreciation (bird watching, photography, etc.), sport hunting, and sport fishing. Or they may wish to have a wetland complex intermingled with housing lots and commercial offices to enhance the diversity and attractiveness of the landscape and purify stormwater runoff. Others may wish to harvest fur-bearing mammals, crayfish, and lumber from their tract to supplement other income. Each of these goals will require creating a diverse, complex wetland system. Conversely, someone interested in bank protection or water purification may be able to attain their objectives with relatively simple, early successional stage systems.

At this stage, the most important facet is to refine, condense, and consolidate vague, general thoughts into numerical values that can be plugged into project planning. Obviously, some objectives will require highly detailed descriptions, whereas others can be accomplished with one or two sentences. For example, a housing developer wishing to add value to and modify runoff from a tract that will have 100 single-family housing units might include created wetlands. In this case, the objective is to provide the maximum diversity of structure and texture to increase visual stimulation and to intercept most runoff. He might specify that the wetlands will border each lot at some point, include various water depths, and have maximum irregularity in the shorelines. Other objectives would include supporting transitional, shallow water, emergent, submergent, and floating leaved herbaceous plants, as well as shrubs and trees.

Form objectives would include square meters of water surface, meters of shoreline length, and the exact location would be graphically depicted to insure the wetland borders all tracts and that maximum shoreline irregularity is attained. Additional descriptions would include water depths from 0.2 m above to 1.0 m below normal water levels with locations for various depths graphically shown, along with size, location, and elevations of islands, plant species, and exact planting locations for each, planting densities and composition, and areal extent of each type after 3 or 5 or more years. His functional objective would include typical water quality parameters — contaminant levels — for storm runoff from the tract along with a measurement of the aesthetics. Each of these is a numerical value that can be described in bid specifications and construction contracts and measured for evaluation at some future time even though quantifying and numerically evaluating the overall aesthetics of the system may be less precise.

Not surprisingly, objectives for wetlands to be used for nature appreciation or education will have similar goals since both functional values are enhanced by increasing complexity and diversity. Conversely, a system that will provide shoreline protection may only need one or two species, but the objectives should include amount and degree of protection as functional goals and the type, planting locations and densities, and the extent of coverage of vegetation as form goals. At the extreme, some successful water purification wetlands consist of rectangular cells with a single emergent plant; but here too, objectives should specify rates for wastewater application and projected removal efficiencies or the desired discharge quality as a functional goal and the size and configuration of cells, plant species and planting density and water level elevations as supporting form goals (Figure 6-3).

Most planners will begin with a single objective in mind but as they explore the available functional values, they are likely to include others. If budgetary limits do not constrain these additions, will the nature of wetland systems preclude some objectives? Or must a sequence of priorities be established to insure that the original goal is not lost? Obviously a wetland with a larger number of functions will provide more benefits and have greater value to the planner and the community. The number of functions and derivative benefits is controlled by the wetland structure required for each function, the type,

Figure 6-3. A constructed wetland providing wastewater treatment and because of its attractiveness to wildlife, important recreational benefits for Arcata, California. (Photo by Robert Gearheart.)

size, and location of the wetland required to produce those functions and benefits, and ultimately the adequacy of financial resources in accordance with site characteristics. Can a single created wetland provide more than one functional value? Of course! Can it provide all known functional values for a single population of users? Probably not, simply because one location may foster one benefit but negate another. For example, a groundwater recharge system must obviously be located in an area with highly permeable substrates, and that system is unlikely to provide appropriate water purification without jeopardizing groundwater quality. A flood water storage/buffering system must be located upstream from the community it is to protect and, without added pumping costs, using that system for water purification of the town's wastewaters is impractical. The latter objective will be less expensive if the wetland is located downstream from the community. But either of these systems could support recreational, commercial production, shoreline protection, aesthetic, educational or research benefits, in addition to the prime goal. The ability to provide the additional values is dependent on the objectives defined and the size, diversity, and complexity of the created wetland. As a general rule, increased numbers of functional benefits are directly related to increasing size, complexity, and diversity within the system.

A large (>5000 ha) diverse wetland could easily support the entire panoply of functional values, depending on its location with respect to other features in the landscape. Conversely, a very small but diverse system (<0.5 ha) might provide some flood retention, water purification, research, and aesthetic benefits,

Figure 6-4. A large, diverse wetland supports many different sizes and types of plants and animals and can also serve flood amelioration and water quality improvement functions.

but its value for recreation or commercial products would be limited because of minimum area requirements for many animals and the small quantity of products (timber, crayfish, etc.) would make management impractical. Anything smaller is unlikely to provide any significant functional values.

DETERMINING FUNCTIONAL VALUES

Since wetland functional values are governed by principles of hydrology, chemistry, and ecology, it is possible to relate expected functional values with diversity and complexity; but we need to define diversity and complexity first. In ecology, diversity is defined as the number of individuals related to the number of different species represented by those individuals in a given area. For example, a system with 10,000 individuals of 10 different species would have much lower diversity than a system with 10,000 individuals of 1000 different species (Figure 6-4). Diverse systems contain fewer numbers of individuals representing larger numbers of species (i.e., 5 individuals in each of 100 different species of butterflies). Conversely, a system with low diversity might have 250 individuals in only 2 species, but it would have the same total number of butterflies and could have the same quantity of biomass.

Complexity is typically related and to some extent a corollary of diversity. A complex ecosystem has many different species at each different level, whereas a simple ecosystem has only a few species at each level and/or few levels. For example, penguins (tertiary consumers) feeding on small fish (secondary consumers) feeding on krill (primary consumers) feeding on plankton (producers) in the Antarctic seas represent a simple system since each level

Figure 6-5. Because most animals are opportunists, few are solely restricted to one trophic level. Snapping turtles feed extensively on plants and invertebrates but also prey on birds and mammals.

has few species even though a number of trophic levels are present. This sequence is referred to as a food chain, that is, energy in the form of food flows up a simple, straightforward pathway to the top consumer level.

Conversely, a complex system might have red and gray fox, raccoons, opossums, and bears (tertiary consumers) feeding on many different species of mice (secondary consumers) feeding on many different types of insects (primary consumers) feeding on many different types of plants (producers). In addition, these carnivores (tertiary consumers) will feed directly on insects (primary consumers) and on plants (producers). Consequently, the number of components in each trophic level is much greater, as is the number of inter-actions and pathways between trophic levels. Energy flow may follow any of a number of different pathways from producers to the top consumer level (Figure 6-5).

Two important corollaries result from these principles. In the Antarctic example, the disappearance of one component (a single species of krill) could cause collapse of the entire system since the fish and penguins dependent on krill would starve and plankton populations may increase unchecked until resources are depleted and mass die-offs occur. However, if one species of insect or one species of mouse were lost in the more complex system, the top consumers would merely shift to feeding on other species of mice, insects, or plants. The complex system has the ability to adjust to and survive disturbances (perturbations) that might destroy the simple system because the complex system has alternate pathways and considerable redundancy that create system resiliency.

Whereas the simple system might suffer major changes and may never recover, the complex system is likely to continue with only minor adjustments following a significant disturbance. Consequently, the complex system exhibits considerable stability or resiliency; that is, it remains relatively unchanged or returns to the original state after experiencing a disturbance. The simple system is likely to exhibit wide fluctuations in types and sizes of populations and it may never fully recover from a disturbance. It lacks resiliency or stability because it lacks the alternate pathways and redundancy incorporated in the complex system.

The geographic and climatic differences apparent in the examples of simple and complex systems are not coincidental. The simple system occurs in a harsh environment (the Antarctic), whereas the complex system is present in temperate regions. Because only a few types of plants and animals have been able to adapt to environmental extremes, unusually hot or cold, wet or dry regions often have relatively simple systems with low diversity and low complexity. Even though these simple systems have tremendous numbers of a few types of plants and animals and may be highly productive, they are much more sensitive to slight disturbances than complex systems that may have lower numbers and lower productivity.

Now we can address the original thesis — increasing complexity and diversity are directly related to the ability of a wetland system to provide more than one functional value, as well as the ability of the system to withstand disturbing factors. If we compare the various functional values with the foregoing discussion in mind, it quickly becomes apparent that certain functions and benefits will require certain levels of diversity and complexity. In fact, we can establish a continuum describing complexity and diversity in terms of three broad categories of anticipated functional values:

water purification < hydrologic buffering < life support

with rapidly increasing diversity and complexity from left to right. If we apply the same approach to commonly used benefits we find

ground water recharge < water purification < flood reduction < shoreline protection < bank stabilization < base flow augmentation < commercial products < research < education < aesthetics < sport fishing < sport hunting < nature appreciation

with diversity and complexity increasing along the gradient. This scaling is based on the minimum required to produce that function and doubtless a few of these benefits might be reversed; for example, the differences between water purification and flood reduction are not great, but the differences between the complexity and diversity of wetlands generating groundwater recharge functional values compared to sport hunting functional values are substantial.

More important is the general rule that a higher system (more diverse and complex) has the ability to also support the functions of a lower system (homogeneous and simple), and the higher systems are inherently more stable and able to withstand disturbances. Both concepts are important to developing and defining the objectives for creating a wetland system. In the first case, the more diverse and more complex system will support the higher objectives as well as the functions of systems lower on the scale. Secondly, the higher system is more capable of withstanding disturbance than the lower system and is less likely to require intensive management to insure continued operation.

A corollary of the fact that simple systems occur in harsh environments, complexity and diversity also serve as indicators of system health and well-being. For example, if we establish a complex system to provide life support and flood reduction benefits, but over time the system becomes simple and homogeneous, we should suspect that some environmental factor is more extreme than anticipated. It may have received more extreme or more frequent flooding than planned or some other factor may be detrimentally impacting (constraining) the system and failure to implement remedial measures could result in system failure with loss of flood reduction values. Typically the life support function will begin to fail long before the hydrology function and failure of the former will provide an early warning to modify the system or influencing factors well prior to failure of the flood reduction function.

Similarly, a relatively complex system created for water purification and nature appreciation that loses diversity and complexity is likely receiving excessive wastewater loading. Harsh environmental conditions restrict the types and variety of plants and animals that can survive with subsequent loss of the nature appreciation benefits although the lower benefit (water purification) is still supported. Of course, environmental extremes may force the system to a simpler and simpler state, but that condition may continue to provide water purification until some disturbance causes system failure since simple systems lack the resiliency and stability of more complex systems.

It should be obvious from the foregoing that establishing the maximum diversity and complexity within the constraints of budgets and site characteristics is advantageous regardless of the principle objective for creating a wetland system. The structure of the more diverse and more complex system will provide a greater range of functional values and benefits, and has substantially more stability which translates into less need for direct management to obtain the desired benefits. Initial cost savings from creating simple systems may be vastly overshadowed by time and expense entailed in required management over the lifetime of the project. The more diverse and more complex systems will not only better serve the principle objective, but they will provide ancillary benefits and reduce long-term operating costs.

CHAPTER 7

ADVICE AND ASSISTANCE

It should be obvious from the foregoing chapters that very few individuals will have expertise in the many different disciplines needed to successfully create a wetland. Many failures have been due to lack of professional knowledge in a critical area the most important of which was often wetland ecology and more specifically wetland creation/restoration methods. Hopefully this small effort will help overcome the latter.

THE TEAM APPROACH

Even an experienced wetland ecologist will need the expertise of other professionals on certain aspects of the wetland project. Consequently, most wetland developers assemble a team with the requisite expertise and wetland experience to conduct the project. At a minimum, fields represented should include:

1. wetland ecology — overall development of project goals and objectives and methods to achieve them, startup and operating management, likely to be the most conversant with other needed disciplines;
2. hydrology — evaluation of existing surface and groundwater and projections on future hydrology within and without the new wetland system;
3. soils science — evaluation of existing soil conditions and suitability for wetland creation including construction materials;
4. wetland botany and plant culture — selection and establishment of wetland vegetation; and,
5. civil engineering — site surveys, preparation of construction drawings, bid invitations and supervision of grading, structural and electrical contractors.

Other representatives could include agronomy, archeology, endangered species, fisheries, ornithology, wildlife ecology, geology, landscape architecture, recreation, environmental education, and others depending on the project goals. Though critical at certain junctures, the total required involvement of these disciplines will be much less than the team members and many developers obtain their services through short-term contracts. Nor will the principle team members need to devote full time on a project — most will be principally concerned with certain aspects in one or more phases of the total activity.

139

TECHNICAL ASSISTANCE

Several federal, state, and even some local governmental agencies and many non-government organizations (NGOs) have information, literature, wetland areas, and staff or member expertise that is invaluable to wetland developers. Agencies and organizations with responsibilities or interests in increasing wetland resources should be contacted for local expertise on wetland restoration or creation, possible cooperative ventures, or potential funding. Due to the diversity and productivity of wetlands, some aspect of wetland ecology intersects the goals or objectives of most natural resource interest groups. Local chapters, regional, or national offices may be interested in developing cooperative projects or long-term management agreements. Objectives in contacting local agencies and NGOs include:

1. technical advice and assistance;
2. inspection and observation of existing regional wetlands;
3. cooperative development programs;
4. financial assistance; and,
5. long-term management.

Regardless of the amount of effort expended in assembling information on wetlands from literature sources, national organizations, or agencies, its value will be overshadowed by the opportunity to visit and observe various types of local wetlands and to talk with the individuals responsible for managing them. Assembling and absorbing information from books, journals, and pamphlets is merely preparation for understanding the complexity and diversity confronting a first-time observer. Of course, a basic understanding is needed to productively discuss construction and operating methods with the responsible staff. Most wetland managers are poorly paid and truly engaged in a labor of love. Not surprisingly, they are not reluctant to share the wealth of knowledge gained through hands-on daily experience with someone else interested in wetlands.

Reviewing the types of wetlands naturally occurring in your region will guide you in selecting the type to be constructed. If many different types of marshes, bogs, meadows, and swamps are naturally present, your options are nearly unlimited; but if only one type — probably a marsh — is present, that is indicative of the probability of success if you attempt to construct a marsh as compared to a bog or swamp. The natural wetlands are likely located in a refuge or wildlife management area and, since it will serve as a model for the system to be constructed, spend some time observing and studying the system. Request permission to accompany the managers in their daily activities and explore in-depth how and why certain procedures have been successful and others less satisfactory in that wetland system. Remember that the true mother lode of knowledge on wetland management still resides on the system because most managers are too busy working with the system and handling the mountain of bureaucratic paperwork to commit their knowledge to formal reports, much less journal articles. Though rarely encountered in the field, beware the attitude that wetlands must be protected at all costs and not disturbed for any reason. That rare individual simply does not understand wetland ecology. Conversely, you may discover that the local wetland manager will offer assistance with your project. If you are so fortunate, remember that he/she likely is already overworked so try not to add too much to his/her workload.

Nothing will replace time spent studying the natural system, reviewing progress or annual reports on it, and discussing creation and management techniques with the managers. This is an opportunity to compare goals and objectives with an actual wetlands that should be very similar to the proposed system. Do not be surprised if project goals or plans are altered after examination of an existing system. Many "failed" attempts at wetlands creation could have been avoided if the developers had assembled and used the information available from this source.

The area manager will also be aware of researchers from laboratories or universities that have investigated various aspects of model wetlands. Contact them for reports and, if possible, additional technical assistance in planning and constructing the new system.

Obtain copies of the excellent references entitled "The Ecology of..." from the Fish and Wildlife Service for each type of wetland in your area. Various reports provide thorough descriptions of the community and supporting geology/soils and hydrology along with distributional maps, successional patterns, management guidelines, and occasionally economic benefits and values and listings of knowledgeable individuals. Combined coverage includes most freshwater and saltwater wetlands of the U.S.

A number of state and federal agencies and local and regional NGOs are charged with or interested in expanding the wetland resource base. Their staff in the local or state office are knowledgeable on wetlands and may have considerable information on regional systems. Some have responsibility for providing technical assistance to landowners or others interested in restoring or creating wetland. Others may be enticed into cooperative programs to create more wetlands and a few may even have limited funds for assisting in restoration or creation projects.

Perhaps more importantly, many agencies and NGOs have permanence and their participation may be crucial to the long-term viability of the new system. Although it may be as limited as changing water levels once or twice a year or as demanding as active management of specific plant or animal populations, most wetlands require some management (Figure 7-1). Without it, successional changes are likely to occur that substantially alter the system from the form that was planned and constructed. Depending on the affiliation and capabilities of the wetland developer, soliciting participation in the planning

Figure 7-1. The National Wildlife Refuge System hosts many types of wetlands varying with regional locations that will serve as examples of the types of wetland that can be created and maintained in that region.

and construction phases and developing a long-term management agreement with an agency or NGO could be critical to maintaining the wetland and fulfilling project goals.

Developers that hope to construct a wetland and walk away are naive and careless with their resources. In the early years, most new systems will require considerable manipulation to achieve hydrological and biological objectives. A well-established system should require much less active management, but the requirement is unlikely to disappear. Consequently, a long-term commitment by the developer or a cooperating organization is essential to project success.

In addition, many of these agencies and NGOs have valuable information on potential sites that should be evaluated by wetland developers. Details on securing specific information to assist in site selection and evaluation are presented in Chapter 8.

Initial contacts for technical or financial assistance should start with a phone call to the local office of the state or federal wildlife agency or the Natural Resource Conservation Service. Numbers are generally listed in phone directories under "government offices." Individuals in these offices normally have daily contact with colleagues in other agencies and with the local NGOs, and can provide appropriate referrals to each.

The Conservation Directory published annually by the National Wildlife Federation (NWF) lists local, state, and regional offices of virtually all

Figure 7-2. A large number of our refuges and wildlife management areas were established to provide habitats for ducks and geese.

conservation organizations (government and non-government), university natural resource departments for the U.S. and Canada, as well as offices of foreign conservation agencies, pertinent periodicals, directories, and other information sources. If your local library does not have a copy, contact the NWF at 1412 Sixteenth Street, N.W., Washington, D.C., 20036-2266, (202)797-6800.

A number of federal agencies support research or management on wetlands and many have local or regional offices with experienced staff familiar with local conditions. Within the U.S. Department of the Interior, the Fish and Wildlife Service (FWS) is responsible for the National Wildlife Refuges. The vast majority of these refuges are wetlands because waterfowl hunters financially and politically supported creation of the system and establishment of many individual refuges (Figure 7-2). Larger refuges have permanent staff with valuable experience in hands-on day-to-day restoration and management of wetlands. In addition, the FWS operates wildlife research laboratories in Laurel, MD, Jamestown, ND, Lafayette, LA, and Denver and Ft. Collins, CO, with emphasis on wetland research at the Jamestown, Laurel, and Lafayette facilities.

The Geological Survey has conducted significant research on wetland hydrology with staff from regional centers in Reston, VA, and Denver, CO as well as from the Geological Survey office in each state. Other Interior offices with interest or responsibility for wetlands and experienced staff include the Bureau of Land Management, the Bureau of Reclamation, and the National Park Service.

In the Department of Agriculture, the Natural Resource Conservation Service (NRCS) has staff in every state and many counties with considerable experience in watershed management and constructing and managing ponds for various uses, depending on the region of the country. Recent legislation has altered the direction of the agency and staff with wetland interest and experience are actively implementing technical aspects of the Sodbuster and Swampbuster provisions of the Food Security Act and the Conservation Reserve Program administered by the Agricultural Stabilization and Conservation Service (ASCS). In addition, the Plant Materials Centers have conducted extensive research on a variety of species and cultivars that are useful in creating and restoring wetlands. The Forest Service (FS) manages extensive wetlands, especially in the southeast, and its staff at the Forest Experiment Stations in Asheville, NC, and New Orleans, LA, have researched forested wetlands for many years.

All branches of the military in the Department of Defense employ professional natural resource managers on individual bases and, since some bases have extensive wetlands, many managers have considerable experience in wetland management. In addition, the Waterways Experiment Station (WES) of the U.S. Army Corps of Engineers has investigated methods to create and restore wetlands on Corps projects throughout the U.S. for many years. The new Wetlands Research Program at WES has lead wetland research by the COE and supervises considerable contract research. With recent changes in regulations on 404 permits, the Corps has added wetland specialists in many District offices.

For a number of years, wetlands specialists in the Environmental Protection Agency (EPA) concentrated on the agency's oversight responsibility for 404 permit applications. More recently, the new Office of Wetlands Protection has become the Office of Wetlands, Ocean and Watersheds (OWOW). Staff at the Washington level and in the regional offices are actively involved with a variety or protection, restoration, mitigation and creation projects and the laboratory in Corvallis, OR, conducts research on wetlands. The Tennessee Valley Authority manages extensive wetlands and the Smithsonian Institution conducts wetland research.

Almost all of the large federal agencies have counterparts in the governmental structure of the states with experienced staff working in the same areas on a local or statewide basis. State wildlife agencies frequently operate wildlife management areas and refuges adjacent to, but occasionally spatially separate from, the national refuges with similar experienced personnel (Figure 7-3). The waterfowl management group within the state wildlife agency is an excellent contact point. Agriculture, forestry, environmental protection, and natural resource management agencies of the states have staff with extensive experience in local and regional wetland matters.

Counterparts to federal and state agencies in the U.S. are responsible for similar programs in Canada. For example, the Canadian Wildlife Service and the provincial wildlife agencies have many years of involvement with and staff

Figure 7-3. Many large wetland complexes are jointly owned and managed by Federal and state agencies as at Reelfoot Lake.

that are highly experienced in wetland management, and cooperative research at the laboratory at Delta, Manitoba laid the groundwork for much of our information on prairie marshes.

Not surprisingly, the history of wetland concern, management, and research in non-governmental organizations (NGOs) is at least as long and, in many cases, predates the involvement of governmental organizations. Of these, Ducks Unlimited (DU) was established by waterfowl hunters as a means to use U.S. funding to acquire and manage wetlands in Canada and, for many years, its activities were largely restricted to Canadian marshes. However, recent policy changes have created the "MARSH" program, which coopera- tively funds wetland projects in the U.S. DU has a large staff throughout Canada and the U.S. and some in Mexico, many of which are former employees of state wildlife agencies or the FWS. To implement DU's acquisition and management programs, their technical staff has developed substantial expertise in hands-on wetland restoration and creation and their fund-raising counter- parts are without parallel.

The Nature Conservancy was established to acquire and protect unique natural areas and, over the years, they have developed innovative methods to finance acquisition and accept donations of land with tax or debt write-off benefits. Some areas are managed by the Conservancy but, generally, the lands are sold or otherwise transferred to a governmental organization for long-term management. Many Conservancy lands provide ideal benchmarks for compar- ison with other wetlands and opportunities to observe and study systems that have been little impacted by human activities. In addition, the Conservancy

Figure 7-4. Wetland protection has been a high priority of the Nature Conservancy due to the large number of unique species occurring in some wetlands and the diverse, productive nature of others.

sponsored many of the state Natural Heritage programs that are the repositories for information on the location of unique species including those legally protected as threatened or endangered (Figure 7-4).

The National Audubon Society was largely a bird-watching NGO, but it has become a very broad environmental advocacy group. It manages significant wetlands including Corkscrew Swamp in southwest Florida and maintains an active research group in Tavernier, FL. Local chapters are distributed throughout the country (Figure 7-5).

The National Wildlife Federation (NWF) is the largest conservation organization in the world and many of its members are dedicated to wetland preservation or restoration. As with many NGOs, NWF has a national office and state associations composed of local clubs at the grass-roots level.

The National Wildlife Refuge Association is a small NGO but its membership largely consists of retired employees of the FWS, many of which spent most of their careers managing wetlands on refuges throughout the country.

The plethora of other NGOs renders individual descriptions impractical; over 500 are listed in a recent issue of The Conservation Directory and that does not include the many state birding clubs or ornithological societies, state wildflower groups, local and state herpetological groups, and even new wetland groups such as The Wetlands Conservancy of Tualatin, OR. Many have specialized interests unrelated to wetlands and others have narrow purposes closely tied to wetlands; still others have very broad environmental purposes,

Figure 7-5. Birds are highly visible components of wetland wildlife and bird watching organizations support wetland protection and creation.

some are local or regional and others have international scope. Local and regional interest groups with broad conservation basis often have members and projects on wetlands.

Do not overlook some of the highly specialized, even single-species NGOs. For example, The Trumpeter Swan Society (TTSS), headquartered in Maple Plain, MN was organized to promote restoration and management of a single bird, but most of its members have extensive wetland experience because swans are dependent on high-quality marshes. Consequently, the interests and policies of TTSS support restoration and management of wetlands across the nation (Figure 7-6). The Raptor Research Foundation (RRF) has members managing and investigating wetlands because many raptors — osprey, bald eagles, peregrine falcons, marsh harriers, short-eared owls, snail kites, etc. — are dependent on wetland habitats.

Professional natural resource associations such as The Wildlife Society, The Society of Wetland Scientists, and The Ecological Society of America, etc. have broad-based memberships with expertise in all aspects of wetland ecology. Many professional societies have regional or state chapters that are excellent sources for local expertise and some chapters are active in local preservation or restoration projects.

As knowledge of the functional values of wetlands spreads, the members of almost all NGOs concerned with natural resources realize that wetlands influence their special interest and most are becoming involved with wetlands or wetland issues to some degree.

Figure 7-6. Trumpeter swans need high quality marshes and consequently The Trumpeter Swan Society is supportive of wetlands creation, restoration, and protection.

FINANCIAL ASSISTANCE

Modern society recognizes that creating wetlands is an expensive undertaking that not only benefits the individual landowner or project leader but will likely also provide a variety of benefits to society as a whole. Consequently, a variety of federal, state, provincial and even local public and private programs may be able to provide financial assistance for a project. While the following discussion concentrates on U.S. programs, counterparts are present in Canada and other countries. In most instances, local offices of federal and state agencies (listed in local telephone directories) should be the first contacts in exploring financial assistance since many programs are administered by local committees.

The U.S. Department of Agriculture, Agricultural Stabilization and Conservation (ASCS), through its Agricultural Conservation Program, provides cost-share funding for approved practices that address point and nonpoint source pollution programs. Approved practices include permanent vegetative cover, erosion control, wildlife enhancement, and developing or restoring existing shallow water areas. Eligible practices must result in long-term and community-wide benefits and be those a landowner is not likely to undertake without financial assistance.

The Conservation Reserve Program (CRP), also administered by the ASCS, encourages landowners to enroll highly erodible lands, lands contributing to serious water quality problems, and converted wetlands farmed prior to December 23, 1985. Enrollment periods are typically for 10 years but may be up to 30 years and landowners receive annual rental payments. In addition CRP provides cost-share incentives for vegetative, water quality, and conservation measures.

Through the Water Quality Incentives Program, the ASCS provides technical assistance, cost-sharing assistance, and annual incentive payments to protect water resources including water quality, wetland protection, and wildlife benefits. The Wetlands Reserve Program (WRP) secures long-term easements and provides technical and financial incentives to restore wetlands. Eligible lands include restorable farmed wetlands, prior-converted cropland (December 23, 1985), adjacent, functionally related uplands, and riparian areas linking wetlands. WRP lands typically have a permanent easement but these lands can be used for other purposes so long as that use does not degrade or reduce the wetland values.

The Natural Resource Conservation Service (NRCS formerly SCS) provides financial and technical assistance with the Watershed Protection and Flood Prevention, the Resource Conservation and Development, and the Great Plains Conservation Programs. Activities covered include watershed protection, improving fish and wildlife resources, resource development and environmental protection, water quality, and wetland protection.

The Farmers Home Administration provides loan reductions in exchange for long-term (>50 years) easements on wetlands that may then be managed by the Fish and Wildlife Service and state agencies.

U.S. Department of Agriculture, Forest Service administers the Forest Incentives (FIP), Forest Stewardship (FSP), and Stewardship Incentives (SIP) Programs. FSP provides technical assistance and SIP provides cost-share assistance to help landowners enhance and protect timber, fish, and wildlife habitat, water quality, wetlands, and esthetic and recreational values. FIP provides technical and cost-share assistance for tree planting, forest stand improvement, and special forestry practices.

The North American Wetlands Conservation Act administered by the U.S. Department of Interior, Fish and Wildlife Service provides funding for wetlands conservation projects including acquisition, restoration and/or enhancement, and wetlands creation. These grants require a minimum of one to one matching from any non-federal source and generally require a minimum 10-year agreement. The Service's Partners for Wildlife program offers technical and financial assistance for restoring degraded or converted wetlands and certain upland habitats. Cost-sharing is not required for small projects but is if the Service's contribution exceeds $10,000 and cooperative agreements must be for more than 10 years. The North American Waterfowl Management Plan targets specific regions of the country and forms partnerships with landowners for technical and financial assistance to protect, restore and enhance wetlands important to waterfowl and other wetland-dependent species.

On the private side, Ducks Unlimited's MARSH program provides matching funds to public agencies and private conservation groups for projects with significant waterfowl benefits. DU's many regional offices throughout the U.S.

and Canada should be the first point of contact but if none is available, contact the Director of Habitat Development, National Headquarters, Memphis, Tennessee.

The variety of smaller or local and regional public and private programs that now offer financial assistance for wetland resources is much too large to attempt to review. As with seeking technical assistance, planners should explore the potential for financial assistance from many of the same organizations discussed earlier.

SUMMARY

With the rapidly growing interest and information base on wetland ecology, few project planners will have difficulty assembling abundant information, considerable expertise, and perhaps cooperative funding to assist them in planning, designing, and constructing wetlands. If necessary expertise is not forthcoming from another organization, employing or contracting with a consulting wetland ecologist is strongly recommended to avoid the many pitfalls that have caused project failures. Contacts and agreements developed during early project phases will not only substantially increase the chances of success, but may be important in long-term viability of the new system. Of course, few developers will refuse financial assistance unless the conditions are too onerous and the agreements covering financial assistance may well include long-term management aspects that are critical to any project.

CHAPTER 8

SITE SELECTION AND EVALUATION

INTRODUCTION

Project success is dependent on availability of a suitable site or the ability to overcome the drawbacks of a less than optimal site. The siting process provides information to identify, compare, and estimate costs of developing alternative sites at an early stage of the project. Not only will it enhance preparation of plans and drawings, but equally important is its ability to assist developers in avoiding impractical or very costly sites or at least select one with full knowledge of the consequences. Finally, site evaluation initiates the process of developing a project budget since much of the information assembled during siting relates to costs for land, construction needs, or operating parameters.

Wetland design, construction, and operation are facilitated by identifying potential problems and opportunities early in the project with careful site evaluation. Site selection and evaluation is a systematic, reiterative process of collecting and analyzing information, modifying plans, identifying, collecting and analyzing more data, modifying plans, and repeating the cycle until the initial concept is polished into a preliminary plan that becomes the basis for the design. Siting summarizes information on topographical, geological, hydrological, soils, land ownership and use, climatic, biological, and regulatory factors that may influence construction, operation, management and possible impacts of the proposed system. It provides baseline data for evaluating alternative sites, choosing compatible designs, assessing important components, drafting concept drawings, and outlining construction and operating plans.

Much of the discussion in this chapter is directed toward siting a created wetland, but most factors pertinent to created wetland sites are also relevant to evaluating the potential for restoring a wetland site. The most important single difference is the site of a restored wetland already exists and comparative evaluation of sites is meaningless or, at most, a comparison of different sites with damaged, impacted, or otherwise modified wetland systems to select the site with the highest probability for success.

This does not mean that projects to restore wetlands will not need to conduct site investigations and evaluations (Figure 8-1). On the contrary, project planners will need to acquire much the same information as those working with created wetlands, but the objectives may be slightly different.

151

Figure 8-1. Agricultural fields occupying previous wetland sites in eastern Arkansas would quickly revert to early successional wetlands if drainage ditches were plugged.

To restore wetlands, one must first determine what type of wetland previously existed on the site. What was the form or structure and what functions were performed by the previous wetlands? What were the hydrologic and soil conditions that supported a wetland complex on this site? What was the composition of the plant and animal communities? Few will be fortunate to have accurate quantitative, historical data on either the form or the function of the previous wetlands. Most likely, casual descriptions and investigations of nearby healthy systems must be relied upon, in which case, detailed, extensive investigations of a neighboring system could be more lengthy and costly than site evaluations for a created system.

In restoring wetlands, the most important factor is likely to be understanding the hydrology since drainage and/or filling are most often the degradation factors (Figure 8-2). In that case, similar site investigations should be undertaken as with created wetland projects before seemingly simple but possibly incorrect solutions — plugging the drains — are implemented. For example, after drainage, accelerated sediment deposition may have erased former channels and/or elevated the substrate causing higher level and longer term flooding that has damaged the original forest. Destroying drain pipes or tile and plugging ditches may be successful in isolated wetlands, but if the subject system is part of a complex, most of which has been drained, interrupting drainage devices may not restore previous water levels since neighboring wetlands may have supplied water to the site through groundwater interconnections. In peat systems, drainage may have permitted decomposition and compaction that

Figure 8-2. Simply plugging the ditch may not restore this bottomland hardwood swamp because increased erosion from surrounding croplands has modified channels and excessive flooding has damaged the forest.

drastically reduced the organic content, soil chemistry and hydraulic conductivity of the substrate. Exotic pests, for example *Melaleuca* or purple loosestrife, may now dominate the system and restoration will require intensive weed control efforts.

Excepting soils and climate, information on most factors identified below will be just as critical in a restoration project as a creation project. Presumably, the climate has not drastically changed though local modifications could have resulted from regional development and presumably the soils have not been completely altered. However, complete drainage, intensive agriculture, and possibly heavy erosion over a long period of time could have been more destructive than filling and paving for a parking lot. At least in the latter case, the original soil, and seedbank, are probably beneath the fill and asphalt and may start recovery after removal of paving and fill material and re-flooding the site.

GENERAL CONSIDERATIONS

The extent of site evaluation will vary with type and magnitude of the proposed wetland, but many important components of siting are common to virtually all development or land alteration projects. Wetland planners must concern themselves with the same regulatory, land use, topographic, geologic, and hydrologic issues that face housing or industrial developers.

Types of information that must be assembled and analyzed include:

1. current land ownership, use, and availability;
2. historical land use;
3. topography and geology;
4. hydrology;
5. soils;
6. climate and weather;
7. biology;
8. regulations.

LAND USE AND AVAILABILITY

One of most important categories to investigate is land ownership, use, and availability. This includes title searches on the proposed site and adjacent lands to identify ownership, easements, rights-of-way, covenants, water rights, liens, and other encumbrances. Though importance varies with different regions, subsurface (mineral) rights and legal claims need to be determined as well as surface ownership since activation of a 100-year-old deep-mining claim might disrupt site hydrology and a new surface mine could substantially alter topography of the new wetland. In arid regions, an outstanding stream water right or extensive groundwater irrigation activities may nullify a planned water source, eliminating further consideration of that site.

Flowage easements on lands adjacent to reservoirs and rivers usually grant the right to inundate the land and may preclude deposition of fill that would reduce reservoir storage capacity. Watershed covenants may prohibit any activity deemed detrimental to the quantity and quality of runoff waters. Intricate webs of utility (water, sewer, gas, and electric) easements and rights-of-way crisscross the country, touching a large proportion of the landscape. Depending on the type of utility, wetland construction and operation may or may not be compatible with dedicated uses. For example, proposing to impound waters above a gas pipeline would not be favorably received since leak detection and repair would be impaired. Conversely, building small ponds below a large transmission line, without impacting tower footings, might be acceptable to the utility but could be unwise since attracting large birds to the complex of wires might increase wire-strike mortality.

Ownership and availability (i.e., willing sellers or lengthy and costly legal maneuvers) may be the most crucial factor in selecting a site. Historical as well as present land use may favor or eliminate a site because of acquisition or construction costs. Adjacent land use is also important since it could detrimentally impact functioning of a wetland or the wetland may have detrimental impacts on current or planned uses of neighboring lands. Intensive agriculture with sediment and chemical-laden runoff adjacent to the site could impair wetland functions and foreshorten its useful life. Conversely, increased humidity and bird populations near a major airport could endanger the health and safety of airline travelers; and building a "swamp" near an expensive housing development or office complex was not thought to improve property values, though public attitudes are changing. Some recent development billboards now advertise "marsh views". In addition, identifying undesirable land uses, toxic dumps, underground storage tanks, etc. could alert planners to potential costs of complex litigation and cleanup liability encumbering a potential site. Hazardous material lists now include a plethora of formerly "benign" substances and many older farm places have underground fuel tanks.

Developers should plan to acquire surface rights through fee purchase or long-term leases and easements. Fee simple acquisition is preferable to easements since small differences in cost may easily be overwhelmed by future complications. Unfortunately, easement rights are worth little when negotiating to obtain outstanding rights, that is, acquiring fee simple at a later date. In most cases, easements will cost nearly as much as fee simple and later costs for remaining rights will be about the same, resulting in purchasing the land twice.

Depending on the projected operating period and nature of other outstanding rights, mineral rights, water rights, and easements may also need to be obtained. Utility rights-of-way generally grant the right of passage, guarantee access for maintenance, and forbid incompatible uses. In some cases, discussing and negotiating planned uses may develop mutually agreeable solutions. In others, acquisition of easements and costs for relocating pipes or lines may need to be included in project cost estimates or an alternate site evaluated. Within reasonable limits, free title to project lands is generally worth the initial price in order to reduce future conflicts and save additional purchase costs at a later date, possibly in an adversarial environment.

Planners should also examine access for site inventory personnel, construction equipment, operating staff and equipment, and for utilities (electricity, phone, etc.). If public access is not available, corridors will need to be acquired in fee simple for people and equipment, though easements may be used for utilities. Distance and costs of extending utility pipes or lines will also be needed in preparing project cost estimates. Some can be expensive. For example, local cooperatives in Tennessee often only require purchase of materials at a cost of about $5,000 per mile whereas the cooperatives in southeastern Arizona add on labor costs, indirects, and overhead and the total may be as high as $24,000 per mile for residential electrical service. If the planned wetland will require large electrical pumps, more substantial wiring, transformers, and connectors could elevate the costs.

TOPOGRAPHY AND GEOLOGY

Site topography — elevation differences and spatial relationships — influences construction costs, erosion potential, drainage patterns, access, and overall feasibility. Shallow wetlands require relatively flat lands but few sites of any size are level. Since earth moving to create level to very gently sloping terrain is second only to land costs in most projects, accurate, detailed contour mapping is essential to project design and planning. During site comparisons, published maps at 1-meter or 5-foot contour intervals are adequate, but mapping to 1-foot contour intervals or less will be needed for final design.

Necessary evaluations of site geology include nature and depth to bedrock, potential construction materials, and other subsurface characteristics. Sites with shallow bedrock can dramatically increase construction costs because large quantities of rock may need to be broken up, removed, and replaced with imported fill. Costs and construction feasibility vary with type of bedrock because some types may be broken with equipment, whereas hard, continuous formations will require blasting. Erosion of exposed or near-surface bedrock creates parent material for soil formation and soil properties in the watershed can often be estimated from knowing the type and composition of underlying rock formations.

Underground limestone formations are often revealed by characteristic surface features typical of karst geology, the kettle and dome patterns, sinkholes, and caves or caverns. Most were formed by dissolution, often subsurface, of limestone and, in moderate to wet climates, this process is on-going. If the proposed site has limestone strata or karst topography, detailed seismographic surveys will be needed. Sudden appearance of an open sinkhole or land subsidence in a new wetland renders water control difficult, if not impossible, and is costly to repair. In arid limestone regions, crevices or channels may have formed under previously wet climates that could similarly jeopardize project goals. Soils formed from limestone parent materials also tend to be highly permeable and difficult to modify to seal pond bottoms.

Most of our land has been explored in more or less detail and much has been actively mined or drilled. Failing to identify old mine shafts, tunnels, air shafts, and bore holes could similarly cause sudden drainage of the new system, while uncapping a "dryhole" in an oil field might suddenly introduce volumes of brackish water or undesirable gases.

SOILS

Soils and parent materials should be evaluated for class and composition, distribution, and depth. Parent or subsurface materials form dike and dam foundation and will be used as construction materials, and topsoil becomes the substrate for plant growth. Important soil attributes include proportions of silt, sand, clay, gravel and organic material, texture and particle size, permeability and drainage potential, erodibility, and chemistry. Wetland plants are similar to other plants in that sandy loam soils provide optimal growing conditions. Evaluations should include potential borrow areas adjacent to the site if fill, dam coring, or bottom lining materials will be needed.

HYDROLOGY

Evaluating site hydrology includes understanding surface and groundwater location, quantity, and quality, along with surface and subsurface flow patterns, connections, and seasonal changes. Obviously, the amount and type of surface water in the drainage basin influence the size, nature, and operation of the proposed wetland; but subsurface waters and direct connections with surface waters may also have considerable impact in some circumstances. Maintaining necessary water levels will be difficult with inadequate runoff and alternate supplies may need to be tapped, at least during dry periods. High sediment loads and agricultural, industrial, or mining pollutants in potential water supplies will need to be analyzed and compensated for if the new system is to provide safe haven for wildlife and recreation opportunities for area citizens. Conversely, clean water and poor soils could require fertilizing after planting to improve nutrient supplies for biological growth. If the wetland is planned as a treatment system, volumes of flow and type and concentrations of contaminants should be well documented for at least one wet and one dry year to provide parameters for system design.

Flooding and accelerated erosion potentials within the basin must be determined so that appropriate protective measures can be designed to accommodate expected runoff. Wetlands near streams or rivers will be susceptible to flood damage to vegetation and physical structures unless protective measures are included in the design. Since permanent streams are often connected with or at the same levels as groundwater, sites near streams may or may not be suitable, depending on the use of the new wetland. All streams are not "gaining" streams; some are "losing" and depth to groundwater increases rapidly with distance from the stream (Figure 8-3). Locating a wetland in the valley of a losing stream may become an exercise in frustration over inability to seal the bottom. Similarly, depending on groundwater for water supplies may or may not be appropriate since little control over the basic management mechanism (hydrology) will be possible and falling groundwater levels could jeopardize the continuance of the wetland.

Springs, sinkholes, and other significant connections between surface and subsurface waters must be identified and considered during design, lest construction activities disrupt normal flows within or downstream of the site. Emerging waters or near-surface water may impact dike and water control locations, hinder construction activities, and influence system operations. It may be necessary to design temporary diversions during construction or permanent structures to include or exclude these sources from future pools. Though pumping from wells will drastically increase operating costs, subsurface waters may provide supplementary sources in critical drought periods.

Groundwater descriptions should include depth, quality, isolated or perched water tables and flow patterns. Depending on surface water supplies and type of wetland, a groundwater discharge or recharge area might be advantageous. Discharges through springs or seeps could supplement surface waters during operation, though hampering construction. Sites with dry streambeds, fissured metamorphic or igneous rock, and karst geology should be avoided if the wetlands are to be used for wastewater treatment. Known recharge areas of any type could be difficult to seal and water losses might be intolerable, especially during dry periods. Of course, if recharge is the design function, then the wetland should be sited over a recharge area.

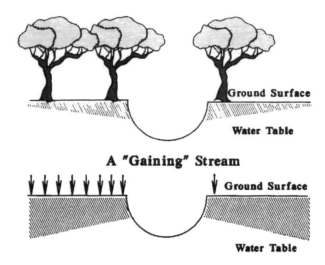

A "Gaining" Stream

A "Losing" Stream

Figure 8-3. In wet climates, stream water levels often depict groundwater elevations but in much of the west, streams carry water down from the mountains into arid plains where water tables may be much deeper than stream water levels.

Locations and type of existing and planned use of surface and groundwater at potential sites should be determined to avoid interrupting current activities or to estimate acquisition or mitigation costs. Potential adverse impacts from other users, as well as to other users, should be estimated during site evaluation. Existing impoundments, especially large reservoirs, in some regions cause substantial increases in groundwater supplies, while ditching and draining cause reductions in most areas. Both can provide indications of groundwater status.

CLIMATE AND WEATHER
Weather patterns and climatic regimes restrict choice of wetland types, dictate biological components, and influence operating procedures. Daily and seasonal precipitation patterns (including frequency, intensity, and duration of storm events) are determinant factors in amounts and timing of runoff crucial to most wetlands. Temperatures can also have important effects on amounts and pattern of runoff. Daily and seasonal air temperatures and, indirectly, water temperatures affect basic chemical and biochemical processes that are the basis of all life forms. Prevailing wind velocity (direction and speed) affects orientation of ponds and perimeter planning during design because of wave action impacts to vegetation and dikes, and water loss from evapo-transpiration during operation. Regional and local relative humidity and incident solar radiation influence and interact with precipitation and temperature, thus controlling many biological and physical processes.

Climatic data are from standard observations at airports, NOAA weather recording stations, Coast Guard facilities, volunteer observer stations and a few other specific sites. While it provides a broad overview of past and expected weather in the region, numerous topographic and biological factors influence generation, location, and diversity of weather patterns affecting a specific site. Generally, the more distant and the more varied the topography between the recording station and the candidate site, the more variation in actual weather is to be expected. Climatological Data, an historical summary for various areas of the country as well as a variety of other climate information for the U.S. is available from the U.S. National Oceanic and Atmospheric Administration, Asheville, NC (see Appendix C).

Hills or ridges, mountain ranges and valleys, large water bodies, upsloping though relatively level terrain, presence and type of vegetation, and north- or south-facing slopes impact precipitation, humidity, solar radiation, and wind currents, creating micro-climates that might be quite different from regional patterns (Figure 8-4). North-facing slopes often have growing conditions typical of more northerly regions, while the opposite is true on the other side of the ridge. Lake effect precipitation and temperatures are well known to residents of western New York, as are sea and land breezes to coastal inhabitants. Up-sloping terrain perpendicular to prevailing winds causes increased precipitation on the windward side and reduced rainfall on the downwind side. The striking differences are evident to a casual observer crossing the divide in the Cascade Mountains and other western ranges. Mountain valleys often generate daily wind patterns with up-slope currents during the day and down-slope, cold air drainage at night. Presence and type of vegetative cover reduce wind action and air temperatures, and increase relative humidity in local areas. Patchwork vegetative cover modifying solar heating results in thermals — rising columns of air over bare ground and descending columns over dense vegetation. Since thermals are the genesis and progenitors of thunderstorms, vegetation patterns also influence local precipitation patterns.

Length of growing season is a generalized parameter that manifests the effects of many climatological variables influencing types of plants or animals that can survive on the site. It obviously will affect selection of plant species to be used, as well as operating constraints for some types of wetland. However, general values represent regional conditions that may need interpolation for describing a specific area or site.

BIOLOGY
Biological factors in site evaluation fall into four categories:

1. animals or plants that would be damaged or detrimentally impacted by construction of the wetland;
2. those that would be detrimentally impacted by project operation;
3. those that may detrimentally impact the new wetland; and,
4. those that facilitate data collection on other aspects of the site.

Figure 8-4. Cooler, wetter environments on the north and east sides of the hills surrounding these potholes support woody plants whereas the warm, dry southern slopes are grass covered.

Creating a wetland on a terrestrial site will obviously be detrimental to the previous inhabitants and, if any species of plant or animal has unusual aesthetic, unique, or commercial value, that loss must be factored into site evaluation and selection. Earth moving to restore an impacted wetland is more likely to impact a unique species since so many wetland species have become rare or isolated because of past wetland drainage.

New or additional wetland habitats may alter local or even continental distribution patterns for wetlands and some terrestrial species. For example, creation of new habitats during the last 40 to 50 years in mid-latitude states — New York, Illinois, Indiana, Ohio, Kansas, Maryland, and Virginia — has dramatically altered migration and wintering patterns of continental waterfowl populations with few wintering in previously used habitats in southern states. Changes in local distribution patterns may bring various species into conflict with other land uses or human activities. In general, restoration or creation has a long way to go before it can replace the original wetland resource base, and most new systems will have overall positive impacts to wetland wildlife.

Presence or proximity to existing wetlands may be important for planting materials or could be setting the stage for invasion of undesirable weeds. Creating a new wetland near an existing system that is heavily infested with purple loosestrife (*Lythrum salicaria*), salt cedar (*Tamarix*), or *Melaleuca* will increase operating costs of the new system. In addition to location within breeding grounds or wintering areas, location within a migratory flyway might be important if waterfowl and other bird populations are an objective.

Figure 8-5. Slight differences in elevation permit loblolly pine and eastern cedar to occur in the heart of this South Carolina swamp. Planning for elevation differences can add considerable diversity to a created wetland.

Presence of plants or animals is perhaps the best indicator of site suitability since they reflect the culmination of present conditions of soil, water, climate, and land use (Figure 8-5). Although information on hydrology, soils, climate, etc. is important, the significant features of the new wetland are the biotic components. Not surprisingly, they often have similar requirements to survive and grow on the site as do the present occupants. Length of growing season, availability of soil moisture, permeability of soils, fertility of soils, and past and present uses can all be interpreted from biological inventories of native or cultivated species. For example, a soybean field in the Mississippi Delta quite likely occupies a previous bottomland hardwoods site with moderately fertile, hydric soils.

If only sedges and wet grasses are found in a high mountain valley, attempting to create a cattail or bulrush marsh is likely to fail because of the short growing season. Bladderwort and pitcherplants indicate reduced nitrogen, chironomids indicate low dissolved oxygen or other pollutants, lichens reflect air quality, blue gramma grass suggests native prairie, and of course wetland species suggest hydric soils and at least remnant wetland systems. *Scirpus fluviatilis* and *Ruppia maritima* are found in brackish waters unfit for establishing *Nuphar* or *Pontederia*, and cottonwood, willow, or salt cedar do not necessarily indicate a good site for cypress or gum. Birds and mammals tend to be less useful because of their mobility, but amphibians, insects, and other invertebrates (especially aquatic types) may be valuable indicators.

REGULATIONS

Laws and regulations that may impact project construction and/or operation need to be considered as soon as potential sites have been identified. Though most vary from state to state, many fit within a federal framework and others are familiar to federal regulatory agencies that will be contacted for information on federal laws and regulations. Bear in mind that even though the proposal is to create or restore a wetland and the proponents may be concerned environmentalists, the same laws and regulations governing dredge and fill or other degradation activities must be complied with by those proposing to build wetlands. Some wetland proponents have the impression that creating or restoring wetlands is an environmentally sound activity and therefore immune from environmental regulations. Fortunately, this is not true. Though most will be impatient with the regulatory review process and some may be affronted or at a loss to prepare an adequate justification for their proposal, wetland developers are not infallible and they are just as likely to alter flood patterns, impact an endangered species, or damage a cultural site as someone proposing a housing development or marina.

The most important regulations cover earth moving (dredge and fill) in rivers, streams, and wetlands, protection of endangered species habitats, protection of wetlands and floodplain management, and preservation of cultural and historic resources. Specific federal laws include:

Sections 401 and 404, Clean Water Act
Executive Order 11988 — Floodplain Management
Executive Order 11990 — Protection of Wetlands
Sections 9 and 10, Rivers and Harbors Act
National Environmental Policy Act
Fish and Wildlife Coordination Act
Endangered Species Act
National Historic Preservation Act
Food Security Act
Clean Air Act

State or local laws and requirements are often a subset of the federal frameworks, though in some instances, local regulations may be considerably more stringent. In addition, most states have regulations governing impounded waters over some minimum size that should be examined for pertinence and many states now have specific wetland protection regulations. The state clearing house generally coordinates review of projects by various state agencies and can provide information on applicability of state regulations.

Low-lying or poorly drained lands are often optimal for creating or restoring wetland since the site may have supported a wetland system in the past. However, construction activities in areas with hydric soils will likely entail dredge and fill actions that are covered by Section 404 of the Clean Water Act and will require appropriate reviews and permits. If a favored site is wet, developers should initiate coordination with the U.S. Army Corps of Engineers at an early stage. Executive Orders 11988 (Floodplain Management) and 11990 (Protection of Wetlands) are applied by Federal Agencies during review and permitting processes under Section 404, Section 10 of the Rivers and Harbors Act, or other federal legislation. Neither Executive Order has legal standing (other than for federal agencies) on its own, but both are guidelines for the interpretation and application of federal laws. Consequently, either or both could be germane through the 404 process if the chosen site has or had wetland characteristics or is in a floodplain.

Since many threatened or endangered species are wetland types, any restoration project could easily fall under the purview of the Endangered Species Act. Alternatively, changing terrestrial habitats to wetland in a creation project could detrimentally impact a rare terrestrial species. Most state offices have a Natural Heritage Program sponsored or initiated by The Nature Conservancy, that will have detailed distribution maps for federally listed threatened or endangered species. Their database often also includes information on species under consideration for federal listing and on species that may have protection under state threatened and endangered species laws or regulations. The nearest office of the U.S. Fish and Wildlife Service has similar information and has statutory responsibility for administering the Act. In most cases, early contact will eliminate any further coordination or action, but in some instances, detailed surveys may be necessary.

Stream bottoms, river valleys, lake and wetland shorelines were favored habitations for past human cultures probably because of the same attributes that make them preferred sites for wetland creation. If the site was previously a wetland, especially a bog, it may have well preserved artifacts due to the anoxic and often low pH conditions of bog waters and substrates. Unfortunately, much of our cultural and historical resource has been lost to past development, but recent federal and state laws and regulations provide fairly comprehensive protection today. The state historic preservation officer can assist project planners in identifying potential cultural sites in the area and advise on inspections or surveys as well as protective measures that may need to be employed.

SPECIFIC CONSIDERATIONS

Locating wetlands to provide recreational or commercial marketing benefits has less constraints than more restricted uses. A constructed wetland (wastewater treatment) site is often predetermined by the location of the wastewater source, that is, an industrial site, a municipal treatment plant, or an acid drainage seep. Since the wastewater source can only be relocated with costly piping or pumping, siting the treatment system is usually limited to evaluating a limited geographic area, which may or may not be an optimal site.

Site evaluation for constructed wetlands should include considerably more data collection on adjacent or downstream lands and evaluation of potential impacts from the constructed wetland. Detailed chemical descriptions of receiving waters and estimates for expected discharge quality should be obtained to evaluate and monitor potential impacts. An early meeting with adjacent landowners to thoroughly explain proposed activities, expected impacts, and project goals is highly recommended.

Planners of flood storage/buffering wetlands, ground water recharge systems, or education facilities are likely to have similar but not as exacting constraints. Hydrologic buffering wetlands must be located in the floodplain, but there may be ample latitude within the valley. Information on nearby and downstream lands will be important in evaluating potential impacts to these areas during operation. Groundwater recharge systems may also have more latitude than wastewater treatment wetlands, depending upon the location and extent of the pervious layers. Many recreational systems offer considerable siting flexibility, but educational systems generally need to be near or on school facilities and site selection will be limited.

DATA COLLECTION

Much of the information needed to evaluate prospective sites is available from local, state, and federal agencies. The first step is to contact these offices for their information on your site(s). After completing this phase, additional steps may be necessary as outlined below:

1. compilation of available information — office;
2. interpretation and evaluation of available information — office;
3. ground or aerial inspection of site(s);
4. preliminary data collection on site(s): surface water quantity and quality, subsurface water parameters from available wells, soil classifications and profiles, subsurface formations, and biological communities;
5. extensive or detailed surveys on the above categories as needed; and,
6. comparative evaluation of data, estimating feasibility and costs for development and operation at candidate site(s), and preparation of preliminary budget.

Data collection efforts for site evaluations must be reasonably adjusted to the magnitude and complexity of the project. Years of effort and thousands of dollars for siting to build a 10-acre system could rarely be justified. Conversely, a low-budget, cursory review for a 1000-acre or multi-million dollar wetland may result in inordinate construction or operating costs when serious problems are discovered later; or the system simply may not function as planned. Most of the information is likely to be available for the asking and should be examined regardless of project size. However, detailed site surveys can be costly and should not be initiated unless necessary to fill in crucial gaps in available information.

Information management systems should be considered in the earliest stages of data assembly. If planners have some form of geographic information system (GIS) support, data should be entered into this system as it is assembled even though data entry may seem costly. The benefits of using a GIS for displaying information during comparative evaluations of sites far outweighs the cost. With the proliferation of personal computers, a variety of evaluation and management simulating programs have been developed that are useful in siting, as well as modeling development/management options. One of the oldest is HEP (Habitat Evaluation Procedures) developed by the U.S. Fish and Wildlife Service. Contact the Division of Ecological Services in Washington, D.C. for information on availability of software and manuals. Two others developed by the U.S. Fish and Wildlife Service, National Ecology Center in Ft. Collins, CO are micro-HSI (Habitat Suitability Index) and HMEM (Habitat Management Evaluation Method). A method for assessment of wetlands functional values, Wetland Evaluation Technique (WET), developed by the Federal Highway Administration has been adapted for PC use by the U.S. Corps of Engineers, Waterways Experiment Station, Vicksburg, MS. The Fish and Wildlife Service has recently developed a computerized ranking system to facilitate land acquisition prioritizing. Contact the regional office of the U.S. F&WS for information on LAPS (Land Acquisition Priority System). Finally, WETLANDS is an electronic data base of state wetland protection programs and contacts developed by the Council of State Governments, Box 11910, Lexington, KY 40578.

Though complexity and cost of data entry into a data management system may seem high, the site selection process will be substantially improved. In addition, most organizations store current information in an electronic form and many have or are in the process of transposing historical records. Inquire about the availability of the data on diskette as you contact each organization. Lastly, creating the information baseline as a management tool for operation of the new wetland system at the chosen site may be just as important as data management in the siting process.

SOURCES AND EXAMINATION OF AVAILABLE INFORMATION

Topographic, geologic, geologic hazard, mineral deposit/exploration, and hydrologic maps and text available from the U.S. Geological Survey (U.S.G.S.) should be examined for relief, elevations, presence and type of waterways, sinkholes or other subsidence indicators, cultural facilities (buildings, roads, etc.), utility lines, mines and wells, abandoned mine shafts, springs, and location and nature of groundwater (Appendix C). Degree of detail varies considerably with the type of information and the area of interest, which is the reason that planners should contact the local U.S.G.S. office for advice rather than simply order the maps or descriptions from the central office. Mapped contours on the common 15-min quad sheet at 1:24,000 scale may not be adequate in mountainous terrain since contour intervals could be 40 feet or more depending on the amount of relief on an individual map.

Much of the U.S. has been mapped for the U.S. Fish and Wildlife Service's National Wetlands Inventory (NWI). NWI maps depict location, extent, and type of wetlands in overlay or composite formats at scales of 1:24,000 to 1:250,000. They are available from a number of state or regional offices of geological survey, wetlands, and map distribution centers (Appendix C). The NWI has also digitized over 30% of the NWI database and information on data products is available — (813) 893-3873. In addition, MicroImages, Inc. (201 North 8th Street, Suite 15, Lincoln, NE 68508; (402) 477-9554) has digitized maps with wetlands and typical cartographic features available for some areas.

Soil surveys, aerial photography, and a variety of information on growing seasons, plant requirements, fertilizer recommendations, and pond construction can be obtained from the district office of the NRCS or ASCS. Specialists in these offices are experienced in almost any aspect of agriculture and forestry, much of which pertains to building and operating a wetland system. Frequently, they are also locally knowledgeable about groundwater conditions and, to some extent, subsurface conditions including bedrock.

Older, perhaps sequential, aerial photography can often be obtained from the ASCS, the U.S. Geological Survey, a local mapping agency such as the Tennessee Valley Authority in the Tennessee Valley region, various other federal and state agencies, or an aerial photography service. Try to determine if any agency or organization may have studied the region for some purpose and then ask if they kept the old prints or negatives. If sequential photography can be obtained, it will prove invaluable in determining site history. Factors such as biological communities, frequency and extent of flooding, drainage, erosion, sedimentation, land use, abandoned or reclaimed mines, buried landfills, waste disposal sites, pipelines, and property ownership (field boundaries from fence or tree lines) can often be detected and/or followed in a sequence of aerial photos.

Careful interpretation can even identify cultural and archaeological resources such as evidence of ephemeral use of an area by historical inhabitants. For example, the Plains Indians collected small rocks to weigh down the fringe of their tipis and then rolled the rocks off when the camp was moved. Even today those clusters of rock rings marking an old camp are clearly seen in aerial photos or low altitude flights over remnants of the native short-grass prairie. Not only do these rock patterns reveal cultural resources, but the fact that circular patterns are still present also indicates the land continuously supported native prairie. Many other parcels may have prairie today but were cultivated at some time in the past and the rock rings were either disrupted or rocks were deliberately cleared to facilitate use of farm machinery.

Large-scale surface geological formations are much more evident from aerial photos than from ground inspections and even subsurface formations may have surface representations. For example, boundaries of cedar glades, a unique terrestrial system in Kentucky, Tennessee, and Alabama caused by near surface limestone formations, are revealed in springtime aerial photos or overflights by the distinctive greenish foliage of prairie clover (*Petalostemon gattingeri*).

Available imagery may go well beyond the expected black and white or color photography. Depending on the area, organizations with different interests and responsibilities may have obtained infra-red, ultra-violet, vertical or side-looking radar imagery, magnitometer surveys, and a variety of other specialized data types. Most can be very useful, but some may need to be interpreted by specialists. For example, plant stress from poor soils, inadequate water, disease, or insect infestation is easily detected by the pale or light reddish color in IR photos because healthy vegetation tends to have dark shades of red. Different species also cause different shades, but considerable experience may be needed to correctly interpret the variations. Understanding magnitometer survey data or side-looking radar images is a bit more complicated. On the other hand, dominant species composition of a forest is easily determined from color aerial photos made during the fall color season and wetland maps (delineation and classification) for the NWI are prepared by photo interpretive specialists from high-altitude photography with occasional ground truthing.

Flood hazard maps available from the Federal Emergency Management Agency depict expected flood frequency in terms of the area inundated by 50-year, 100-year, or 500-year floods and can be useful in identifying flood ways, flood plains, and flood-proofing requirements.

All in all, the proportion of information available for most sites from maps, aerial photos, surveys, and textual descriptions is likely to be much greater than will be obtained from typical on-site surveys. Only if planners initiate intensive, detailed and expensive surveys of all factors important in siting are they likely to obtain an equivalent amount of useful siting information. Much of the available information is free or provided at nominal cost and can be reviewed, categorized, and evaluated with only the cost for the evaluator. Consequently, the majority of siting effort should be placed on office collection and interpretation of available information. Other than a low altitude overflight, new aerial photography or a walk over, on-the-ground surveys should only be planned and initiated for major information gaps in critical siting factors.

After office comparisons have been completed, low altitude (500 to 2000 m above ground level) overflights will expedite verification of assembled data for alternative sites, especially if the sites are widely separated. Flights should be scheduled in early morning hours on sunny, calm days to avoid thermal turbulence and because low-level sun angles and subsequent shadows improve topographic perspectives. Depending on factors of interest, presence or absence of tree leaves and other foliage may obstruct or enhance site evaluations and flights should be planned accordingly. Choice of altitude is a compromise between scale and fine detail and time over a specific area since relative speed increases with lower altitudes reducing the available time to study specific details. Generally, a high-altitude pass followed by one or more overflights at lower altitudes is the best combination.

Other types of useful information include county and regional maps, drilling records from water, oil, and gas wells, stream water and biological surveys, permit records from mineral or petroleum explorations and highway construction, environmental impact statements, and zoning regulations. Even patterns and names on road maps can be useful. Straight roads in England are striking evidence of old Roman roads, in contrast with typical serpentine roads of other periods. Turnpikes, tollways, and ferry roads are often names that reveal historical routes that may have associated cultural resources as do names including fort, mill, or mill pond. Additional information sources include government agencies, universities, local builders, surveyors and contractors, utility companies, and naturalist or sporting clubs and societies.

After completing assembly and evaluation of data collected from the above largely remote sensing methods, permission to inspect the site should be obtained and a careful walkover carried out with specialists as needed. Most likely, information will be scanty or absent on one or more factors that can be supplemented by a simple field examination by an experienced professional. Trips to the site are also opportunities to determine access and discuss site history with local residents, either owners or neighbors, as well as verifying other information collected to date.

Most likely, detailed topographic mapping will be necessary at some stage in the process since planning will require 1-foot or smaller contour information. Other on-site surveys that may be useful include auguring, test pits, percolating and compaction testing, along with limited surveys of vegetation, soils, hydrology, cultural resources, and geologic formations such as rock outcrops. Current and expected use of the site and neighboring lands may often be identified during site inspections and interviews with residents.

During these contacts, project goals should be carefully and fully explained to gauge public attitudes and reactions, or to modify attitudes if necessary and feasible, before major commitments have been made. Though wetlands are virtuous in many circles today, remnants of the "dark and dismal swamp" persist, as do unpleasant experiences with some aspects of wetlands. Concerns over snakes, mosquitos, odors, depressed property values, aesthetics, or crop depredation must be addressed frankly and patiently even though many derive from misinformation.

At this point, planners should have enough information so that significant deficiencies, if any, are obvious. Detailed contour mapping, perhaps laboratory analysis of soils, test pits, or shallow drilling for groundwater, bedrock information, and borrow materials are the most likely surveys that will need to be conducted. However, some sites may potentially have a listed rare or endangered species, significant archaeological or historic resources, subterranean caverns, or unusual groundwater distributions, and comprehensive surveys must be initiated. Generally, the responsible regulatory agency will be able to informally advise on local or regional expertise that should be obtained if all other factors favor that site.

Extent and detail of these surveys, as with other means of information collection, must be representative of complexity and magnitude of the project. Minimally, the siting process will have assembled the information needed to identify hazards to be avoided and to develop plans and cost estimates for project construction and operation during the planning stage. If additional information seems necessary, astute planners will obtain it since an overly conservative design may be much more costly to build and operate than the apparent cost savings from limited site investigations.

CHAPTER 9

PROJECT PLANNING

DESIGN SELECTION

Perhaps the most enjoyable part of this process is designing the new wetland. This is the time to release the creativity of your imagination, to translate ideas into drawings, and to revise ideas and drawings to accomplish your objectives, all without exceeding available funds. But once a site is selected, the properties of the site may accommodate your wishes or they may place serious constraints on your innovative concepts.

By determining the functional values sought from the new wetland, you have already narrowed the choices of wetland types. Though certain kinds are more suitable for certain objectives, the site you have selected or are forced to use may further restrict the usable options. Previously unbridled design innovations may need to be tempered or even completely revised after a careful review of the properties of your site. Or constraints imposed by site properties may preclude building and operating the type of wetland that will provide the functional values listed in your objectives. Three characteristics — water supply, topography, and land costs — are perhaps the most important factors to consider since any one of them may eliminate some designs or sites. The need for an adequate water supply is obvious. Topography may be less evident, but it is no less important. Relief, or the differences in elevations, and spatial relationships, the locations and proximity of high and low areas, are important determinants of the potential size, shape, and depth of the new system. Of course, seasonal fluctuations in runoff or possible alternative supplies and the seasonal needs of various wetland types, soil and parent material type and condition, property boundaries and adjacent land uses, and types of natural wetlands in the region must also be evaluated before selecting a specific design. But topography determines the amount of earth moving required and grading and land costs often represent 50–90% of project costs.

Examining natural wetlands is important because the created wetland must closely imitate natural systems adapted to that region if it is to succeed without excessive operating and maintenance costs. Since we hope to create a system that will be largely self-maintaining, we are not likely to succeed if we attempt to build a wetland type that is adapted to entirely different climatic, hydrologic, and edaphic conditions. Regardless of the functional benefit expected from the new wetland (the objectives), the new system should mimic natural wetlands as closely as possible. To attempt otherwise may be frustrating and expensive.

AVAILABLE WATER SUPPLY

Adequate supplies of water at the appropriate time of year are crucial to sustaining a wet system. Do normal precipitation in the region and runoff from the specific watershed match the size and shape of the new system? How large a watershed is needed to supply adequate runoff for the preferred type or size or shape? Will water inputs at least equal exports during a critical portion of the year for the chosen wetland type? Must the surface area be reduced to cut down on water losses, or should another watershed be investigated? Answers to most of these questions are developed from examining precipitation, evaporation, and evapo-transpiration values for the region and a particular site. Understanding these factors in conjunction with variables influencing amounts of runoff provides valuable insight into the feasibility of creating a functional, self-maintaining wetland. Not surprisingly, wetlands in arid regions need runoff from a larger area of watershed to maintain adequate water levels during the driest part of the year and, even then, may lose depth or dry out during drought years. Conversely, a small wetland fed by a large watershed in a wet climate will not lack for water but may require elaborate and costly water control structures and emergency spillways for flood protection.

Although a variety of expressions have been developed to estimate available water supplies, the basic water budget concept is simply:

Change in Stored Volume = (Inflow + Precipitation) — (Outflow + Evapotranspiration)

Inflow can be in the form of surface flows and subsurface or groundwater flows and similarly for outflow.

Potential water sources and some methods of exploiting them include:

1. Precipitation: obviously cool, wet climates with high P/E ratios can support shallow basin wetlands with only precipitation as a source of water;
2. Surface runoff: watershed runoff collection is commonly used for many ponds and wetlands and much of the following discussion relates to this source;
3. On-line stream flow: placing a dike across a stream and impounding stream flow;
4. Off-line stream flow: creating the wetland adjacent to a stream with an excavated channel, and water control structure, leading to the wetland or in a few extreme cases, pumping from the stream into the wetland;
5. Lake shore: modifying the shoreline of a lake or other water body to use overflow waters from the lake;
6. Lake or reservoir: elevating the substrate in a lake, reservoir or river by depositing soil materials within the water body or by modifying the lake bottom to create the required shallow water conditions; many island type

and/or aquatic bed wetlands have been created by judicious deposition of dredge spoil materials.

7. Seepage or spring-flow: creating a basin or impoundment downstream from a seep or spring;
8. Ground water — passive: excavating below the upper surface of the water table as has happened in many mining situations but this also includes shallow excavations to intercept flow-through groundwater supplies;
9. Ground water — active: though few wetland developers could afford the high costs of pumping groundwater to maintain a wetland, some have been incidentally created adjacent to and in irrigation holding ponds by agricultural interests.
10. Combinations of any of the above: obviously precipitation inputs will be combined with other sources in most wetlands but arid regions may require more than one source.

The first step is obtaining average monthly values for precipitation and evaporation losses in the district and then estimating evapo-transpiration losses from the types of wetland under consideration. Precipitation and evaporation losses, commonly combined as the precipitation/evaporation ratio (P/E), may be obtained from the nearest office of the U.S. Weather Bureau, the Natural Resource Conservation Service and many extension agents, or from climatological data compiled by the National Climatological Center in Asheville, NC. Precipitation values are derived from long-term averages of actual precipitation records. Evaporation values may also be developed from historical observations or they may be computations of the quantity of water expected to evaporate from an open water surface under prevailing conditions of air temperature, relative humidity, and wind speed over a period of time. In either case, they are commonly referred to as lake or "pan evaporation." In simplest terms, the annual or monthly P/E ratio predicts the amount of water that would remain in an open-topped shallow container, a pan, after precipitation inputs are reduced by evaporation losses during the specified interval.

Though this ratio is valuable information, it may need slight modification since a wetland may lose water to percolation into underlying soils and because vegetation is not present in deep-water lakes or the abstract pan. Plants complicate the analysis by transpiring — losing water from cell surfaces directly exposed to the atmosphere to exchange oxygen, carbon dioxide, and other gases. Conversely, plants obstruct air flow near the water's surface (Figure 9-1). In windless regions or under still conditions, the atmospheric layer nearest the water surface becomes saturated and little additional evaporation occurs. But under windy conditions and turbulent flow, the boundary air layer is replaced continuously. If the relative humidity is high, evaporation rates may not increase, as is often the case on a still night. On a hot, dry day, continuous renewal of the saturated boundary layer with dry air may dramatically increase evaporative losses. Clearly, anything that obstructs or influences air flow patterns

Figure 9-1.   Evapo-transpiration losses from dense emergent stands are generally lower than evaporative losses from open water surfaces because of plant influences on the micro-climate near the water surface within a stand of vegetation.

near the water's surface will influence evaporation rates and the effect will be greater in hot, dry weather than in cool, wet periods. Plants, both herbaceous and woody species, along the edges or within the wetland have variable impacts on air flow patterns depending on the density, height, and areal coverage of the stand.

If the selected wetland type is primarily shallow open water, standard P/E ratios fairly approximate water balances. However, in large, dense stands of tall plants, cattail, bulrush, reeds, or woody species, transpiration losses from photosynthetically active plants become significant and determining a site specific P/E ratio becomes important. Though transpiration losses were once thought to be additive to pan evaporation losses, recent measurements have shown that combined loss from evaporation and transpiration in a densely growing emergent species (marsh) is approximately 80% of pan evaporation loss in a specific region. Apparently, lower wind speed and exposed water surface coupled with high relative humidity within an emergent stand substantially reduces direct evaporative loss from the water's surface and, even with the addition of transpiration losses, the combined total is lower than from open water surfaces.

Obviously, a P/E ratio can be computed for the entire year, but even a cursory review of factors contributing to this expression suggests that monthly values should be examined. This is especially important when designing a wetlands in drier regions. P/E values change, sometimes dramatically, during different seasons and evaporation losses during the driest part of the year may far outstrip precipitation inputs imperiling water levels in wetlands. Hot, dry months with low P/E ratios may cause dramatic evapo-transpiration losses from wetlands without compensating inputs from runoff. Cool, wet months with positive P/E values will likely support higher water levels because losses are reduced and runoff is increased. However, runoff extremes must then be considered in designing flood protection measures.

Northern climates further complicate the evaluation. High P/E ratios during cold months do not simply translate into increased runoff because of evaporative losses (snow sublimation) during the winter and differential runoff of meltwater in spring. Snowfall from November through January may largely disappear through sublimation in the Northern Plains; significant pothole-filling runoff results from snowstorms in February and March. Large amounts of runoff may be generated by snow melting over frozen ground but much potential runoff is lost to infiltration when snow melts over unfrozen ground. Neither condition is unusual and both may occur in different parts of the watershed depending largely on air temperatures and snow cover. Extreme cold prior to the accumulation of an insulating snow blanket may freeze soil moisture deep into the underlying parent materials. In spring, deeper layers melt later than surface layers and melt water runs off-site rather than percolating into the soil; runoff will be much lower following winters with substantial snow cover prior to extreme cold.

In cold and windy regions such as the Northern Plains, blowing and drifting snow accumulates in depressions and standing vegetation and downwind of fence lines, tree lines, hills, buildings and any other obstruction. Cold, relatively dry conditions render the snow susceptible to blowing and drifting for many days after the storm has passed. For example, the south and east slopes in a watershed will often have substantially more snow than the north or west facing slopes because the prevailing winter winds are strongly northwesterly. The differences can be as extreme as bare ground on the upwind side of the hill and 2- to 5-m snow banks on the downwind side. Huge snow banks projecting out on the downwind side of ridge tops in the Cascades and the northern Rocky Mountains represent a similar phenomenon.

To further complicate water estimates, wetland vegetation entraps substantial quantities of wind-blown snow. Tall marsh plants (1 to 2 m) are frequently almost buried in snow drifts if high winds accompany or follow a significant snowstorm. In intensively farmed areas, it is not unusual to find almost all the snow in a few remaining pothole marshes and very little on bare, fallow fields after a day or two of high winds (Figure 9-2). The remainder is, of course, piled up in the farmer's yard. On the other hand, northern forest vegetation intercepts fallen snow and a considerable proportion may sublimate without ever reaching the ground.

In the U.S., annual P/E ratios are high — precipitation is greater than evaporation losses — east of 96 or 97° west latitude (roughly the Mississippi River) and in Northwest and low in the Great Plains, Intermountain West, and the Southwest Deserts. However, higher elevations (mountains) and coastal

Figure 9-2. Streaks of windblown snow, accumulations in pothole marshes and snowbanks downwind of fence and treelines reveal the prevailing northwesterly winds in the Northern Plains. A snowflake rarely melts where it falls.

regions (ocean and lake) in both sections of the country receive more precipitation, cooler air temperatures reduce evaporative losses, and the P/E ratio is higher than surrounding areas. To some extent, forests are indicative of regional P/E ratios since most natural forests occur in regions with moderate to high P/E values. Riparian forests supported by high water tables and/or local micro-climates (higher P/E) in western river valleys are evident but not unequivocal exceptions since this woody vegetation often taps subsurface water supplies.

Generally, less than 5 acres of watershed is needed for each acre-foot of water in a farm pond east of the Mississippi River. Since shallow wetlands have greater surface area to volume ratios, additional watershed acreage will be needed. For example, a 20-acre wetland with average depths of 0.5 feet would have 10 acre-feet of storage and nominally need a 50-acre watershed; but a conservative design would increase that by 50% for a total of 75 acres. At lower elevations in the West, watershed size for farm ponds ranges from 20 acres in the Plains to 140 acres in the Southwest. These are gross approximations for typical conditions. Actual runoff amounts are strongly affected by amount, intensity, and duration of rainfall as well as watershed soils, topography, and vegetation cover. Methods for developing runoff estimates with precipitation descriptors and watershed characteristics are presented in the section on emergency spillway design in Chapter 10.

However, these methods were developed to estimate watershed area needed to supply runoff that will sustain an open water pond. If the new wetland is located in a humid region (P/E ratio >>1), is densely vegetated, has an impermeable liner, and has precise discharge control, somewhat less area will be needed because the annual evapo-transpiration rate from the wetland will be approximately 80% of the evaporation rate from a comparable amount of open-water surface in a pond.

After evaluating available information, visit potential sites during both wet and dry periods of the year and check on stream monitoring data with the local offices of USGS, NRCS, COE or TVA. If your project is large but the streams aren't monitored, it may be worth establishing monitoring stations to measure runoff and streamflows at various points in the watershed. Data collected over one year can then be compared with average meteorological data to develop fairly precise estimates of available water supplies.

Surface and subsurface water quality should be measured as well as quantity. Contaminants may contribute nutrients and enhance productivity or they may decrease productivity, be toxic to plants and animals, bioaccumulate, or limit uses of desired products. For example, crayfish, turtles, or frogs might accumulate potential toxics above levels approved for human consumption and the product is unmarketable. Pollutants may be carried into the site by surface or underground water flows or even be generated on site if underground storage tanks, pipelines, or transmission corridors are present. If the site was a wetland it may have been used for dumping hazardous or even toxic materials.

Many common water quality parameters can be quickly measured in the field with electronic probes and other modern equipment or even with the old standby Hack kits. Some probes measure 4 to 5 parameters of interest in a few seconds. Proper procedures for collection, transport and storage of water samples vary with the parameter of interest, nature of the sample and times involved. Simply dipping a flask of water and mailing it off to the lab is not appropriate if you are interested in dissolved oxygen, biochemical oxygen demand (BOD), bacterial counts, or any volatile compound. Follow sampling and handling guidelines carefully or find an experienced technician to collect the samples for you.

Due to the widespread requirement for monitoring discharge waters under NPDES permit regulations, laboratories staffed and equipped for water quality analysis are generally close at hand. Analysis costs for common contaminants, including heavy metals, are relatively inexpensive but analyses for pesticides and complex organics are much more costly. Unfortunately the complex substances are the most likely to cause problems and if there is any reason to suspect their presence, it would be false economy to forego that testing.

If adequate supplies of water for the new wetland are questionable, tall, dense emergents over most of the area is preferable to large, open waters if either type will meet project objectives. If not, consider including a zone of dense emergents or a shelterbelt on the upwind perimeter to reduce wind speed and evaporation losses from open-water areas. Remember, however, that in arid regions the roots of screening trees will grow towards water, perhaps into the wetland. Commonly occurring willows and cottonwoods are poor choices for wind breaks since they do not conserve water as well as mesquite (*Prosopis*), acacia (*Acacia*), or locust (*Robinia*).

Irrigation return flows; that is, water that has been used to irrigate cropfields or pasture may be a tempting source of normal operating or emergency supply, but should not be used under any circumstances in any region where the P/E ratio is less than 1. Farmers in a few areas, Georgia, Florida, etc., with humid climates have recently begun to use irrigation during the driest part of the year. In those areas (P/E ratio >1) return flows may be usable, after chemical analyses for fertilizers and pesticides, for temporary or emergency makeup water only. In dry climates, basically west of the Mississippi River, soil salt contents tend to be high and many areas have elevated concentrations of selenium. Typically, irrigated fields have a layer of salts 4 to 20 cm below the surface caused by long-term flooding for irrigation under high evaporation but low percolation conditions. Runoff from these fields contains high salt and often high selenium concentrations that will be deposited in the wetland, bioaccumulated, and impact fish and wildlife. Marshes in a National Wildlife Refuge in California (Kesterson) are no longer usable, in fact dewatered, because of selenium concentrations caused by using irrigation water return flows for many years.

Groundwater has been successfully used to create wetlands in some locations where the P/E ratio is high and groundwater supplies adequate. But attempts in more arid regions of the western U.S. have not always been successful. Nor would it be wise to depend on groundwater in irrigation regions because irrigation is slowly drawing down the water table in most. However, surface mining frequently goes well below the upper limit of the water table assuring future water supplies for ponds and marshes. Reclamation guidelines for coal surface-mined lands have included creation of ponds and wetlands for wildlife habitat for many years but most are dependent on runoff or springs. Current reclamation guidelines generally require "back to contour" restoration eliminating pits and depressions. But pre-act abandoned mine sites occasionally have pits and holes that intercept ground water. In many cases, the pit is too deep for wetland vegetation but shaping the sides can greatly increase the amount of fringe area suitable for wetland plants. Recently many constructed wetlands have been located below seep areas to remediate acid drainage and protect streams and rivers. Phosphate mines also tend to have deep excavations but a number of projects have converted these areas to productive wetlands for a variety of uses. For example, a 440-ha constructed wetland built on clay settling ponds at an old phosphate mine provides high level, polishing wastewater treatment for the city of Lakeland, Florida as well as habitat for eagles, wood storks, and myriads of other wildlife species. At another 20-ha phosphate mine site, ponds were created above and within the water table and monitoring indicated the new wetlands were rapidly approaching the species diversity of nearby natural wetlands after only two years.

More recently, innovative projects in New England and the mid-Atlantic states have created wetlands in old gravel pits converting a landscape eye-sore into a pleasant marsh/forested wetland. Developers often face requirements for mitigating wetland impacts from highway and other projects but regulators increasingly discourage creating replacement wetlands on undisturbed terrestrial sites. However, old gravel pits represent a considerable acreage of degraded lands with little value, especially in glaciated regions. And many, perhaps most, gravel pits intercept fairly reliable ground water supplies to sustain a new wetland (Figures 9-3 and 9-4).

Figure 9-3. An inactive gravel mine site near Brentwood, New Hampshire. (Photo by Photo Hawk, courtesy of Normandeau Associates.)

Figure 9-4. Carefully planned grading and planting transformed the gravel pit into an early stage wetland in only one year. (Photo by Photo Hawk, courtesy of Normandeau Associates.).

SITE TOPOGRAPHY

Site topography will guide, if not dictate, optimal locations for dikes or dams since closing off a narrows often requires less fill material than arbitrary or randomly placed dikes. Placement of dikes and relative elevations above the dikes determine the areal extent and depth of impounded waters and influences water level manipulation capability of the new wetland. Size, shape, and depth of each pond in the new wetland result from the ability of dikes and water controls to impound waters to elevations at or above upstream land surfaces. A broad, flat area will need only a low dike to flood a large area, whereas high dams may be needed in areas with high relief. The short, high dam may be less costly than a long, low dike, but the amount of shallow waters created and operational flexibility must be matched with project objectives. The proportion of shallow (5 to 20 cm) water depths available for marsh plants or for bottomland hardwoods is large relative to the cost in the flat area, but small at the steep site. Conversely, rapid increases or decreases in water depths may be impractical without a large capacity or several water control structures in the flat area, while a large structure may quickly raise or lower water levels in a narrow valley. In the broad, flat area, rapid lowering of water levels to avoid stress on favored trees may not be possible in certain seasons without many control structures.

A deep linear system, protected from wind exposure to reduce evaporative loss, is optimal for a groundwater recharge/infiltration system assuming it is underlain by highly permeable substrates. This narrow pond or lake would also be better for water storage and supply.

Total dirt moving and grading requirements may often be reduced by carefully positioning dikes and dugouts. Construction grading will be a major expense and balancing cut and fill in nearby areas will reduce materials transport within the construction area or off-site. Fill material for dikes should be obtained by removing thin layers of earth from broad expanses rather than forming a deep borrow pit to maximize the extent of shallow-water areas. Excess fill may be used to widen or increase dike height, form parking areas and observation turnouts, or to create nesting islands or hummocks in selected locations. Low gradient mounds later overtopped with shallow water can be used to form islands of shallow water species (plantains or arrowhead) within selected deep-water areas, thereby increasing diversity of the system. Elevated ridges or mounds supporting terrestrial biota increase diversity and, if carefully located, alter distribution and velocity of flow patterns in bottomland hardwoods.

Site topography also influences runoff patterns, especially amounts and patterns of flow. Both spatial distribution and timing of runoff vary with drainage patterns because of elevation differences between different portions of the watershed. If an adequate water supply is uncertain for the size and type of wetland selected, individual ponds or cells may be sized and positioned to receive runoff from different portions of the contributing watershed. Topography within the site and in adjacent areas must also be considered in terms of access for construction equipment and future operation and management needs.

Topography may enhance or negate the functional values expected from the new wetland. For example, a productive, well-developed marsh may receive little use by waterfowl if it is a linear system nestled in a long, narrow ravine or valley. Many waterfowl are reluctant to rest or feed in areas with views that restrict their ability to detect approaching predators. Some diving ducks, grebes, and loons may have inadequate take-off distance and the wetland could become a trap, as happens when they mistake a rain-slick roadway or parking lot for deeper water. Similarly, a linear wetlands perpendicular to the main stem of a river will be less successful for flood water storage since high flows will tend to bypass the wetland rather than topping the river bank and spreading into a system adjacent and parallel to the main channel. A round or elliptical wetland may cause short-circuiting and reduce the effective treatment area in a wastewater treatment wetlands or increase evaporative losses in a groundwater recharge system.

SOILS

Soil type and erodibility and composition of underlying materials will also influence wetland design. Foremost is the permeability of the substrate, since trying to impound water over pervious soils may require excessive compaction or lining. Conversely, locating a groundwater recharge wetlands on impermeable clays is not likely to achieve project objectives. Placing dikes in areas with impermeable clays will reduce costs for transporting materials and reduce or eliminate dike seepage. In some low areas, soil materials carried in over time have created an impermeable bottom layer that is ideal for most wetlands. However, the underlying material might be highly pervious and virtually impossible to seal after the upper layer is broken through. This is often the case in wide valleys because much of the floodplain was, at one time or another, part of an old channel receiving sand and gravel deposits. As the river meandered away, only fine clays were deposited during floods, creating an impervious seal above very porous materials. Very shallow excavations and importation of clay dike materials may avoid penetrating the bottom seal, but there is always a potential for major leaks stemming from future activities.

Designers should seek assistance from a local soils scientist and/or the NRCS to conduct thorough soils testing throughout the project site. Soils analysis should include soil composition and chemistry as well as permeability to obtain an understanding of fertility and growing conditions. Permeability can often be estimated with a rough "perk" test — simply digging a hole, filling it with water and timing the period for complete drainage. A standard "perk" test includes a 6- to 8-in hole, scarification of side walls, presoaking, filling to a specified height (usually 6 in) and careful measurements of the drainage or percolation rate (inches/hour).

Highly erodible soils in the watershed above the new wetland may require a deep retention/sedimentation pond above the wetland to reduce deposition within the system. On the other end, water control structures and spillways will need careful design to reduce potential downstream erosion. A well-developed topsoil on-site should be removed, stockpiled, and later spread in the graded cells to accelerate development of wetland vegetation and substrates. Poor clay or sandy soils should also be saved since they are probably better planting media than parent material, but they may require adding lime, fertilizer, or organic materials to improve growing conditions. Very poor soils may also affect plant survival and propagation, thus impacting the choices of plant species and prolonging the start-up period.

OTHER FACTORS

Unless a long-term cooperative agreement has been developed with adjacent landowners, location of property lines will constrain size and location of the wetland. Unfortunately, property lines rarely follow natural landscape features, but instead are based on arbitrary conventions (the township and range system in much of the U.S.) or historical use patterns (metes and bounds of the eastern states). Dam locations and future water levels must be related to on-site and off-site land elevations so that adjacent lands will not be flooded during normal operation, periods of high runoff, or when beavers assume management of water levels. Impounded waters should not be permitted to cause land subsidence, shoreline erosion, or become a "public nuisance" or "deadly attraction" that may jeopardize neighboring children or livestock. Opinions of adjacent landowners should be solicited to identify fears and, if feasible, accommodate desires. Remember, creating a mosquito/alligator habitat in a Florida subdivision or a mallard/blackbird habitat among the small-grain farms of North Dakota will not endear you to your neighbors.

Configuration of the watershed may suggest that a series of small cells located on different but proximal drainages would optimize available water supplies or runoff and roughly approximate the desired total acreage. For example, a small wetland in the mouth of each small ravine or hollow in the upper portion of a watershed may, in the aggregate, create more total wetland acreage at less expense and with lower water requirements than one large system in the lower reaches. The former system may be easier to plan and construct, depending on the relief of the upper watershed; and it will be less susceptible to sedimentation during stormwater runoff. However, if runoff is the sole water source in arid regions, the combined runoff from much of a large drainage may be needed to sustain water levels during the dry period and the lower site would be favored.

SPECIFIC DESIGNS

SIZING AND CONFIGURATION

Natural wetlands and created wetlands vary from small backyard systems to thousands or millions of hectares. Size and shape of created wetlands are determined by site characteristics, desired functional values and funding available for land acquisition, construction, and long-term maintenance. A simple *Iris/Sagittaria* marsh of only 20 to 200 m² may purify discharges from the household septic tank and add a focal point of floral diversity to a homeowner's backyard, but only small forms of wildlife are likely to find adequate habitat. Conversely, providing significant groundwater recharge or flood water storage capacity may require hundreds or thousands of hectares as would a wetland supporting populations of larger wildlife species.

As with so many other design factors, size and configuration are largely dependent upon the target functions, though in practice, designers tend to make the size fit the budget available for land costs and construction. Unfortunately, few consider long-term maintenance requirements or establish a funding base to cover these costs. In addition, little consideration is given to ensuring that the planned size is adequate to support all of the components that enhance self-regulatory processes and reduce maintenance requirements. Current attempts to reduce size and complexity of wetlands constructed for wastewater treatment to the barest essentials are quite likely to lose an important attribute — self-maintenance — that is an important component in comparative cost analyses. Commonly described wetland creation failures are little larger than mud puddles — often 0.2 ha or less. Attempts to mitigate the loss of 0.2 ha on the fringe of a large natural wetland by creating an isolated 0.2 ha wetland on a terrestrial site are doomed to failure. But aggregating many 0.2-ha mitigation needs into a 25- or 50-ha system through mitigation banking shows considerable promise.

Wetlands designed for wildlife habitat must provide food and shelter — the essential ingredients for life — in adequate amounts and locations in a large enough area to provide living space for the target species. Small animals tend to have small home ranges requiring less area to support a viable population. For example, a few frogs, salamanders, toads, small fish, turtles, and small plants (submergents such as *Potamogeton, Ceratophyllum, Myriophyllum,* floating types such as *Lemna, Azolla,* and small species of *Brasenia, Peltandra,* and *Sagittaria*) could thrive in a backyard or school yard system and even songbirds would visit occasionally (Figure 9-5). Along with myriads of macro- and micro-invertebrates and plants, this small but complex system could function as a laboratory for formal classes, for informal visits by neighborhood children, or simply as a tranquil spot of nature in over-developed suburbia. At the smallest extreme, it would not be much different from the goldfish/lily ponds fashionable around the turn of the century. However, it would likely require quite a bit of attention to maintain the original mix and, without a zone of emergent/wet meadow plants or shrubs, even small mammals (mice or voles) are unlikely to find homes.

Attracting or providing homes for larger animals will require considerably more area and seasonal aspects become important considerations (Figure 9-6). Depending on the season, habitat requirements for a single species may vary

Figure 9-5. Small backyard wetlands can add diversity to over-developed suburbia and support a variety of plants and small animals for educational and recreational uses.

Figure 9-6. Blue-winged teal are typically identified with Northern Plains wetlands but they use many different types of wetlands in different regions of the country for breeding, migrating, or wintering including this parrot feather bed in a South Carolina swamp.

substantially. During spring and early summer, many waterfowl seek the isolation of small ponds; mallards or wood ducks may be found in marshes or ponds, some temporary, of a hectare or less. During migration or on the wintering grounds, ducks and geese congregate in large flocks that tend to avoid small wetlands, probably because of limited protection from predators but perhaps also because the large flocks simply need larger physical space. However, in many instances, waterfowl loaf almost shoulder to shoulder in a small portion of a large bay or marsh. Though they actively use only part of the total area available, they rarely frequent wetlands that are only as large as the area used in the larger system. In some cases, the small portion used changes and, over time, much of the larger system may receive some use. However, in others, only a small percentage of the total marsh or swamp is visited by the flocks. Undoubtedly, security from predators is a factor in this behavior, but other elements may be important.

Feeding behavior and food habits vary considerably, though only a few wetland species are as restricted as some terrestrial species. Wading birds tend to eat anything that walks, crawls, or swims as long as it is small enough to swallow. On the other hand, snail kites (*Rostrhamus sociabilis*) live almost entirely on one type of snail (*Pomacea*). In general, animals with highly specialized food habits will require a much larger tract to produce adequate food supplies for a self-sustaining population than will species with generalized food habits (Figure 9-7). The latter may be able to forage on a wide variety of food items including plants, insects, herps, mammals, and birds. Snapping turtles (*Chelydra serpentina*) (large and generalized) occur in almost any wet environment east of the Rockies, from small ponds and even wastewater lagoons to extensive marshes and swamps; bog turtles (*Clemmys muhlenbergi*) (small and specialized) are restricted to a few bogs with narrowly defined water chemistry and plant species.

Project goals may go beyond simply the presence or absence to setting production targets; that is, the goal may be to raise or produce a certain number of one or more types of plants or animals during a given time period (Figure 9-8). In that case, planners must review scientific literature on the desired species and determine typical production rates per unit area of wetland for that species or group of species and size the new wetland accordingly. Depending on the species, time considerations and scheduling may also become important. A harvestable crop of crayfish can be raised in less than a year on relatively small areas, but commercially valuable stands of cypress or cherrybark oak require large areas and many growing seasons. In general, much information is available on production rates and management methods for economic products. Products as varied as bullfrogs, muskrats, crayfish, alligators, turtles, many fish, rice, bulrush, reeds, timber, and recently, many herbaceous wetland plants have been commercially managed. Start by contacting the local agricultural extension office, the offices of NRCS, federal and state forestry and wildlife, and nurseries in Appendix C. Then spend some time in

Figure 9-7. A viable raccoon population needs a sizable wetlands to provide adequate supplies of food and cover for all its members.

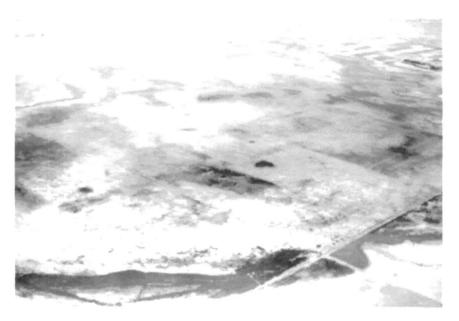

Figure 9-8. Larger wetlands, simply because of their size, have the capability to support many more of the important functional values than do small systems. The diagonal line running from far left to the intersection at the lower right is one mile long.

the local college library. Of course, if your objective is crayfish production in Nebraska or reed for roof thatching in Ohio it might be well to contact experts in Louisiana or near Fertos Lake in Hungary.

Sizing wetlands for flood storage or hydrologic buffering is much less exact since our understanding of this function is limited. However, storage capacity may simply be determined from the water volumes, peak runoff periods, and planned retention times for average storm flows of any given period. For example, flood storage wetlands upstream from an urban area may be designed to store and slowly release excess flows caused by a 10-year 24-hour storm or a 100-year 24-hour storm. Total volumes and peak discharge rates may be estimated from anticipated rainfall and watershed characteristics to determine the requisite storage capacity. Wetland size is then predicted from the water depths attainable and the area needed to store flood waters for a given time period. For example, if the predicted storm runoff volume is 1,000,000 m^3, then the wetlands should be capable of holding 1-m deep water over 1,000,000 m^2 (100 ha), 2 m over 50 ha, or some other combination. Obviously, this wetlands should be located in the restored floodplain along either or both sides of the river. Entry into and exit from the wetlands may be simply by over-topping the river bank, or specific channels, levees, and control structures may be needed.

The basic concept is to simply provide the capacity to store a given volume of water, to reduce the flow velocity by forcing it through dense vegetation, and to gradually release flood waters into the river or stream for an extended period after the storm. Depending on time of year, the storage and release period could range from 2 to 4 weeks or be as short as 3 to 5 days without significant damage to the wetlands. During winter, bottomland hardwoods often withstand inundation for weeks without apparent harm; but flooding for more than 4 to 5 days in spring and early summer will cause stress and mortality. A cattail or reed marsh could tolerate an additional 0.2 to 0.5 m of water for 2 to 3 weeks in the growing season, but longer flooding will cause changes in submergent and wet meadow species and eventually in reed or cattail. High velocity flows may knock down emergent plants but most will simply regrow from roots and rhizomes. Consequently, storage capacity and wetland sizing must be related to the time of year when flood protection is needed and the species used in the flood storage wetland.

The obverse of flooding, but a part of total hydrologic buffering, is the augmentation of low river flows during dry periods; this poorly understood phenomenon is very likely a simple function of storage and release over an extended period. Doubtless, waters draining into a river from a floodplain wetland during intervals between storms tend to maintain consistent river levels, moderating the peaks and troughs that would otherwise occur. However, wetland drainage may vary substantially with hydraulic conductivity of the duff/litter layer, the soil layer, and the underlaying parent materials. Some flow will occur across the surface, but once the water level in the wetland has dropped below the berm or overbank, return flow will be through channels or underground. In many cases, drainage back to the river may be largely sub-terranean due to the highly permeable nature of underlying materials in many floodplains.

Since flood storage wetlands are planned as a "gentle" sort of dam, they should occupy the entire width of the floodplain in the selected portion of the valley. If this is not possible, then levees or dikes must prevent flood waters from escaping around an end or through a non-wetland zone within the flood-plain. They could also be located on one side of the river if the opposite bank is naturally or artificially raised at or slightly above the planned water level ele-vation within the wetland. Since current velocity is highest on the outer radius of a curve, overtopping normally occurs first on a bend and, if only one side of the river is available, an astute planner will select an area for the wetland immediately downstream of a sharp bend with a low bank.

If flood water ingress and egress is broadly across the river bank, shape may not be important. However, if ingress and egress are through channels or old oxbows, then an elliptical, rounded diamond, or V-shape pointing downstream will be most efficient in accomplishing rapid accumulation and gradual releases. In these instances, low dikes or levees may be needed to guide flood waters into and contain them in the wetland, and small control structures will be useful in managing releases.

For wastewater treatment, wetland size and configuration are determined by the volume of flow and the influent and effluent concentration of pollutants. Required effective treatment area is related to mass quantities of organic mate-rials, metallic ions, or concentration of microorganisms. Conversely, retention times or hydraulic loading may be used to estimate required volumes and, indi-rectly, treatment areas. Detailed guidelines are presented in Chapter 15.

Lastly, the new wetland should be designed to blend in with other features of the landscape, including those on adjacent lands under other ownership. Within the constraints outlined above, shape and dimensions of the new system will be determined by the principle functions. However, dams across small drainages tend to form tear-drop shapes and excavated ponds/wetlands tend to be rectangular because of earth moving equipment procedures. Neither adds much and may even detract from the aesthetics of the landscape; but adding a small water body (a pond/wetland) can significantly improve landscape aes-thetics with a little foresight and planning. Regular shapes (angular, rect-angular, or round) should be avoided if at all possible. Smooth, rounded, elliptical, or sinusoidal curves and irregular lines form pleasing shapes that mimic and merge with natural features of the landscape, thus augmenting visual attractiveness and habitat diversity.

If regular configurations must be used, appearance of a rectangular exca-vation may be improved by selective grading of shallow areas along the bound-aries to create irregular shorelines and increase edge effect. Spoil islands, unequal and uneven location of deep and shallow water regions, patchwork pat-terns of different plants and different trees, plants with different heights and growth form, slow and fast growing trees, careful location of observation points and vistas, plants or trees with colorful fruits or foliage, showy flowering species, smoothly contoured dikes, and screened control structures can all reduce the obtrusive, "engineered" appearance of a new wetland.

SEALING AND LINING

Most natural wetlands occur in topographical lows because water runs downhill and accumulates in depressions. However, all depressions do not support wetland ecosystems. Many basins are directly connected to subterranean channels (karst sinkholes) and others lack an impervious substrate, allowing potential water supplies to infiltrate to ground water. Without ponding or flooding during the growing season, these sites support, at best, transitional and, more commonly, terrestrial ecosystems rather than wetland. Even though adequate water, energy, and nutrients are carried off adjacent uplands, the uninterrupted connection to ground water permits passage of these critical elements through the system and into ground waters. Conversely, depressions with wetlands have a barrier to reduce or prevent water losses to the underlying materials and, consequently, undergo extended periods of inundation required to exclude terrestrial plants and animals. Only a few exceptions exist wherein inflows (surface and/or subsurface) balance percolation losses and other losses and the basin nurtures a wetland ecosystem without having an impermeable lining. Most natural wetlands are perched above an impervious layer that reduces or prevents water loss to underlying strata, as occurs in upland environments.

A few planners may be fortunate in having a continuous, reliable source of clean water from a spring or permanent stream to offset subsurface losses, and insuring the new wetland has an impervious layer may be less critical. Others may compute a balance of inflows from seepage or perhaps wastewater flows and losses from all factors and conclude that moderate or large infiltration losses are tolerable. In the first case, through flow could result in the loss of important nutrients and energy and the new wetland may be less productive than one that continuously cycles nutrients. In addition, even with relatively clean sources, introduction of unmodified compounds originating in uplands or within the wetland to groundwater may detrimentally impact groundwater quality. Conversely, if the reliable water supply is any type of wastewater, isolation from groundwater is imperative in order to protect groundwater, even if adequate water is available to sustain a leaky wetland complex. Though water quality near the discharge end of the complex may be high and not perilous, insuring that leakage only occurs at the lower end may not be practical, and partial protection is justifiably discouraged by regulatory agencies.

Since an impervious liner is so important, designers must evaluate composition of soils on site and determine the hydraulic conductivity with percolation tests. Soils in many otherwise suitable sites may lack adequate clay content or the particle size is too large to produce a natural seal. If the conductivity is greater than 10^{-6} or 10^{-7} cm/s, sealing or lining must be considered, especially in arid climates, wastewater treatment systems, or other situations where loss from infiltration could jeopardize water level maintenance during extended drought periods or impact ground water.

Compaction of in situ soil materials is the simplest and least costly method of sealing the bottom, but the clay content must be greater than 10%, and a wide range of silt, sands, and other small particle sizes should make up the majority of the soil. During construction, the bottom is tilled to a 30- to 40-cm depth and then rolled with a sheepsfoot roller under optimum moisture conditions to create a dense tight layer of the same depth. If insufficient clay is present, more suitable material (clay content >20%) may be imported from a borrow area and layered across the bottom and up the sides of the dikes to operating water level elevations and compacted as above.

If permeability is high or borrowed clay unavailable, soda ash or bentonite clay may be obtained through the nearest farmers supply or direct from the source. Bentonite is typically applied at 5 to 15 kg/m^2 (1 to 3 lbs/ ft^2) and soda ash at 0.5 to 1.0 kg/m^2 (0.10 to 0.20 lb/ft^2), depending upon the results of laboratory analysis of native soil materials. Bentonite is used if in situ materials have inadequate clay and are mostly coarse-grained particles. Very fine-grained clay soils may also permit substantial seepage, in which case soda ash (a chemical agent) is used as a dispersing agent to rearrange clay particles and seal the bottom. Both are layered uniformly across the bottom and up the dikes, tilled in to 40- to 60-cm depths, and compacted with a sheepsfoot roller before flooding. Bentonite is a colloidal clay that swells many times its original volume when wet, but shrinks again upon drying. Consequently, it must be kept wet after application and should not be used if significant water level fluctuation is expected or complete drying will occur. The compacted soda ash layer must be protected against penetration by roots, animals, or from erosion by covering it with a 40- to 60-cm layer of soil or rock near inlets and spillways. Bentonite typically retails for $200.00 per ton and soda ash for $350.00 per ton at farm suppliers, but substantial discounts on large quantities can usually be obtained directly from the source. However, if large amounts are needed and/or the source is distant, materials cost plus freight charges may become prohibitively expensive.

If bentonite sources are not nearby and/or the required application rate becomes greater than 10 kg/m^2, comparative costs of synthetic materials become more attractive. Geotextiles are extensively used for a variety of sealing purposes and generally available in most areas, though relatively expensive to purchase and costly to install. Depending on thickness and composition, most synthetics are susceptible to damage from root growth and must be carefully placed below a 40- to 60-cm layer of soil materials. Proper installation also requires placing a layer of fine sand to cushion the synthetic, and applying special adhesives to weld long strips together to form impermeable seals. Most synthetics are also vulnerable to UV radiation damage and must be covered with at least 10 to 15 cm of soil wherever used. Since installation is critical to obtaining a good seal, it should not be undertaken by the inexperienced.

Irrespective of the method used to create an impermeable liner, 40 to 60 cm of soil should be placed above the liner to support planted vegetation. Roots of few if any wetland plants will penetrate deeper than 40 cm unless severe prolonged drought conditions occur. Providing adequate substrate for root growth decreases the potential for root penetration of the liner and subsequent leakage.

Most of these methods could be used in constructing smaller wetlands, but attempting to seal hundreds or thousands of hectares in a large system is usually impractical unless a bentonite mine is adjacent to the site. If the soils on the site of a large system are pervious, it is generally better to find another site since even compaction of in situ materials over a large area could quickly become the most expensive element in project costs. However, regardless of the means, the created wetland should have a relatively impervious liner to preserve water supplies, foster nutrient cycling, and protect groundwaters.

ACCESS: OVERLOOKS AND BOARDWALKS

Lakes and impounded waters, even without the added features of complex, diverse biological communities in wetlands, tend to attract human visitors. This is especially true with society's current interest in wetlands. Consequently, visitor accommodations should be included in the wetland design unless system function or integrity precludes other uses. Encouraging ancillary uses will often improve public support for creating wetland for many other purposes. For example, at a recent workshop on constructed wetlands for wastewater treatment in Cannon Beach, OR, one town citizen explained that over two thirds of the community attended the ribbon cutting ceremony for the town's new wastewater treatment wetland. When I asked treatment plant operators in the audience, "How many residents of their towns knew the location of their wastewater treatment plant?", most said almost none and one did not think the mayor or council could find his. In Cannon Beach, many residents, tourists, and a few elk enjoy the quiet footpaths through a large wetland a few blocks from Main Street; and they support the operation and budget needs during critical periods.

Public information and education programs should be considered in planning, design, and operating stages. Certainly, public information campaigns and outreach programs will be vital to acceptance and long-term success of conserving wetlands and should be a major element of any planning effort. Encouraging public visits to view the processes at work will enhance a project's acceptance, win over valuable allies, and garner broader public support for creating wetlands.

Part of this effort should solicit support from local sportsmen's organizations; environmental groups; naturalist, birdwatching, and wildflower societies; and other associations and individuals interested in wetlands and wildlife. Even constructed wetlands for wastewater treatment may prove to be excellent areas for hunting, birdwatching, and other outdoor recreation. For example, a 1988 issue of Birder's World, a national birdwatching tabloid, has a 5-page article on California's Arcata Marsh and Wildlife Sanctuary. Entitled

"Birding Hot Spots," the story describes the history of the town's wastewater treatment project and over 200 species of birds and thousands of waterfowl and shorebirds using the marsh. Minot, North Dakota's constructed wetland system was a major field trip for participants at the national meeting of the American Birding Society in 1993. User surveys of visitors to the Arcata wetland rank walking and isolation as important values. Widening the support circles to include average citizens will improve acceptance of wetlands.

Visitor accommodations must provide reasonable protection from possible dangers that may be encountered or the owner may fall victim to one of our litigatious society members. However, "reasonable precautions" may be differentially interpreted and even the careless that manage to drive off a dike into the pool may be successful in some courtrooms. Generally, preventing access to known hazards, correcting hazards forthwith, or taking reasonable precautions will avoid most problems that could result in legal liabilities. But if any action or lack of action can be construed as "gross negligence, willful or wanton conduct" or any condition could be construed as dangerous or perilous, either correct it or close the area. Laws, courts, and juries vary widely and a prudent manager will seek legal counsel and purchase adequate insurance.

Visitor facilities range from occasionally mowed footpaths to elaborate visitor centers with creative displays, multimedia theaters, and full-time interpretive staff. The latter may be found at a few National Parks (Everglades) and Wildlife Refuges (Okeefenokee) or nature centers owned and operated by conservation organizations (Audubon, Nature Conservancy), utilities (Florida Power and Light), private entities (Cypress Gardens near Charleston, SC), and local and state governments (Myrtle Beach, SC). Footpaths with or without interpretive signs, and roadways with turnouts or parking at viewing points will provide basic access for most general-purpose ancillary uses. Both should be carefully located on higher ground or dikes to avoid impacting wetland hydrology or biological communities. Obviously these facilities should be accessible to handicapped persons.

Most wetlands are flat with few hills or mounds that could furnish an overlook or vantage point; but appreciation for wetlands is hindered if the viewpoint is limited to human eye level. In created systems, dikes and water control structures may elevate newcomers above the tops of marsh vegetation, but rarely to treetop levels. Getting down close to wetland plants and animals and rising well above the system aid in understanding the fascinating complexity on the one hand and grasping the incredible diversity on the other. Carefully planned boardwalks furnish the up-close perspective, while mounds or short towers are useful in visualizing the overall system.

Boardwalks are justly popular, providing access within the wetland while protecting the visitor and the wetland from intimate contact. Location and construction of boardwalks must be prudently undertaken lest either harm the structure and/or function of the wetland system. Boardwalks are typically

Figure 9-9. Boardwalks need not be elaborate or expensive and are essential for visitor access in most wetlands.

framed across short pilings driven into the substrate and made of treated lumber (Figure 9-9). Occasionally, cypress or other naturally resistant wood is used, but creosote or chemically pressure-treated lumber is often readily available. Precautions are necessary in handling treated materials since in a few instances, CCA (copper, chromium, arsenic)-treated lumber has caused mortality to wildlife in wet situations. Since the purpose is to expose the visitor to wetland diversity, boardwalks should transect as many different types of habitats as feasible, but should not follow boundaries between zones lest they become barriers to animal movements. Linear or angular patterns are easier to construct but much less likely to blend with the surroundings than shallow curves, loops, or circles. Attention can be focused on points of interest or vistas by widening the walk, adding a bench, or putting in a short tower (Figures 9-10 and 9-11).

In marshes or bogs, overlooks need only be a few feet above the surface to furnish an educational vantage point. Building up a small portion of a dike or forming a mound or low hill with excess spoil material along a border may be slightly more expensive initially, but will require little maintenance and cost over the years. In addition, mounds are less likely to become safety hazards through accident or neglect and, consequently, have lower potential liabilities.

In forested wetlands, a tower may be necessary to achieve a similar perspective since the viewpoint must be at or above treetop level. Though costly to construct and maintain, the benefit to visitor understanding and appreciation of large wetlands is commensurate with the expense. Unused fire towers are popular with visitors on many National Wildlife Refuges because of the improved perspective from the elevated viewpoint. Unfortunately, few staff or visitors have an opportunity to observe wetlands from light aircraft at low altitudes, and providing a tower is the next best alternative.

Signs are important to visitors and to project planners and managers. Signs should not only be preclusive, but they should be informative. If closure or caution is needed, explain the requirement in terms of wetland benefits, functions, and values, or perhaps visitor safety. Signs, small displays, and kiosks also provide opportunities for education on wetland ecology and explanation of project objectives (Figure 9-12).

Regardless of the level of visitor accommodations, it is imperative that facilities are included for access control since visitation during certain periods may impact the principle functions or endanger the visitor. Depending on available surveillance, signed and locked gates may be adequate or sturdy steel posts and bars may be required. If vandalism is experienced, the U.S. Forest Service steel pipe design places the padlock inside part of the gate piping. In any case, signs should explain the reasons for and the duration of anticipated closure and specific areas that are closed or open.

Figure 9-10. Adding small benches encourages visitors to pause a minute while natural vegetation screening allows wildlife to resume their normal activities.

Figure 9-11.   This boardwalk provides visitor access as well as supporting a pipe carrying highly treated wastewater for polishing in a Carolina Bay near Myrtle Beach, South Carolina.

Figure 9-12.   Educational signs can enhance visitor understanding and enjoyment of a wetland developing support for specific projects as well as other wetlands.

CHAPTER 10

MAJOR STRUCTURES

DIKES, DAMS, AND BERMS

Restoring wetlands may require little dike construction, but creating a new system will likely require building one or more dams or berms if elevations are relatively flat. In any case, these structures function to modify site hydrology by restricting water flow to create a pool of standing or slow moving water that supports the biotic communities (Figure 10-1).

In general, dikes will be fairly low. If heights are greater than 2 m or failure could cause serious damage downstream, planners should employ full engineering design procedures or should secure the assistance of an experienced engineer to insure dam integrity and safety under anticipated conditions (Figure 10-2). With lower dikes or berms, caution is necessary, but design and construction are less complicated and feasible for less experienced planners as long as catastrophic failure would not cause legal liabilities.

The foundation for the dike should consist of impermeable clay material or bedrock without crevices, seams, or fissures. If the location has sand or gravel, it may be usable, but the assistance of the local NRCS engineer or another engineer experienced with local conditions should be obtained. Bedrock should be carefully inspected to insure that excessive seepage will not occur through seams or cracks and that bonding can be obtained with dike materials.

If available dike material is not impermeable clay, a clay core should be installed in the dike, insuring that the core is well bonded to an adequate foundation. Slope on the sides should be a minimum of 3:1 with top width adequate for vehicular travel. On a very small system with low dikes, widths may be 1 to 1.5 m to at least provide space for mowing and foot paths. Larger systems need top widths of at least 3 m to accommodate vehicles of operating personnel and mowing equipment. Dikes used by visitors should not be less than 4 m, and preferably 5 m wide. Depending on side slopes, a planned and constructed dike with a 3-m top width will likely have a usable top width of 2.0 to 2.5 m due to edge slumping and unstable conditions. However, as top width increases, width of the base increases proportionately and a considerable amount of wetland area may be lost to dike networks. For example, a 2-m high dike with top width of 3 m and 3:1 slopes will have a 15-m base.

Muskrats and nutria often burrow into dikes and can cause dike failure unless preventive measures are included in the designs or problems are identified and corrected quickly. The latter, in some areas, could become an expensive

197

Figure 10-1. Dikes in Bear River marsh not only regulate water levels in upstream pools but also protect freshwater marshes from salt water influences when water levels rise in Great Salt Lake. Note the stoplog control structures beneath each bridge.

Figure 10-2. Simple berms, often too narrow for foot traffic, have controlled water for centuries in Thai rice fields but frequent maintenance is needed and water level control is only through breaching the dikes.

operating procedure. Wide dikes (4- to 6-m top widths) with 4:1 or greater slopes and corresponding bases rarely have serious muskrat damage. Muskrat tunnels and dens generally do not extend into the dike over 1 to 1.5 m. However, burrowing, in some cases from both sides, into narrow or steep dikes often causes collapse of the surface and may cause total failure. In small systems, installing welded wire vertically in the dike during construction will prevent muskrats from burrowing through the dike and causing failure. Rock rip-rap placed on both surfaces from pool bottom to 0.7-m above normal water elevations will also discourage muskrat burrowing, but it inhibits vegetation and impacts aesthetics.

WATER CONTROL STRUCTURES

Dikes or dams retain and impound water to form the hydrologic environment for the new wetland, but some means of manipulating the water elevations behind each dike is essential to wetland system management (Figure 10-3). Most created wetlands will require deliberate management — generally, water level manipulation — especially during the first 3 to 5 years to encourage vegetation establishment (see Chapter 16). Once the plant and animal communities are well established, little management will be required until undesirable successional changes begin some years later. Typically, these are represented by drastically declining productivity or by invasions of terrestrial species. Constant water levels for many years in most marshes will cause production decreases because essential nutrients are immobilized in reduced states within the substrate and not available to the plants. Complete drying for an extended period during the growing season (drawdown) oxidizes and mobilizes substrate nutrients so that an explosion of growth often follows reflooding.

Terrestrial plants can only invade an older system because water depths are gradually decreasing from accumulation of humic materials, peat, and sediments. Moderately increasing water elevations during the growing season will inhibit or eliminate terrestrial invaders without severely impacting wetland plants. However, over time, the required level may exceed the elevation of the top of the water control structure or dangerously reduce remaining freeboard on the dikes. At this point, complete drying during the growing season and perhaps burning the accumulated peat, under appropriate conditions and with necessary permits, will be needed. In severe cases, dredging of the accumulated materials may be required.

Some wetlands have been designed with only an overflow spillway to prevent excessive water levels that may damage dikes. Many constructed wetlands treating acid drainage in remote areas rely solely on overflow spillways because water control structures are expensive and susceptible to vandalism. A few small constructed wetlands treating livestock waste or row-crop runoff also lack control structures for similar reasons (Figure 10-4). In these cases, normal biological productivity and complexity are subordinated to the primary purpose of wastewater treatment and even invasion of terrestrial plants

Figure 10-3. Water control structures placed in or near the dikes facilitate water level management in ponds and wetlands.

Figure 10-4. An outlet pipe set at the proper elevation provides inexpensive water level control in small systems.

as the depression gradually fills is acceptable. However, management of these systems is only possible with earth-moving equipment and little can be done to provide the critical disturbance element that would retard the inevitable succession to a terrestrial system. At that point, it may be necessary to breach dikes, dewater the system, and dredge out the accumulated deposits, and repair the dike to restart the wetland.

Overflow spillways in small, low-flow systems often consist of a low portion in the dike with dense coverage of water-tolerant, mat-forming grasses, sedges, rushes, or rock riprap. During construction, a porous geotextile fabric or one of the various organic mattes is placed over the spillway to reduce erosion until planted vegetation has become established. Repair of minor erosion may be needed after unusual storms, but little investment is jeopardized and repair costs are insignificant. More rugged protection may be obtained with riprap or a layer of concrete placed within and along the sides of the spillway. The extreme is a fixed elevation, concrete structure that is much more costly to build and provides little additional protection or control though some have been used in larger systems.

Depth and duration of flooding control much of the plant community and, directly and indirectly, the animal community in wetland ecosystems. Consequently, wetland management primarily consists of water level manipulation, and water control structures in the dikes are essential tools for managing created wetland systems. Though a variety of designs are available, keep in mind that the principle use will be for setting a specific elevation to maintain a desired water depth in the upstream pool. In some types of established wetlands, that elevation may not change during the year or over the course of many years. In others, it may be necessary to raise or lower the level during the growing or non-growing seasons to simulate natural hydrologic cycles. Infrequently, the water control will be used to drastically lower and perhaps gradually raise the pool level to foster germination and establishment of wetland plants. It may also be used to completely drain the pool for dike repair or other needed maintenance, or for deep flooding to retard or reverse successional changes. Generally, a certain water depth will be maintained for weeks, months, or even years on end and the choice of water control structure design should reflect expected operation and the principal function. The ideal water control structure will:

1. provide for fairly precise regulation of water elevations;
2. have the capacity to raise water levels to the maximum permissible level with a safe margin of dike freeboard, essentially the elevation of the emergency spillway outlet;
3. have the capacity to completely dewater the pool;
4. allow changes to be easily made;
5. not require changes because of increases or decreases in inflows or from precipitation;

6. consist of simple structures requiring little or no maintenance;
7. not be susceptible to vandalism;
8. not be susceptible to blockage from debris or plant growth; and,
9. inhibit blockage by beaver, muskrat or nutria.

Available designs do not meet all criteria completely, though some are more appropriate to certain circumstances than others. Various types of valves or penstocks are perhaps most commonly used and are most suitable for very large structures with high flow capacities. Valves, whether gate or ball, can be installed in virtually any diameter of pipe and flows are regulated by partially obstructing the opening (reducing the functional diameter of the pipe). Large valves often have stem screws or other mechanical devices to facilitate raising and lowering the gate despite considerable hydraulic pressure. With a given pressure and above very small volumes, valves will provide accurate regulation of flow volumes. At minimal flows, valves are susceptible to clogging that may require frequent opening of the valve to its maximum to flush the blockage.

Valves are designed to regulate **volume** of flow, not **water level elevations**. With experience, it is possible to determine the appropriate setting to balance outflow from a wetland pool with normal inflow and even to learn what degree of adjustment is necessary to compensate for a certain amount of rainfall. However, many mistakes will be made in developing this experience and the water control must be adjusted after every moderate or heavy rain and then again when the increased inflow has passed through the system. Conversely, it will also need to be adjusted, perhaps repeatedly, during a prolonged dry period. Though valves are needed in high flow systems, they require considerably more adjustment than simpler methods because valves regulate volume of flow, not water levels. However, if water depths are much more than 2 m in the lower portion of the pool, other methods may be impractical or even dangerous to adjust and valves should be used. Larger systems with valves generally have substantial concrete control structures and developers should obtain competent engineering assistance in design and construction. Planners, especially of large systems, must also check federal, state and local laws and regulations for building requirements and seek professional assistance since structure failure could have serious legal implications.

STOPLOG STRUCTURES

Two types of water controls designed to regulate water elevations have been in use in smaller systems for many years. Perhaps the oldest and most widely used is the "flashboard" or "stoplog" type depicted in Figure 10-5.

The control structure is normally built of reinforced concrete within the dike at the lowest end of the pool and the floor of the opening is located at or below the bottom the wetland pool. A metal channel ("U" shaped) must be installed vertically in the stoplog slots because wood and metal logs will not seal against concrete and considerable leakage will occur. Height of the opening (<2 m) is scaled to the maximum depth desired within safe limits of dike

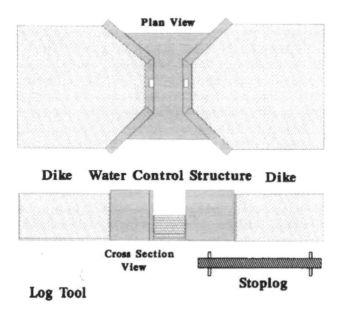

Figure 10-5. Stoplog water control structures of concrete provide reliable means to precisely regulate water levels for increased flow volumes in large wetlands.

freeboard and, commonly, the top elevation is the same as the elevation of the emergency spillway. Width of the opening is determined by the cross-sectional area required to completely drain the wetland pool in a specified period — usually 3 to 5 days. The opening width should also accommodate increased flows projected from runoff following a 1-year 24-hour storm event. Accommodation of greater flows is incorporated in the width and configuration of the emergency spillway. If a width of more than 1.5 to 1.7 m is needed, 2 or more banks of slots to hold logs with concrete islands (dividers) may be used. Greater widths in each opening are not practical since 4 x 4 or 6 x 6 logs will not retain their shape against water pressure over greater lengths. Alternatively, more than one individual structure may be placed in the dike if a larger total opening width is required.

Concrete wingwalls on upper and lower sides of the dike protect the dike from erosion and reduce seepage around the control structure. If the substrate of the pool is not impermeable, a vertical concrete wall may be poured below and perpendicular to the floor of the opening and parallel to the dike to prevent seepage and/or erosion below the control structure. The anti-seep wall is in turn embedded in the clay coring of the dike.

Water level regulation is achieved by placing "logs" or "boards" in the control slots to the desired elevation. Logs are commonly made of treated timber, metal, concrete, or PVC. One flashboard or log designed to fit in the structure slots is shown in the foreground of Figure 10-5. Concrete reinforcing rod pins through each end of the log are made to engage a simple fork of similar material to raise or lower each log. Normal inflows raise the pool level above that point and the excess simply flows over the top board. Excessive inflows from heavy rain elevate pool levels, and larger volumes flow out since the depth of flow across the log is commensurately greater. Conversely, reduced inflows to the system cause the pool level to fall below the top board, suspending outflow until pool levels are again raised by rainfall or increased inflows. Consequently, adjustment and/or readjustment is unnecessary and operation is minimal.

Major water level adjustment is simple on the one hand and could be more difficult on the other. Large increases in depth are obtained by simply placing more logs in the slots to reach the desired level. However, major reductions may require stepwise removal of boards since additional boards beyond the first 2 to 3 will have substantial horizontal and vertical water pressure holding them in the slots, in addition to the weight of the water soaked boards. If all of the boards must be removed concurrently, a block and tackle or tractor-mounted front end loader may be the only safe, practical method of extracting the boards. Otherwise, 2 to 3 boards may be removed and, after the pool level has fallen and pressures are reduced, additional boards are removed and the process repeated until the last board is out.

Construction of a stoplog control requires considerable labor in building forms, tying reinforcing rod, and mixing and pouring concrete, and costs are higher than for a control that is fabricated off-site and set into the dike during construction. In addition, transportation of lumber and concrete materials may be expensive or difficult in remote or inaccessible areas.

The stoplog structure is more expensive than the modified versions described below or the swiveling pipe design, but it has greater capacity for large flows and is less susceptible to debris blockage than either the pipe or culvert designs. Though muskrats rarely block this type unless water depths have become very low because of accumulated deposits, beaver are adept at placing chosen materials between the logs and walls and raising subsequent water levels to attain their objectives. If beaver are allowed to continue managing the control structure, over a period of time the large amount of imported material may be so tightly interwoven that extraction without explosives or heavy equipment is a long and arduous task. Human vandalism may be reduced by securing logs with metal straps or mounting bolts and lock fittings adjacent to the slots.

Recently at least two manufacturers have begun producing small stoplog control structures of fiberglass and PVC that employ the same design concepts but overcome the need for fabricating the structure on site. Two models produced by Agri-Drain Corp. are shown in Figure 10-6. These are constructed of 1/2-in. thick extruded PVC sheets, connected at the corners with anodized aluminum bracing and secured with stainless steel screws. Waterproof caulking is used throughout. Stoplogs are also made of 1/2-in. thick PVC with neoprene for sealing between logs and against the aluminum track and include metal hooks for handling. Various combinations of an equal number of stoplogs in 5-in. and 7-in. heights allows for ready adjustability. Upstream and downstream pipes

Figure 10-6. Inline (left) and inlet PVC stoplog control structures are light-weight and inexpensive overcoming the disadvantages of concrete structures for small to medium systems.

are connected with a flexible rubber sewer coupler that accommodates corrugated plastic tubing, PVC pipe, corrugated metal pipe, and various other pipes.

Both models are available with pipe diameters ranging from 4 in. to 18 in. and structure heights from 2 to 14 ft. The inline structure collects upstream water in a straight pipe passes it up and over the stoplogs and discharges through a downstream pipe. It is essentially a water control mechanism inserted into a pipe. The inlet type collects upstream water over the top of the stoplogs and then discharges through a downstream pipe. It is similar to the flashboard culvert described in following paragraphs. The inlet structure must obviously be installed upstream from the dike and should be well anchored but the inline structure can be located almost anywhere along a pipe from the upstream to the downstream side of the dike (Figure 10-7). Placing it in the dike would provide support but if it is free standing above or below the dike, the inline type should also be anchored. Prudent planners will include anti-seep collars with both models.

The smallest inline version — 4-in. pipe and 2-ft high — is priced at $230 FOB Adair, IA; a similar inlet model is $175. The largest inline — 18-in. pipe and 14-ft high — costs $910 and the inlet model is $720. Most wetland developers will need a mid range model, perhaps 12-in. pipe size and 6-ft structure height. The inline version is $460 and the inlet type is $325.

These structures are not expensive and much less complicated than forming a concrete structure on site. In addition, two husky men can handle even the largest. The most commonly used versions can be transported in the back of a pickup truck or with an all-terrain vehicle and handled by two workers. If beaver are likely to occupy the wetland, even placing screening or bars over

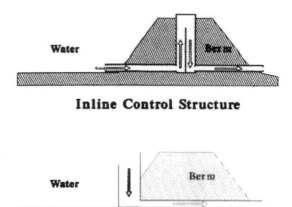

Inline Control Structure

Inlet Control Structure

Figure 10-7.   The inline stoplog structure may be installed in the dike for anchoring and easier access and can be made more beaver and vandal proof.

the opening are not likely to prevent problems with the inlet model. But the inline type, used with fairly long pipes (>4 m) on both ends, will prevent beaver access to the stoplogs. If short pipes are used, beaver will detect the leak, search for a source near the sound of running water and likely plug the pipes. But they aren't nearly as likely to trace back the length of a long pipe and stumble on the opening. Locking the cover will also reduce vandalism opportunities.

A flashboard fitting modifying the simple metal pipe riser (flashboard or stoplog culvert, pipe drop inlet) provides considerable flexibility in water level manipulation and is generally available at the local metal culvert fabricating plant (Figures 10-8 and 10-9). Flashboard fittings are rarely placed on culverts with more than 0.7- to 0.8-m diameters, so this structure is not suitable for large systems or systems with greatly fluctuating inflows. Asphalt and other coatings are normally available if system waters are corrosive, as in some acid drainage treatment wetlands; an anti-seep collar should be included in the purchase specifications. The horizontal pipe must be placed in a trench below the floor of the wetlands pool so that the bottom of the flashboard opening is at the level of the pool floor to insure complete drainage capability.

Operation of the flashboard culvert is similar to the stoplog control structure in that boards are simply placed in the slots to the desired height and subsequent water depth in the wetland pool. Since the boards are shorter and thinner, extracting all the boards at one time is generally feasible. However, lifting water-logged boards from the unstable platform of a small boat, if the structure is in deep water, may require considerable strength and dexterity. Operation and maintenance is minimal, though the smaller size opening is more likely to have problems from debris blockage, in which case a trash rack of steel rod or flat iron built around the pipe may be required. Since the upright pipe with

Figure 10-8. Culvert-type flashboard water control structures are readily available and durable but require equipment for transport and placement.

the flashboard fitting is located in the deeper water of the pond and often some distance out from the dike, this type is less susceptible to vandalism, though the standard design is easily blocked by beavers.

An enclosure around the vertical portion of the flashboard culvert has been quite successful in preventing beaver manipulation of water levels. Several of these designs have been in use in beaver-inhabited wetlands for over 10 years without obstruction. The enclosure simply consists of a larger diameter culvert or, in some cases, a metal 55-gal drum placed over the top of the upright section of the flashboard culvert. Holes are cut in the sides near the bottom and 3- to 5-m lengths of 10- to 20-cm plastic pipe are securely fitted in the holes. A metal cover secured with chain or metal strapping prevents beaver access through the top of the enclosure.

Beaver apparently detect leaks that need plugging, primarily by the sound of water flowing and perhaps secondarily by sensing flows. The principle underlying the design of the enclosure is to prevent beaver access to the flashboard fitting. Although beaver doubtless hear the water flowing over the boards and splashing into the pipe below, the enclosure prevents them from reaching and blocking the source of the sound — the flowing water. Since it

Figure 10-9. Culvert structures will maintain precise water elevations and are easily adjusted but they are susceptible to debris blockage and beaver management.

is necessary to avoid creating rushing water sounds or to create an obvious flow pattern at the distal ends of the plastic pipe, the pipe diameter must be large enough, and 10- to 20-cm pipes should be used to prevent beaver from detecting and blocking flow into the pipes. If the pipes are too short, beaver may accidentally block the ends in attempting to stop the water flow over the flashboards within the enclosure.

This anti-beaver enclosure has been successfully used for many years in created wetlands and has even been installed in beaver dams. However, given the adaptability of the beaver, some individuals will doubtless learn to manage this design as they have most others.

SWIVELING PIPE STRUCTURES

The swiveling pipe concept is the least expensive and perhaps the most simple structure providing excellent capacity for water level regulation (Figure 10-10). Constructed of plastic pipe, it is inexpensive, readily available, and easily transported to and assembled in remote locations. The slotted or perforated collector pipe on the upstream side of the dike is often placed in a trench and covered with coarse gravel or rock, hence, the discharge pipe must also lie in a trench below the dike. Alternatively, the collector side could consist of a short riser pipe or a flush-mounted, hooded inlet pipe. Either of the latter two will require a trash rack and perhaps anti-vortex baffling on the end of the pipe. The swiveling section on the lower side is an L-shaped piece of slightly larger diameter pipe fitted over the smaller pipe with a lubricated

Figure 10-10. Swiveling elbow-type water control structures are inexpensive and suitable for small flow systems.

O-ring fitting. The pipe section beneath or through the dike should have an anti-seep collar, as for metal culverts or corrugated pipes.

In operation, the elbow or L-shaped pipe is pivoted up or down by inserting a short length of wood board into the end of the pipe to increase leverage and carefully rotating the pipe. Raising or lowering the pipe outlet establishes the water level in the wetland pool behind the dike. In the vertical position, it will maintain the greatest depth, whereas horizontally it will drain the pool if the pipe joint is correctly placed at or below the floor of the pond. Depending on the tightness of the fitting, friction is normally adequate to hold the L-shaped pipe at the desired angle and outlet elevation though, occasionally, it may need to be fastened with a short length of chain. A concrete pad or rock riprap should be placed around and below the discharge pipe to reduce erosion and, if flow measurement is needed, the pipe should discharge into a small well with a weir plate fitting at one end.

The swiveling pipe structure is inexpensive; pipe, fittings, and cements are readily available and easily assembled, and operation is minimal. However, this type is limited to small flow systems since pipe diameters much greater than 25 to 30 cm are impractical to rotate and it should not be used where inflow is expected to vary significantly. Generally, the discharge is equivalent to the diameter of the pipe and additional inflow with greater depths and higher hydraulic pressures only increases the velocity of outflow. Increased velocity translates into increased volumes, but the capacity to accommodate substantial increases is obviously limited. Since the inlet structure is below the pool water surface, it is not susceptible to debris blockage though accumulated humic materials and other deposits may eventually clog the slotting or perforations. Beaver can easily block a horizontal collector pipe within the pond, but may be frustrated by using an anti-beaver enclosure over a short pipe riser. The swiveling pipe is the most vulnerable to vandalism since inappropriate pipe rotation is not difficult and plastic pipe can be easily damaged unless the swiveling discharge pipe is located in an enclosed concrete well.

Pipes used with the PVC stoplog controls or the swiveling pipe design or for any other purpose, should have screening on any open end to exclude small turtles. Pond and painted turtles becoming lodged in pipes have caused repeated maintenance problems in many wetland systems.

ESTIMATING CAPACITY

Flow capacity in stoplog control structures is determined from the following:

$$Q = 3.1 \ L_w H^{3/2}$$

where: Q = the capacity of the structure, m³/s or ft³/s
L_w = the length of the opening, m or ft
H = the depth of flow over the stoplogs, m or ft.

For example, a structure with 1.0-m long stoplogs (actually opening width of 0.8 m) and 20-cm deep water flowing over the logs would have a capacity of:

$$Q = (3.1) \ (0.8) \ (0.894)$$
$$Q = 0.22 \ \text{m}^3/\text{s}$$

In the flashboard culvert, hydraulic head must be taken into account and flow capacity is determined from the following:

$$Q = 0.6A(2gH)^{1/2}$$

where: Q = the culvert capacity, m³/s or ft³/s
A = the area of the culvert opening, m² or ft²
g = 9.8 m/s or 32.2 ft/s
H = hydraulic head over the center of the culvert

A flashboard culvert with 0.5 m culvert diameter and a 1.5-m riser would have a capacity of:

$$A = (3.141) \ (.25)^2$$
$$= 0.196$$

$$Q = (0.6) \ (0.196) \ (5.422)$$
$$Q = 0.639 \ \text{m}^3/\text{s}$$

One method for estimating flow in an open channel is by a modified Manning's Equation:

$$Q = \frac{1}{n}\left(\frac{(wd)^5}{(w+2d)^2}\right)^{1/3} S^{1/2}$$

where: Q = velocity — m³/s
      n = Mannings roughness coefficient
      w = channel width, m
      d = channel depth, m
      S = channel slope — decimal equivalent percent

To calculate Q in ft³/s, replace the 1 with 1.49 and use feet dimensions for w and d. For a smooth concrete channel with a width of 0.75 m, depth of 0.5 m and slope of 0.2%

Q = 1/0.014 { (0.75 × 0.5)⁵ / (0.75 + 2 × 0.5)² }¹/³ 0.002¹/²
Q = 0.43 m³/s

Manning's roughness coefficient, n, varies considerably. For example, n is often given as 0.014 in a smooth rectangular channel, 0.035 in a channel with short grass, 0.1 in taller grass, and can be 0.25 or greater in poorly maintained channels. Again, typical values for a variety of conditions are included in hydraulics texts along with nomographs for quick estimations.

STRUCTURE COMPARISONS

Overall, the concrete stoplog control structure is the most durable and trouble-free, though more expensive to build than the other types. With smaller and less fluctuating flows, the PVC stoplog structures and either of the other designs will suffice, though the stoplog types will accommodate more flow variability and are more amenable to protection from manipulation by beaver or damage from vandals. Valves require excessive adjustment, are expensive, and are vulnerable to beaver and vandals, although they are capable of managing higher flows than the less complex designs.

At the extreme, if the created wetland is located in an area with low base flows but very high inflow from intense storms, a large upstream watershed or high runoff from an un-vegetated or largely impermeable watershed, a combination of designs including the stoplog and/or the flashboard culvert designs might be used along with a valve structure. Unfortunately, storm events are as likely to occur at night as during work hours and valve adjustment may not be accomplished until after water levels in the wetland have risen significantly. Perhaps the least complex solution for construction and operation is one or more stoplog designs coupled with careful design and maintenance of the emergency spillway. High runoff from an unvegetated watershed is also likely to carry substantial quantities of suspended materials and the designer should consider including an upstream retention/sediment pond to avoid high sediment accumulations that would have serious impacts on the wetland. In this case, water controls in the retention pond may regulate and stabilize flows into the wetland, thereby reducing the need for large-capacity structures in the wetland.

TABLE 10-1
COMPARISONS OF WATER CONTROL DEVICES

Attribute	Stoplog Culvert	Flashboard Pipe	Swivel	Valve
Flow Capacity	Mod.	Low	Low	High
Durability	High	Mod.	Low	Mod.
Water Level Regulation	High	High	High	Low
Adjustment Requirements	Low	Low	Low	High
Ease of Adjustment	Mod.	Mod.	High	High
Debris Blockage	Low	Mod.	High	Low
Vandalism Vulnerability	Low	Low	High	High
Beaver Vulnerability	Mod.	Low (w/encl.)	High	Mod.
Construction Cost	High	Mod.	Low	High
Complexity	Low	Low	Low	High
Material Availability	Mod.	Mod.	High	Low
Materials Transport	Low	Mod.	High	High

If beaver problems are anticipated in a large wetland or one with high flows, two or more manufactured inline stoplog structures or flashboard culverts with anti-beaver enclosures will provide flow capacity as well as protection from beaver management of water levels. Flashboard culverts or concrete stoplog structures are more costly and transporting materials to remote sites may be more difficult, but their longevity, limited need for adjustment, and resistance to blockage by debris or beaver make them advantageous at sites with little access control or maintenance capability.

Wetlands created for most purposes will not need inflow control or distribution devices but flow distribution is critical in constructed wetlands designed for wastewater treatment. Details on inlet and flow distribution structures are provided in Chapter 15.

In summary, the choice of design is likely to be strongly influenced by size of the wetland and flow volumes since cost comparisons must incorporate long-term operation and maintenance requirements which often exceed initial construction costs. The more costly water level regulation devices have lower adjustment requirements and reduced maintenance costs because of the design simplicity and durability of the construction materials. In contrast to the concrete stoplog structure in Table 10-1, the PVC stoplog models would have lower durability, higher ease of adjustment, low beaver vulnerability (inline type), low construction cost, and high availability. Reinforced concrete and heavy corrugated metal pipe are more resistant to corrosion/decomposition and to vandalism and beaver impacts than plastic pipe or complex valve structures. In the final analysis, the type selected must have adequate capacity to regulate wetland water elevations with minimal need for adjustment and maintenance for anticipated base flows as well as limited stormwater runoff. The stoplog designs, concrete and PVC, and the flashboard culvert, sized for expected volumes, have the best combination of attributes to satisfy requirements to regulate water levels as the principal means of managing the created wetland.

EMERGENCY SPILLWAYS

Since watershed runoff estimates suffer from inexact weather predictions as well as watershed surface and vegetation characteristics, including the capacity for all possible flows in the selected water control structure(s) is not only costly but impractical. But the need to relieve excess water pressure for dike protection during and after high rainfall or snow melt runoff is crucial to long-term success of the wetland. If the dikes are washed out by excess flows, water level regulation is obviously lost until possibly costly repairs are accomplished. Depending on time of year and duration of water loss, wetland plant and animal communities may suffer serious impacts. Therefore, an emergency spillway must be included in every dike to protect dike integrity and the dependent biological communities.

The function of the emergency spillway is to provide extra discharge capacity in addition to the flow capacity of the water control structure(s) during high runoff events. It should pass excess water over or around the dike so that ponded water levels do not overtop and damage the dike or water control structures, and excess flow is discharged without damaging either side of the dikes, the control devices or the outlet channel below the water control structure. Though its operation may be infrequent and for limited periods, failure due to inadequate design or poor construction and maintenance will jeopardize the wetland system.

Dimensions of the emergency spillway are determined from the size and nature of the contributing watershed and the storage capacity of the wetland in conjunction with projected runoff volumes from standard design storm frequencies and duration. Width, depth, length, and slope also depend upon desired operation and subsequent location of the spillway, as well as soil erodibility and type of grass cover. A bypass spillway that conveys storm runoff around the wetland will reduce sediment deposition within the wetlands but is likely to be more costly to construct because of the length and critical construction requirements. It will also require an additional water control structure to permit normal runoff into the wetland while diverting excess flows through the spillway around the wetland. Consequently, less costly methods commonly locate spillways adjacent to or in the end of a dike, thus exploiting the velocity reduction factors — vegetation roughness and impounded waters — within the wetland system. The design must then provide an adequate cross-sectional area (width and depth) to pass peak discharge volumes from a specified rainfall event without eroding the dike.

Needless to say, the spillway should be constructed of or in erosion-resistant materials, covered with a dense stand of sod-forming grass with a 15- to 25-cm height and have a slope downstream of not more than 3 to 4%. The spillway may simply be a low spot in the dike or a short channel around one end of the dike. In either case, it should be straight or, at most, shallowly curved with grass-covered 3:1 slopes on either side, and the slope should drain towards the downstream side of the dike. The spillway with a shallow slope should be continued fully to and meet the downstream discharge channel at a narrow angle (i.e., the spillway should join the discharge channel at an angle of less than 20°), to reduce potential erosion on the spillway and in the discharge channel. Final grading of the spillway should ensure that depressions are not present in the spillway that would hold water and jeopardize grass cover or integrity of the dike.

Storm runoff is a function of:

1. rainfall amounts and expected frequency;
2. infiltration rates of the watershed soils;
3. land use and vegetative cover conditions; and,
4. slope of the land in the watershed.

The Natural Resource Conservation Service (NRCS) of the U.S. Department of Agriculture and the U.S. Geological Survey have extensive experience in monitoring and predicting runoff volumes from different storm events on small watersheds with different soils, vegetative cover, and slopes throughout the U.S. and its territories. The district office of the NRCS can provide assistance in determining soil types in the targeted watershed and invaluable local information on interpretation of vegetative cover and slopes in estimating storm runoff. Local experience is also useful in interpolating rainfall amounts expected from a 24-hour storm of 10-, 25-, 50-, or 100-year frequency since regional factors (large water bodies, hills, or mountain ranges) can significantly influence local rainfall patterns. NRCS specialists have considerable experience at modifying general guidelines for all aspects of designing farm ponds, including runoff estimation and spillway requirements, in most areas of the U.S. Adequate design needs for protection of a wetland system differ little from comparable requirements for a farm pond. The typical created wetland simply has less depth and greater surface area than most farm ponds. Before final drawings are prepared, the NRCS specialists should be asked to review runoff calculations and emergency spillway dimensions to insure the design is adequate. While in the NRCS office, request a copy of Ponds — Planning,

Design, Construction, SCS Agricultural Handbook No. 590, and study the information on all aspects of pond construction.

Predicted rainfall amounts are obtained from the Rainfall Frequency Atlas of the United States and the Precipitation Frequency Atlas of the Western United States, prepared and distributed by the U.S. Weather Bureau. These atlases include maps depicting the expected amount of rainfall during a 24-hour storm event based on 10-, 25-, 50-year, etc. storm frequencies. In each case, the rainfall amount is that amount expected from a storm that will, on average, occur only once in 10, 25, 50, or more years depending upon the map in question. Conversely, the 25-year storm will only have a 4% probability of occurring in any given year and the 50-year storm will have a 2% chance of occurrence in any year. Rainfall amounts in excess of the amount predicted for a 50-year storm event are only anticipated to occur, on average, once in more than 50 years, or once during the next standard category, every 100 years.

Practical construction considerations and impact severity in the event of failure suggest that wetlands in watersheds of less than 20 acres with average soils, slopes, and vegetative cover should have spillways designed to accommodate the 24-hour 10-year storm event, and those in larger watersheds should be designed for a 24-hour 25-year storm event. As a generality, rainfall from the 10-year storm in the eastern U.S. ranges from 9 in. near the Gulf Coast to 3 in. in northwestern North Dakota and comparable values for a 25-year storm are, respectively, 11 in. and 3.5 in. For example, values for Tennessee are approximately 5 in. for a 10-year storm, close to 6 in. for a 25-year storm, and 6.5 in. for a 50-year storm. Since rainfall amounts are not greatly different between storm categories, a conservative designer would use the next higher category to insure the adequacy of his spillway dimensions.

The NRCS has developed a series of curves and equations to estimate volumes and peak discharge rates from four hydrologic soil groups with different infiltration rates, interrelated with nine land use and vegetative cover classes, as shown in Table 10-2. Soils in group A have high infiltration rates and those in group D have low infiltration with high runoff potentials. Runoff curve numbers obtained from Table 10-2 may then be multiplied by the percentage of the watershed in each category to obtain an average runoff curve number for that watershed. This is related to expected rainfall, in Table 10-3, to determine the volume of runoff from a watershed of given size using the conversion factor of 325,851 gallons per acre-foot.

For example, a 200-acre moderately sloping watershed with 110 acres in good pasture on soil group A and 90 acres in poor woods on soil group C would have an average total runoff of 11.5 million gallons after a 5-in. storm event.

$$0.55 \times 60 = 33.00$$
$$0.45 \times 77 = 34.65$$
$$= 67.65$$

TABLE 10-2
RUNOFF CURVE NUMBERS INTERRELATING SOIL
INFILTRATION CAPACITY, LAND USE,
AND VEGETATIVE COVER[a]

	Hydrologic Soil Group			
Land Use and Cover	A	B	C	D
Cultivated				
Without soil & water cons. treatment	72	81	88	91
With soil & water cons. treatment	62	71	78	81
Pasture				
Poor cover	68	79	86	89
Good cover	39[b]	61	74	80
Meadow	30	58	71	78
Woods, shrub or forest				
Thin, poor cover, no mulch	45	66	77	83
Good cover, mulch and humus	25	55	70	77

[a]   Adapted from *Ponds — Planing, Design, Construction*; SCS Handbook No.
     590.
[b]   Use a runoff curve number of 60 if the table value is <60.

TABLE 10-3
DETERMINATION OF RUNOFF DEPTH[a]

Runoff Curve	Equation Constants		
Numbers	A	B	C
CN 60	−7.698	−169.37	−23.63
CN 70	−16.75	−538.52	−33.54
CN 80	−36.89	−1948.69	−53.94
CN 90	−128.11	−18550.96	−145.54

where: $I = A + B/(C + X)$
          I = runoff depth in inches
A, B, & C = constants
          X = rainfall in inches

This expression provides a fairly good approximation for
rainfall amounts between 1.5–12 inches.

[a]   Adapted from tabular data in *Ponds — Planning, Design,
     Construction*; SCS Handbook No. 590

TABLE 10-4
DETERMINATION OF PEAK RATES
OF DISCHARGE[a]

Runoff Curve	Equation Constants	
Numbers	A	B
Flat Slopes		
CN 60	−8.41	3.53
CN 70	−11.99	4.61
CN 80	−15.45	5.89
CN 90	−22.19	7.76
Moderate Slopes		
CN 60	−16.25	6.19
CN 70	−21.86	7.92
CN 80	−29.15	9.90
CN 90	−47.28	12.97
Steep Slopes		
CN 60	−28.71	10.47
CN 70	−42.92	13.55
CN 80	−55.82	16.43
CN 90	−72.06	20.13

where: $P = A + B\ X$

P = peak discharge rate (cfs/inch of runoff)

$A\ \&\ B$ = constants

X = drainage area in acres

This expression provides a fair approximation for 30 — 1000 acre watersheds.

[a] Adapted from graphic data in *Ponds — Planning, Design, Construction*; SCS Handbook No. 590

From Table 10-3, 5 in. of rain on a 200-acre watershed with a runoff curve number of 67.65 or 70 will produce

$$I = A + \frac{B}{(C + X)}$$
$$I = -16.75 + (-538.52/(-33.54 + X))$$
$$I = -16.75 + (-538.52/(-33.54 + 5))$$
$$I = 2.11 \text{ in. of runoff}$$

Then,

200 acres × 2.11 in. = 422 acre-in.

422 acre-in. / 12 = 35.17 acre-ft

35.17 acre-ft × 325,581 gal/acre-ft = 11,450,684 gal

or approximately 11.5 million gal

For perspective, if the new wetland has a water surface area of 35 acres, this storm would raise the water level slightly more than 1 foot if no discharge occurred.

In addition to the total runoff, we need to estimate the peak discharge rate in order to determine the dimensions for the emergency spillway as follows. From Table 10-4,

$$P = A + B \quad X$$

where: P   = peak discharge rate in cubic feet per second per inch of runoff
          A and B = constants
          X   = drainage area in acres

For a curve number of 70 and moderately sloping watershed,

P   = -21.86 + 7.92   200
P   =   90.15 cfs/in of runoff depth
and Q =   90.15 cfs/in. × 1.39 in. = 125.31 ft³/s

at the peak discharge rate during this storm.

For a 50-ft spillway with a 3 to 4% slope and good grass cover of 6 to 10 in. on erosion-resistant materials and maximum velocity of 5 ft/s, the dimensions of the spillway should be (Table 10-5):

W = Q / q = 125.31 / 3 = 41.77 ft wide, and
depth of flow (Hp) for a 50-ft long (L) spillway will be 1.4 ft

Since the minimum freeboard on the dike should be 1 ft,. the emergency spillway should be built 42 ft wide and the spillway surface should be 2.4 ft below the top of the dike (deep). However, since the spillway may need to function prior to good grass cover establishment, the calculated width is an approximation and to compensate for rounding errors, a conservative design would add 10 to 20% to each dimension, or

42 × 1.1 = 46 ft wide and
1.4 × 1.1 = 1.5 ft deep

Location of the spillway is also important since it must convey excess waters around or over the dike without damaging the end, top, upstream, or downstream surface of the dike or the discharge channel. In some cases, a wingwall or short stub dike may be needed to guide stormwaters around the end of the dike and in others rock riprap may be needed on portions of the wingwall, spillway, or dike surface.

TABLE 10-5
DETERMINATION OF DISCHARGE
AND FLOW DEPTH[a]

Maximum Velocity (V) ft/s	Discharge (q) cfs/s/ft	Flow Depth (Hp) Spillway Length (L, in ft)	
		50	100
4	2	1.2	1.4
5	3	1.4	1.6
6	4	1.6	1.8

[a] Adapted from *Ponds — Planing, Design, Construction*; SCS Handbook No. 590.

The example calculations, while developed specifically for excavated spillway estimates, also provide a fair approximation for natural spillways. In addition, they tend toward the conservative side, but most designers will do likewise since proper functioning of the emergency spillway is crucial to the security of the dikes and the continued well-being of the wetland communities.

CHAPTER 11

PLANS AND CONSTRUCTION

PREPARATION OF PLANS

Duplicate copies of the detailed site topographic maps are handy for developing initial plans. Soil types, subsurface formations, hydrologic characteristics, cultural artifacts, utility rights-of-way, biotic components, potential borrow areas, and any other factors identified in the site surveys that could be useful in selecting locations for dikes, water control structures, visitor facilities, and access can be included or prepared as overlays. Including everything on one map will likely clutter the map, so only factors relevant to locations of construction activities should be included.

With the ease of duplication today, numerous copies of the site map facilitate sketching in potential locations for various structures without having to erase and redraw every mistake. In addition, sketches of alternative locations can be prepared from which preliminary calculations of acreage needs, runoff estimates, cut and fill requirements, etc. can be developed for comparative purposes. Much of the initial work is likely to be done with basic instruments — scaled rulers, compass, or dividers — even in those instances where planners have access to sophisticated computer programs. It is simply much faster to sketch out the first rough plans.

Exponential developments in computer technology now provide the capability and speed to do some rough sketching with a computer assisted drawing (CAD) program. Speed and memory capacity of newer computers and creation of many smaller CAD programs for layman designing has fairly well relegated drafting instruments to the closet. For example, I often prepare concept drawings on a laptop computer with only a grid overlay and the topographic survey map even in regions without or with incompatible electrical power. A wire-drawing CAD, DANCAD (see Appendix C) has minimal memory requirements along with user friendly features, a short learning curve, and substantial drawing, scaling, and printing flexibility. Prepare a grid overlay by drawing the lines in the CAD program and print it with a laser printer on transparency film and simply place the grid overlay on the survey map to locate the coordinates of the contour lines or any other feature. Then use the coordinates to transfer the map features to a drawing in the CAD program. Be careful that your grid overlay units are compatible with the survey map and whatever units your concept drawing will have, i.e., consistently use either metric units or English units. Converting from one system to the other is cumbersome and provides another opportunity for error.

More powerful CAD programs are now available into which the topographic survey information is fed and fairly rapid drawings can be generated directly on electronic representations of contour maps. Since elevations are included in the survey information, depicting potential depths and area of inundation is easily accomplished. A few advanced programs have the ability to create a drawing, calculate the amount of cut and fill required for specific designs, and store each drawing and calculations as a separate file so that comparative evaluations can be printed out for study and discussion purposes. Rough sketching is a bit slower with these, but much more information is available for later comparisons. Some require very powerful computers (work stations) and have lengthy learning curves.

The ultimate tools are advanced CAD programs, geographic information system (GIS) capability and site surveys employing global position system (GPS) technology. This combination can create accurate, detailed drawings and overlays depicting virtually every aspect of the site and construction plans. But it will likely require expansion of the project team to include the technical expertise to effectively use these technologies. As with many other aspects, project size and cost influence the level of detail needed and large expensive projects should employ the most advanced technologies.

Regardless of the means employed, planners should explore all possible alternatives for developing the selected type of wetland within each of the potential sites under evaluation. Because of the many possible interactions of project goals, site characteristics, feasibility, and costs, it is not infrequent that a seemingly unusual design furnishes an optimal combination of compromises; and of course, each draft design will represent a different set of compromises since the ideal site can rarely be found.

Once initial sketches have been prepared and evaluated and a site selected, concept plans should be developed, preferably with one of the many available CAD programs. Though this step can also be accomplished with rudimentary drawing devices, developing the concept plans with a CAD program greatly facilitates modification, evaluation, and duplication. Concept plans should include locations and dimensions of all structures, type and location of construction activities, and location of hazardous or cautionary areas such as utility rights-of-way, subterranean caverns, etc. With computer drawing tools, depiction of perspectives of the completed system from different viewing angles can be easily developed to assist in refining overall aesthetics. Depending upon the program(s) used and printing capabilities, the concept plan may be expanded and done in more detail to prepare construction drawings and specifications.

Construction drawings and specifications should have sufficient detail so the contractor can readily understand what is expected of him and the developer has sound basis for legal recourse if the final product does not meet contract specifications. Generally, well prepared, clear, and detailed plans and specifications will enhance the probability of the final product resembling the planned system; but investment in preparation of drawings should be proportionate to overall project complexity and cost.

Minimally, drawings and specifications must include clearing and grubbing limits, final grades, utilities, borrow areas, type of structures, dimensions, dike and structure materials, bottom permeability, valves or other water controls, areas or vegetation to remain undisturbed, erosion control measures, and planting requirements including sodding or seeding and mulching dikes, spillways, and other disturbed areas. A more detailed plan might include composition and type of valve or even a specific manufacturer's model number, spacing of reinforcing rod in concrete, composition and grade/thickness of piping, etc.

Ideally, construction plans will include:

1. boundaries of construction activities, including clearing and grubbing limits;
2. access for construction equipment and transportation corridors;
3. locations of cautionary or hazardous areas;
4. utility rights-of-way and contacts;
5. quantities, location, and dimensions of borrow areas;
6. areas or vegetation that should not be disturbed;
7. erosion control measures to be taken during construction and revegetation methods during final stages;
8. locations, dimensions, and materials specifications for structures;
9. locations, length, top and base widths, elevations, upstream and downstream slopes, permeability and coring for dikes or berms and spillways;
10. type, size, location, materials, and elevations of water control structures;
11. pond bottom and side permeability specifications and methods to attain required permeabilities, including liners and liner installation if needed;
12. elevations, slopes, and contours of pond bottoms and permissible tolerances;
13. elevations, dimensions, composition, grades/thickness, manufacturer, and/or model for piping and valves or other water control structures;
14. dimensions and specifications on lighting, switches, wiring, outlets, pumps, and other electrical facilities;
15. type and method of placement of sand, gravel, rock, or rock riprap;
16. species, sources of supply, planting spacing, planting dates, and expected survival of wetland vegetation;
17. seeding, fertilizing, mulching and liming, or sodding of dikes, berms, spillways, and any other disturbed areas;
18. provisions for on-site construction supervision;
19. methods for determining permeabilities and other contract specifications; and,
20. types, sizes, and numbers of construction equipment.

Planners should check and recheck and have a competent colleague verify elevations of pool bottoms and area, dike tops, piping, and controls or valves in the drawings. Discovering during construction, or even later, that system operation will require water to flow uphill can be embarrassing and costly. Simple arithmetic errors easily result in that unusual phenomena. Planners should also compile a list of materials, check availability and prices, and develop cost estimates for each phase of the project to assist them in evaluating construction bids.

Cost estimates should include:

1. contour mapping surveys and construction staking;
2. preparation of construction drawings and specifications;
3. preparation and distribution of bid invitations and advertisements;
4. site preparation: clearing, grubbing, and dewatering if needed;
5. categorized construction activities for major units
   (i.e., dikes, water controls, roadways, spillways, visitor facilities, etc.);
   a. materials;
   b. equipment;
   c. labor;
   d. supervision;
   e. overhead percentages;
6. planting wetlands vegetation; and,
7. revegetating disturbed areas.

PREPARATION FOR CONSTRUCTION

Construction drawings and specifications will become core elements in the contract for constructing the wetland. The contract should also include construction and planting periods and any restrictions on either, type of earth-moving or other construction equipment, provisions for site access and security, completion dates and penalties or early completion bonuses, performance bonding requirements, methods for testing permeabilities, methods and dates for determining plant survival and replanting if necessary, start-up and acceptance procedures, and any other aspect that could be confused or result in a later disagreement. As with construction drawings, larger, more complex (hence, more costly) projects generally require more carefully thought out and detailed contracts. However, even in the largest projects, a certain amount of good faith and understanding between the developer and contractor is usually necessary since including every minute detail in the contract is virtually impossible. Of course, payment arrangements at the completion of certain phases or at final completion and acceptance should be included.

Vegetation planting requirements are new to many designers and contractors and there is a strong tendency to write performance specifications for plants similar to those for concrete structures, electrical wiring, or plumbing.

Unfortunately, rigid specifications for biological factors rarely reflect real world situations. Review the admonitions in Chapter 13. Your objectives should have been defined in terms of functional values, i.e., performance of the new system. Be very careful of over specifying the plant community (form-structure) in the invitation to bid or contract. Requiring 90% coverage of species A and B after 90 days may or may not be the most cost-effective method of achieving your objectives. Regardless of the plant species specified and planted correctly and even if water levels are managed properly, numerous other species are likely to appear in the first few years. And one or more may become dominant, resulting in limited coverage or even the absence of species A or B. Has the project failed? Was the planting done improperly? Should you require replanting? Before you try to answer these questions examine the important question. Is the project producing the functional value(s)? Is it accomplishing your functional objectives? Which is more important? Having 90% coverage of species A and B after 90 days or having a diverse, complex plant community that supports the desired function? Keep that in mind while preparing bid specifications and negotiating with the planting contractor. Remember the objective is to perform a function, not necessarily to have a specific form (structure). The art of wetland plant culture is too imprecise to warrant the degree of specifics commonly used to describe grading and physical structures. Doing so enormously increases planting costs since the contractor must cover any number of re-plantings in his estimate.

Furthermore, there have been many cases were the planting subcontractor established a healthy, vigorously growing stand of the desired species only to have the contractor or operator apply inappropriate water level management that inhibited or devastated the new growth. Planting subcontractors need to be very careful of agreeing to contract conditions specifying X% survival after Y days unless the start-up and normal water management conditions are clearly spelled out in the contract. And designers and bid writers need to recognize that many conditions beyond the control of the planting contractor may substantially alter growing conditions and subsequent composition of the plant community.

The most important factor influencing growing conditions is system operation. Far too many designers and operators, especially in constructed wetlands for wastewater treatment, think that more is better. That wetland plants need deep (>20 cm) water or that increasing water depth increases retention time and improves system performance. Increasing water depth in sewage treatment lagoons increases retention time and usually improves performance. But constructed wetlands are not lagoons and beyond a minimal level, retention time has little influence on performance. Furthermore, wetland plants, especially new plantings, do not do well in deep water. Older established stands may survive in deep water but the stress will inhibit if not eliminate reproduction and expansion of coverage. And over time, coverage is likely to decrease because of stress and mortality. Planting contractors can not be held responsible for poor

Figure 11-1. Sedges thrive in the shallow waters of this Wisconsin marsh but if water levels were raised, they would lose out to the floating-leaved water lilies occupying the deeper waters in the roadside borrow ditch.

survival and growth if operating conditions, especially water depths, are deleterious to plant growth. And planting contractors need to insure that planned operating conditions are appropriate for the specified plants and are included in the contract.

By now it should be apparent to the reader that I don't believe it possible to over emphasize the critical nature of proper water level management especially when the designers and/or managers lack wetland experience or even biological training (Figure 11-1). Though not as much a problem today, 5 to 10 years ago few nurseries had experience with and even fewer specialized in wetland plants. I recall a contract horticulturalist that couldn't understand why cattail didn't survive planting in almost 1.0-m water depths at one project. "It's a wetland plant isn't it?" Fortunately, a glance through Appendix C will reveal a substantial increase in firms specializing in growing and planting wetland species. Water management problems today are more likely to be caused by improper design and/or operation. Of course, operating plans should be included in the contract so that prospective bidders can evaluate growing conditions for the new plants.

Once the contract has been developed, the invitation to bid is abstracted from the contract and advertised as required of public agencies and/or through local news media and contractors association newsletters. Invitations should be sent to potential contractors known to planners or identified by the local highway department, NRCS office, or trade associations. Advertisements should contain a brief description of the project, bid deadlines and procedures, expected completion dates, information on obtaining the invitation to bid, and a contact for further information. Invitations to bid should include a conceptual drawing and general specifications, unusual requirements, deadlines, construction dates, penalties, and bidding forms. Breaking down the project into discrete components and requiring bids for each component on bidding forms will facilitate evaluation and comparison of bids received. In large or complex projects, bid invitations often include complete construction drawings and the proposed contract.

Since wetland construction is likely to be unusual for most prospective contractors, the invitation to bid and advertisements should include a time, date, and location for a pre-bid site inspection and conference. Although wetland construction is essentially the same as building a shallow lagoon, many potential bidders will have reservations and will likely increase bid amounts, in some cases astronomically, as insurance. A thorough explanation of project goals, construction expectations, and a site walk-over will allay many fears and promote realistic bidding.

Prior to the pre-bid conference, the information on construction plans must be transposed to the site through staking so that potential contractors can readily determine the expected work. Hopefully, basic stakes and elevations are still present from contour mapping and available for a baseline. Placing construction stakes locates individual construction activities and delineates the elevations, grades, and lines specified in the plans. Boundaries, depth of cut, elevation of fill, slope angle and position, location and elevation of piping and structures, and all of the other information on construction plans should be transferred to the site through clear and ample stakes. Staking must clearly show contractors what, where, and how much is to be done to facilitate bid preparation and to insure the finished product adheres to construction plans and specifications.

Area staking consists of marking access, boundaries of the construction area, clearing and grubbing limits, the proposed water level, and borrow areas. In addition to boundary stakes, borrow areas must have cut stakes to indicate excavation depths to insure that proper fill materials are obtained and not unsuitable materials that may underlie the borrow area.

Dikes are located by placing stakes along the centerline and fill and slope stakes upstream and downstream at the point of intersection with the ground surface. Spillways are staked along the centerline with cut and slope stakes at the intersections with ground surface. Locations, dimensions, and elevations of water control structures, visitor facilities, roadways, utilities, and other structures should also be clearly marked so that construction workers can readily determine project components. Planners also benefit from seeing the project staked out on site. More than one set of plans has been revised after the site was staked for construction.

Despite careful and thorough site inspection and evaluation, planning and preparation of drawings and specifications, some eventuality is likely to be overlooked. Consequently, completion of invitations to bid, construction plans, a contract and construction staking will only mark the completion of the major portion of project planning. Quite likely, construction conditions and activities will require revisions in the original plan, which is the reason that on-site supervision by competent personnel and/or ready access to project planners is critical during the construction phase.

CONSTRUCTION

Planning activities grade into construction activities as planning, construction staking, the pre-bid conference, and contract award(s) follow one another. Probable delays before equipment actually moves on site include minimum advertisement periods for public agencies, contractor's scheduling on other jobs, and inclement weather. Since the final construction activity is planting vegetation or improving growing conditions for native vegetation, and both should be accomplished during spring or early summer, project schedules and construction time frames must be carefully established and clearly understood by planners and contractors. Generally, construction is most rapid in dry seasons, typically late summer and fall, in which case aquatic planting should be delayed until the following spring. In the interim, flooding the system may be useful to check grades, elevations, sealing and operation of piping and water control structures, electrical accessories, and other facilities. Identified errors or problems can then be corrected before vegetation is placed in the pools the following spring. If bentonite was used, the pools should be flooded until just prior to planting to avoid cracks and leaks. Of course, dikes and spillways and any other disturbed areas should be vegetated immediately following construction.

Construction activities include clearing and grubbing, excavation, grading, transporting and placing fill, compacting, placing sand, gravel, or rock riprap, installing liners, placing and tilling in sealing substances, disposing of waste or excess fill, building or installing water control structures and piping, installing electrical facilities and other utilities, planting wetland vegetation, seeding, and mulching or sodding disturbed areas,

Schedules for various phases of construction should be agreed upon prior to any activities so that moving equipment in and starting work do not interfere with other planned activities on site and nearby site disturbance is kept to a minimum. Dry weather, generally late summer and fall, is typically optimal for construction and the job may be finished quickly. During wet seasons, contractors may only be able to work for 2 to 3 days and then be idle for days or weeks, dragging out the period when the site is disturbed; erosion may be high and construction activities disrupt other site work. Scheduling should also include consideration of neighboring landowners and land uses to cause as little disturbance to them as possible. Even though it may seem advantageous to have heavy equipment working at night to complete the job, disrupting the neighbor's relaxation or sleep will lay the seeds for future problems.

Prior to construction planners may need to:

1. investigate and, if necessary, divert or pump water from the site;
2. mark areas that should not be disturbed;
3. mark any trees with flagging that should be left or limbed or anything other than clearing;
4. identify locations of silt barriers (with fence or straw bales) needed, and insure that contractors comply with regulations and any special requirements;
5. discuss equipment types and numbers with contractors to insure expeditious work activities; improperly sized equipment will increase construction time and costs and may inhibit accurate construction.

For example, top width of the dikes will be at least as wide as the bulldozer blade because it is difficult for the operator to build anything less. If the site is wet, at least two dozers or other machines should be used since one will often be needed to extract the other after it mires down. Obtaining a unit off-site each time the dozer is stuck will cause needless delays. If the site is very wet, a dragline using supporting mats is generally more efficient than dozers and scrapers or backhoes. Trackhoes can create more feet of dike in a day but they generally have less reaching ability than draglines.

The system planner or a construction supervisor familiar with all aspects of the project (including design objectives and management plans) should be on-site or at least monitor activities on a daily basis. Invariably, some aspect has been overlooked or is not anticipated, and modifications will be necessary. If someone knowledgeable is not available, contractors may stop work activities until advised, or independently modify the plans. In the first case, valuable time will be lost; but in the second, major errors could occur that may be costly to correct and contractors may be unwilling to absorb the extra costs.

Specific construction activities vary substantially with site conditions, type of wetland, and construction equipment and work force. However, construction supervision is primarily insuring that the contractor adheres to construction drawings and specifications and developing and agreeing to modifications as necessary to accommodate unforeseen circumstances. Modifications should be approved by the system planner or someone familiar with the technical specifications, future management plans, and requirements before adoption. Though construction supervision is as varied as different projects, general guidelines and precautions are applicable to a variety of wetland construction activities.

Topsoil should be removed and stockpiled for later use during the early stages of clearing. If a wetland soil is removed for later use, store it underwater to avoid oxidizing and releasing bound metals or other substances that could detrimentally impact the new system if they were re-dissolved. This also prevents included seeds from drying and dying. All permeable soil materials, organic matter, rocks, trash, or debris should be removed in preparing a solid, impermeable foundation for dikes and other structures. Stream beds must be widened and deepened, and all stones, gravel, or sand removed to the clay foundation so that fill material will bond properly. Natural holes or holes caused by clearing and grubbing should be cleaned out and filled with suitable fill material. In each instance, the objective is to insure that clay fill materials will abut and bond to a clay foundation to insure continuity of the impervious materials.

Waste or spoil materials that will not be needed for fill should be placed in boundary areas, sloped and contoured to blend into the surroundings, and stabilized by seeding or sodding. Excess earth could be used to provide knoll overlooks, islands, wider parking areas and visitor stops, visual and/or auditory screening, or any other imaginative use that will not interfere with the principal functions of the new wetland.

Construction should always follow contract specification as closely as possible. In many wetlands, slight deviations may not be critical; but if the system will be used for wastewater treatment, grading must meet the specifications in the plan within described tolerances to achieve proper functioning of the system. Out-of-tolerance lateral bed slopes may not only cause ponding, but are likely to cause channeling or short-circuiting that reduces the effective treatment area in the cell and depresses performance. Similarly, improper grading along the cell length could make it impossible to set and maintain proper water depths, as well as causing channeling and short-circuiting. Obviously, inability to manage water depths would severely retard establishing or managing the desired plant community and degrade functioning of any type of wetland.

Specifications on dike materials and dimensions are established to reduce or eliminate leakage or seepage and to insure dike integrity under expected water pressures. Failure to follow specifications on materials, top width, base width, or slopes could result in weak or leaky dikes or highly erodible slopes that may be difficult to stabilize. Installing the clay core and culverts or controls and anti-seep collars in the dikes must also be done carefully, with proper fill compaction to prevent seepage.

Meeting permeability specifications is important in all wetlands. Be careful that contractors do not excavate deeper than planned and penetrate an impermeable layer into a permeable layer. Permeability testing should be agreed upon prior to construction, and frequent testing is necessary during construction. Compacting in situ or fill material must be done with proper equipment and only when moisture conditions are satisfactory. If necessary, sprinklers may be used to achieve proper soil moisture conditions before and during compacting. Bentonite or soda ash blankets must be carefully installed and tilled into the bottom following specifications in the designs to insure proper functioning. If synthetic liners are used, installation must follow manufacturers' instructions for bed material, sealing (liner-to-liner and liner-to-piping and controls), and insulating material above the liner. Construction equipment and workers can easily puncture synthetic liners and clay blankets unless special precautions are communicated and followed. Once final grading and compacting has been accomplished or a liner installed, equipment should not be permitted on the cell bottom or dike sides, and foot traffic should be minimized.

If parent or fill materials are very fine clays that easily powder after disturbance, plan to flood the system for a few weeks and dewater it before final grading. Preliminary flooding is also the simplest method to check grade and structure elevations throughout the system. Final grading before flooding may need to be repeated since it is not unusual for settling to cause elevation differences of 15 to 30 cm in cell beds and occasionally on dikes. Piping and control structures may also shift as fine materials settle.

Installing water control structures may be as simple as laying PVC pipes, or as complex as building forms and pouring concrete in place. In either case, correct elevations and adequate support are critical to future management ability. Correct proportions of cement and sand and approved temperatures are important in achieving design strength for concrete structures. Appropriate adhesives, bonding techniques, and curing temperatures are required to insure correct joining of plastic piping. As always, specified elevations for each and every segment of pipe or portion of a concrete structure are especially important. Generally, wetland system operation is dependent on gravity flow that is only possible if correct design elevations are achieved in all aspects of the system.

In as much as vegetation planting requirements vary considerably with region, site conditions, species, and planting season, detailed information is provided in Chapter 13. In general, planting supervision, as with other construction activities, is principally insuring that contractors adhere to contract specifications so that design objectives will be achieved.

Optimal planting conditions for cut materials or seed are created by shallow flooding, followed by dewatering but not complete drying — leaving soft, moist soil conditions. Common pitfalls include improperly storing planting materials — too dry or too warm or too long — careless handling and poor orientation. Planting stock should not be dug more than 2 days before planting and should be stored and transported in a cool, dark, humid environment. Damaged stock may perish or take longer than normal to begin growth and wetland plant roots are as sensitive to damage as any garden plant. Handling tree seedlings requires similar precautions and seedling roots must not be allowed to dry for even a few minutes. Rows must run perpendicular to the direction of flow to improve coverage and reduce channeling, even though it may be easier to operate equipment up and down the long axis of each cell. After planting is completed, flood the area with 1 to 2.5 cm of water, but insure that water depths do not overtop cut stalks or the new plantings may die. As new growth begins, water levels may be slowly raised, but should not overtop the new growth. Water levels in a new forested wetland should be managed to maintain moist but not saturated soils during the growing season and should only rarely saturate soils around new seedlings the first winter.

Figure 11-2. Impounding a small stream and fluctuating the water levels during spring and summer developed the proper conditions for seed germination and growth of the variety of wetland plants in this created marsh. After 15 years it is virtually indistinguishable from nearby natural marshes.

Hand or natural seeding is less expensive, but much less reliable for starting the new plant community. Germination rates of many aquatic plant seeds are less than 5% per year, so large quantities must be collected and distributed. Whether hand or natural seeding is used, any deeper sections should be shallow flooded in late winter and early spring and dewatered at the onset of warm weather to establish warm, moist mud conditions (Figure 11-2). Careful monitoring and regulation of water levels at or just below the bottom of the deeper sections is important in order to maintain the proper soil moisture conditions for germination and sprouting. After the new growth has reached 10 to 12 cm, water levels should be raised to between 2 and 4 cm above the substrate to inhibit or kill terrestrial species, but should not overtop wetland plants.

Trees or shrubs along the perimeters should be installed after a good grass cover has become established in the event that regrading and reseeding should become necessary because of erosion or other problems. Generally, trees or shrubs should not be planted on dikes unless the dike is very large and root growth is not likely to impugn the integrity of the dike. Similarly, fencing for livestock or security should not be installed until after disturbed areas have become well covered with stabilizing vegetation.

All control structures, piping, wiring, pumps, and other electrical facilities, seals, water levels, and flow distribution should be tested for proper operation or elevations before formal acceptance. Plant survival can be estimated by observing the percentage of plants with new and vigorous growth evident and areal coverage. Since some aspects — settling, subsidence, leaks, or seepage — may not become apparent until some time after initial operation, planners should establish a test or start-up period with the contractor and arrangements that may be needed to correct problems. Elevation variances will largely become apparent with initial flooding, as will problems with piping and controls. However, seeps or leaks may be undetected for fairly long periods, and plants showing good growth initially, may die later from any number of factors. Since some of these are controlled by project personnel (i.e., water depths) planners should obtain agreements with contractors on remedial actions and responsibilities.

Any erosion on dikes, spillways, around control structures, or in the upper ends of cells should be immediately filled with clay or soil materials, compacted, and reseeded or resodded as necessary. Before the vegetation provides a protective screen, wave action may erode dikes or earthen fills and temporary log booms or slat fencing may be needed. If seepage or leaks are detected, water levels should be lowered immediately. Depending on the location, adding additional sealing material, bentonite, or soda ash may correct the problem. However, leaks around control structures or at the base of a dike will require excavation of fill materials and replacement with compacted, impervious clays. If a synthetic liner was used, additional panels may need to be added and securely bonded to previous panels.

In summary, proper construction supervision is principally having a knowledgeable person ensure that construction drawings and specifications are followed precisely and developing and documenting modifications as necessary. Since experience levels and work quality may vary substantially between contractors, time and effort required for supervision will also vary, but close attention to details is always a prudent investment. Few contracts fully describe every component and a considerable amount of faith and trust between planners and contractors is essential. However, even the best contractors may misunderstand or misinterpret drawings and specifications, or more likely, encounter unforeseen situations, and readily available advice or directions will save time and expense and possibly, future management capabilities. Supervision also includes frequent and thorough inspections of all aspects of the project during start-up and immediate attention to any problems. In as much as some problems may not be identified until after the contractor has moved off-site, good working relationships between the construction supervisor and the contractor are important in implementing remedial actions before minor faults become major failures.

CHAPTER 12

PLANT SELECTION

INTRODUCTION

The diversity and complexity of natural wetlands are principally the result of interactions of three important factors: (1) hydrology, (2) substrate, and (3) vegetation. The first two strongly influence the vegetation, as do climate and proximity to other wetlands. Since restoring or creating wetland is dependent on duplicating these factors and their interactions, an understanding of the ways they influence vegetation is important to proper selection of species, planting or establishment methods, and operating conditions.

HYDROLOGY

Water depths, frequency and duration of flooding, and water chemistry are the most important factors determining the survival and growth of plants in a wetland system. Depth influences gas exchange between the substrate and the atmosphere, decreasing oxygen contents below deeper waters. Depth also restricts light penetration, even in clear waters, though highly turbid waters are unlikely to permit adequate light penetration to support photosynthesis much below the surface. The vegetation zonation characteristic of most wetlands is largely due to the influence of water depth because certain species are adapted to or require certain depths, whereas others prefer different depths. In fact, a seasoned observer can judge water depths and bottom types in different wetland systems simply by noting the plant species present in each area. Much of the diversity and spatial heterogeneity of natural wetland systems is the result of different elevations and, consequently, water depths within the system.

Frequency, duration, and seasonality of flooding also strongly influence species that are used for developing different wetlands or different areas within the wetland. Different wetland plants withstand various degrees of inundation depending on when and for how long the flooding occurs. Most shrub and tree species and some emergent plants need a period of lower water levels or very limited flooding during the growing season, but can withstand prolonged inundation in the dormant season, generally fall and winter. In dormancy, plant oxygen needs are reduced and long and/or deep flooding may have little impact. However, oxygen limitations during the active growing period may cause stress and eventually mortality in the same plants. Some swamp trees can endure fairly long periods of shallow flooding, but may be damaged by deep flooding; whereas others are vulnerable to any inundation beyond 4 to 5 days during the growing season. Some shrubs and trees can survive extended flooding during one growing season but flooding for two or more will cause mortality.

Far more created wetlands have had problems with too much rather than too little water. This is especially true with many constructed wetlands treating wastewater and apparently is a carry over from lagoon management or the concept of retention time in treating wastewater. Designers and operators erroneously believe that since it's a wetland plant it must have water, in some cases, lots of water. Attempts at planting cattail or bulrush in up to 1-m water depths have routinely failed. In other cases, plantings were correctly made in "soupy" muds and quickly developed vigorous growth but then water levels were raised to 0.8 to 1 m at full operation and plant mortality ensued. Excepting the submersed and some floating-leaved species requiring water to support them upright, most wetland plants thrive in moist soils not much wetter than many agricultural crops. Most wetland plants lack water conservation measures so in dry climates they will use more water than terrestrial species. But they can obtain needed water from moist substrates similar to any terrestrial plant. Though most submergents simply wait out the drought, many emergents thrive during drought years common to precipitation cycles in the Prairie Pothole region. Only prolonged, severe drought allows terrestrial species to invade these basins and eventually they will out compete wetland emergents if the drought continues.

A selection of herbaceous and woody species grouped by ranges or duration of flood endurance is presented in Tables 12-1 through 12-6. This list is not all inclusive, nor will all species survive in all regions of the country or all types of wetlands. Planners should use the list as a guide for comparison with herbs and trees endemic to their region to select species for planting in different depths of the new wetland. Obviously, it would be just as inappropriate to attempt to establish cypress in northern Manitoba as it would to plant black spruce or tamarack in Florida.

Note the headings on Tables 12-2 through 12-6; each includes "...Water Depth Tolerated..." or "...Will Tolerate Flooding...". Wetland plants have developed the ability to tolerate flooding or innudation. That doesn't mean they thrive under those conditions. In fact, with the above exceptions, most will grow and reproduce more vigorously in shallower waters or with less flooding than their placement in a specific category would suggest. Location in any of the categories in these tables simply means they are commonly found growing under those conditions in nature. That means that those species have developed the ability to tolerate that depth or flooding regime. Because excess water will cause stress or mortality of plants with less tolerance, species that tolerate flooding avoid competition from the less tolerant species. We don't find wetland plants growing in terrestrial sites because they can't compete with species that are better adapted to the dryer conditions. If we eliminate terrestrial species (weeding out), wetland plants will thrive in moist terrestrial habitats.

TABLE 12-1
ENVIRONMENTAL REQUIREMENTS OF SELECTED HERBACEOUS WETLAND PLANTS

Species	Soil	pH	Ranges	
			Salinity (ppt)	Depth (cm +&– surface)
Spartina cordgrasses	Sandy loam	5–8	10–30	20 to –50
Calamogrostis reedgrass	Silt	5–8	0–10	20 to –20
Phalaris Reed Canary	Silt loam	6–7.5	0–.4	10 to –10
Juncus rushes	Organic	5–8	0–30	10 to –10
Cyperus nutsedges	Organic, clay	3–8	0–.4	10 to –30
Carex sedges	Organic, clay	5–7.5	0–.4	10 to –50
Typha cattail	Organic, silt, clay	5–10	0–25	10 to –70
Phragmites reeds	Silt, sand, clay, organic	3–8	0–30	30 to –150
Scirpus bulrushes	Organic, clay	4–9	0–32	10 to –100
Potamogeton pondweeds	Organic, silt, clay	4–10	0–20	–5 to –300
Vallisneria tapegrass	Silt, clay	5–8	0–10	–5 to –100
Ruppia widgeongrass	Silt, clay, organic	5–10	0–25	–5 to –100
Nuphar spatterdock	Organic, silt	3–8	0–5	–50 to –200

TABLE 12-2
WATER DEPTHS TOLERATED BY SELECTED HERBACEOUS WETLAND PLANTS

Scientific Name	Common Name
Transitional — seasonally flooded	
Bidens spp.	Beggarticks
Echinochloa crusgalli	Barnyard Grass
Hymenocallis spp.	Spider Lily
Lysimachia spp.	Loosestrife
Hordeum jubatum	Foxtail Barley
Sesbania spp.	Hemp
Polygonum lapathifolium	Pale Smartweed
Iris fulva	Red Iris
Osmunda cinnamomea	Cinnamon Fern

TABLE 12-2 (continued)
WATER DEPTHS TOLERATED BY SELECTED
HERBACEOUS WETLAND PLANTS

Scientific Name	Common Name
Osmunda regalis	Royal Fern
Setaria spp.	Foxtail Grass
Spartina pectinata	Prairie Cordgrass
Panicum virgatum	Switchgrass
Asclepias incarnata	Swamp Milkweed
Distichlis spicata	Saltgrass
Alopecurus arundinaceus	Foxtail
Scolochloa festucacea	Marshgrass
Lysichitum americanum	Yellow Skunk Cabbage
Symplocarpus foetidus	Skunk Cabbage
Hibiscus moscheutos	Swamp Rose Mallow
Hibiscus militaris	Halbeard-leaved R. Mallow

Shallow — seasonally flooded to permanently flooded to 15 cm

Scientific Name	Common Name
Calamagrostis inexpansa	Reedgrass
Leersia oryzoides	Rice Cutgrass
Juncus effusus	Soft Rush
Saururus cernuus	Lizardtail
Carex spp.	Sedge
Eriophorum polystachion	Cotton Grass
Cyperus spp.	Chufa, Sedge
Iris virginicus	Blue Iris
Iris pseudacorus	Yellow Iris
Panicum hemitomon	Maidencane
Dulichium arundinaceum	Three-way Sedge
Beckmannia syzigachne	Sloughgrass
Panicum agrostoides	Panic Grass
Scirpus cyperinus	Woolgrass
Habenaria spp.	Swamp Orchids
Cypripedium spp.	Lady's Slipper
Thelypteris palustris	Marsh Fern
Paspalum spp.	Knotgrass
Hydrocotyle umbellata	Water Pennywort
Caltha leptosepala	Marsh Marigold
Isoetes spp.	Quillwort
Phalaris arundinacea	Reed Canarygrass
Sarracenia spp.	Pitcherplant
Jussiaea repens	Water Primrose
Justicia americana	Water Willow
Polygonum coccineum	Swamp Smartweed
Polygonum pensylvanicum	Pennsylvania Smartweed

Mid Depths — 15 to 50 cm water depths

Scientific Name	Common Name
Polygonum amphibium	Water Smartweed
Cladium jamaicensis	Sawgrass

TABLE 12-2 (continued)
WATER DEPTHS TOLERATED BY SELECTED
HERBACEOUS WETLAND PLANTS

Scientific Name	Common Name
Acorus calamus	Sweet Flag
Calla palustris	Water Arum
Sparganium eurycarpum	Burreed
Zizania aquatica	Wild Rice
Eleocharis spp.	Spikerush
Alisma spp.	Water Plantain
Scirpus americanus	Three-square
Typha latifolia	Wide-leaved Cattail
Typha angustifolia	Narrow-leaved Cattail
Phragmites australis	Giant Reed
Scirpus validus	Bulrush
Scirpus fluviatilis	River Bulrush
Sagittaria latifolia	Arrowhead
Pontederia cordata	Pickerelweed
Glyceria spp.	Mannagrass
Nasturtium officinale	Watercress
Limnobium spongia	Frogbit
Peltandra cordata	Arrow Arum
Menyanthes trifoliata	Buck Bean
Vaccinium macrocarpon	Cranberry

Deep — 50 to 100 cm water depths

Potamogeton pectinatus	Sago Pondweed
Vallisneria americana	Tapegrass
Ranunculus flabellaris	Yellow Water Buttercup
Ranunculus aquatilis	White Water Buttercup
Callitriche spp.	Water Starwort
Ruppia maritima	Widgeongrass
Ceratophyllum demersum	Coontail
Myriophyllum	Milfoil
Utricularia spp.	Bladderwort
Anacharis	Water Weed
Nymphaea odorata	Fragrant White Lily
Nuphar luteum	Spatterdock
Brasenia schreberi	Water Shield
Nelumbo lutea	Water Lotus
Nymphoides aquatica	Water Heart

Floating

Lemna spp.	Duckweed
Azolla spp.	Water Fern
Wolffiella spp.	Wolffiella
Wolffia spp.	Watermeal
Spirodela spp.	Giant Duckweed
Sphagnum spp.	Sphagnum Moss

TABLE 12-3
WOODY PLANTS THAT WILL TOLERATE
FLOODING FOR MORE THAN ONE YEAR

Scientific Name	Common Name
Taxodium distichum	Bald Cypress
Salix nigra	Black Willow
Carya aquatica	Water Hickory
Planera aquatica	Water Elm
Nyssa aquatica	Water Tupelo
Gleditsia aquatica	Water Locust
Forestiera acuminata	Swamp Privet
Fraxinus pennsylvanica	Green Ash
Quercus nuttallii	Nuttall's Oak
Quercus lyrata	Overcup Oak
Ilex decidua	Deciduous Holly
Cephalanthus occidentalis	Buttonbush
Carya illinoensis	Pecan
Pinus serotina	Pond Pine
Cornus stolonifera	Red-osier Dogwood
Salix lasiandra	Pacific Willow
Salix exigua	Narrow-leaf Willow
Salix hookeriana	Hooker Willow
Campsis radicans	Trumpet Vine
Diospyros virginiana	Persimmon
Rosa palustris	Swamp Rose

Bear in mind that the depth/flooding categories in Tables 12-2 through 12-6 were developed from observations of plants growing in natural wetlands, most of which have relatively clean waters. If your water source is not relatively clean or even wastewater, tolerance levels may be much less depending on the level of contaminants.

In natural marshes, various grasses occupy the highest regions that either have very shallow flooding (2 to 5 cm) or limited seasonal flooding (Figure 12-1). Maidencane and reed canary grass occur in the deeper water portions of this wet meadow zone slowly grading into sedges, *Iris*, rushes (*Juncus*), woolgrass (*Scirpus cyperinus*), and spikerushes (*Eleocharis*) with increasing depths. These are, in turn, replaced by arrowhead, sweetflag, pickerelweed, and water plantains further inward. Cattail and giant reed often intergrade with the last zone, but may extend out to depths of 15 to 30 cm with the large bulrushes (*Scirpus validus*), often occurring in 30- to 50-cm depths. Few emergents colonize deeper waters. At this point, submergents (pondweeds, coontail, or tapegrass), become dominant if water chemistry and clarity are suitable (Figure 12-2). In the deepest regions (1 m and more) a few pondweeds (e.g., *P. robbinsii*) may be able to grow if clarity is high; but in turbid or colored waters, only rooted floating species, spatterdock, and water lilies persist.

TABLE 12-4
WOODY PLANTS THAT WILL TOLERATE
FLOODING FOR ONE GROWING SEASON

Scientific Name	Common Name
Kalmia polifolia	Bog Laurel
Ledum groenlandicum	Labrador Tea
Sambucus callicarpa	Elder
Spirea douglasii	Hardhack
Ledum groenlandicum	Labrador Tea
Picea mariana	Black Spruce
Larix laricina	Tamarack
Chamaecyparis thyoides	Atlantic White Cedar
Vaccinium uliginosum	Blueberry
Thuja occidentalis	Northern White Cedar
Populus heterophylla	Swamp Cottonwood
Tsuga heterophylla	Western Hemlock
Picea sitchensis	Sitka Spruce
Toxicodendron vernix	Poison Sumac
Magnolia virginiana	Sweetbay
Persea borbonia	Redbay
Rhus glabra	Smooth Sumac
Populus fremontii	Fremont Poplar
Quercus lobata	Valley Oak
Salix piperi	Dune Willow
Liquidambar styraciflua	Sweetgum
Populus deltoides	Cottonwood
Quercus imbricaria	Shingle Oak
Quercus palustris	Pin Oak
Diosphyros virginiana	Persimmon
Fraxinus americana	White Ash
Acer rubrum	Red Maple
Celtis laevigata	Sugarberry
Celtis occidentalis	Hackberry
Alnus glutinosa	Black Alder

Similar depth and flood duration-dependent zonation occur in wooded wetlands with characteristic tree-shrub complexes present at various elevations and in various flooding regimes. Differing bands of vegetation encircling a shallow lake or other depression are quite similar to those found around marshes or in bogs. But in many river swamps the wettest regions tend to be in old channels, oxbows behind the overbank region, so that a distinct pattern is not as evident as in depressional marshes. In fact, the zone immediately adjacent to the channel, the overbank, is often higher and much drier than other sites because larger particles are deposited nearer the channel during floods. Nevertheless, greater and greater distance from the channel tends to translate into shorter and shorter hydroperiods and eventually the forest

TABLE 12-5
WOODY PLANTS THAT WILL
TOLERATE FLOODING FOR LESS
THAN 30 DAYS DURING THE
GROWING SEASON

Scientific Name	Common Name
Populus grandidentata	Bigtooth Aspen
Tilia americana	Basswood
Carpinus caroliniana	Ironwood
Acer negundo	Box Elder
Smilax spp	Greenbrier
Rhus radicans	Poison Ivy
Vitis riparia	Wild Grape
Parthenocissus quinquefolia	Virginia Creeper
Cornus stolonifera	Red Stem Dogwood
Quercus phellos	Willow Oak
Quercus nigra	Water Oak
Quercus bicolor	Swamp White Oak
Quercus falcata	Spanish Oak
Quercus macrocarpa	Bur Oak
Nyssa sylvatica	Black Gum
Platanus occidentalis	Sycamore
Betula nigra	River Birch
Ilex opaca	American Holly
Crataegus mollis	Hawthorn
Acer negundo	Box Elder
Acer saccharinum	Silver Maple
Alnus rugosa	Hazel Alder
Gleditsia triacanthos	Honey Locust
Ulmus americana	American Elm
Ulmus alata	Winged Elm

becomes indistinguishable from the adjacent upland forest. But the general zones tend to be linear and in some case many kilometers long.

Only a few (cypress, black willow, water hickory, water elm, water tupelo, water locust, swamp privet, green ash, Nuttall's, and overcup oaks) can survive long-term inundation (i.e., flooding during the growing season) for more than one year. A few (cypress, tupelo, willow, and overcup oak) grow in standing water for many years, similar to emergent herbaceous species. However, most wetland trees and shrubs are only adapted to endure short flooding periods during the growing season, perhaps long periods during fall and winter, but not permanent inundation as are cattail or bulrush.

Natural bogs exhibit similar zonation though the causitive factor(s) is slightly different and in fact, the zonation is often more the result of successional changes than directly related to water depths. Classically, various species of bog/marsh emergents and *Sphagnum* moss began growing in the

TABLE 12-6
WOODY PLANTS THAT WILL TOLERATE
FLOODING FOR LESS THAN 5 DAYS
DURING THE GROWING SEASON

Scientific Name	Common Name
Betula populifolia	White Birch
Betula alleghaniensis	Yellow Birch
Betula payrifera	Paper Birch
Acer saccharum	Sugar Maple
Populus tremuloides	Quaking Aspen
Tsuga canadensis	Eastern Hemlock
Picea rubens	Red Spruce
Fagus grandifolia	American Beech
Carya tomentosa	Mockernut Hickory
Carya ovata	Shagbark Hickory
Carya cordiformis	Bitternut Hickory
Carya lacinosa	Shellbark Hickory
Cornus florida	Flowering Dogwood
Gymnocladus dioica	Kentucky Coffee Tree
Juglans nigra	Black Walnut
Prunus serotina	Black Cherry
Quercus velutina	Black Oak
Quercus alba	White Oak
Quercus rubra	Red Oak
Quercus shumardii	Shumard Oak
Sassafras albidum	Sassafras

shallow, shoreline waters of a lake or smaller depression and slowly extend out into the lake. But decomposition of dead plant material is slow due to the cold climate, acidic and low nutrient conditions, and organic matter (peat) sinks to the bottom and begins to accumulate. Over time, peat accumulations on the lake bottom extend up to and may be indistinguishable from the peat immediatly underlying living *Sphagnum* and other plants so that a deep layer of peat now occupies much of the near shore regions. As conditions near the old shoreline become more stable, larger plants are able to colonize and shrub and tree seedlings appear. Taken to the ultimate, this gradual, successional process will fill the depression with peat supporting a forest — more correctly, a forested wetland or northern swamp. It is true that water depth and quality in the near vicinity of each plant influence whether or not that species can survive at a specific site or zone in the bog. And the obvious zonation progressing outward from shore would suggest that water depth is the controlling factor. But the advancing peat layer that supports shrub and tree species is often floating above 5 to 10-m or more of lake water. Water depth directly below a shrub could be quite deep but the shrub or tree is protected by the floating mat of peat and its roots extend horizontally in the upper, better oxygenated layers of moss and peat.

Figure 12-1. Upland grasses occupy dry sites in the foreground with transition species near the water, emergents in shallow reaches, and submergents in the deep zones.

Figure 12-2. Cordgrass occurs in the transition zone on the far right grading into arrowhead and then a dense mat of American pondweed in deeper waters in the background.

Similar progressive peat accumulation and vegetative patterns occur in many pocosins — essentially shrub-bogs of the Carolinas. Typically a depression is invaded by *Sphagnum* that supports herbaceous and eventually woody species and similar peat accumulations eventually fill the basin. In a sense, pocosins can be considered southern bogs though water chemistry may differ and plant species diversity is frequently much higher than in northern bogs. Some Carolina bays share similar soil types, floral and faunal species, and distributional patterns with pocosins but bays develop in unique elliptical basins, often have marsh vegetation and even open water reaches. Pocosins occur in a variety of geological formations with poor drainage.

To repeat, the best guidelines for selecting plant species for the new wetland will come from observing a similar wetland in the vicinity. Bear in mind that new plantings or seedlings will be less able to survive extremes than will the mature individuals in the reference wetland. Seedlings of tolerant tree species planted among mature representatives of less-tolerant tree species have been killed by prolonged flooding during winter that stressed, but did not kill, the less-tolerant species.

The categories in Tables 12-3 through 12-6 (flood tolerance) are defined by maximum flooding tolerance of mature individuals of native species. Nursery stock with different genotypes may be much less tolerant of prolonged or deep flooding. Conversely, shrubs and trees in a wet category will usually thrive in a dryer regime. For example, cypress was planted on reservoir shorelines and in sinkholes in east Tennessee, far out of its normal range. The shoreline stands rarely are flooded and the sinkhole stands are only flooded in very wet years, but robust populations have existed for 50 to 60 years. Relatively complete canopy cover resulting from closely spaced plantings has virtually eliminated invasion by competitive species. At the other extreme, willow oak is frequently planted as a shade tree or ornamental on city streets, even in hilly, well-drained areas, and some of the largest examples occur in infrequently flooded city parks in the Piedmont and Coastal Plain.

Physical and chemical parameters of water also affect survival and growth of wetland plant species. These include water clarity, pH, dissolved nutrients and other compounds, salt concentration, flow velocity, and dissolved oxygen. For submerged species, water clarity is important since light penetration is reduced in turbid or stained waters, thereby limiting photosynthesis. Under these conditions, rooted aquatics with floating leaves (*Nymphaea, Nuphar, Brasenia, Potamogeton nodosus*, and other floating leaved pondweeds) usually can overcome physical limitations on light penetration. In clear but fast flowing waters, species with filiform leaves and extensive root systems (*Vallisneria, P. pectinatus,* and *Scirpus subterminalis*) are resistant to current disturbance often growing on rock strata in fast flowing rivers.

Salt concentrations in the substrate or in the water column strongly influence plant survival by impacting osmotic balance, the determining factor in direction of passage of water and dissolved materials across cell membranes.

Water tends to move towards the region of higher salt concentrations and nonadapted plants in high salt environments are dessicated by saline waters or unable to take up water and nutrients from saline soils. In alkali wetlands of the West or in coastal regions with brackish waters, cordgrass (*Spartina alterniflora*), salt grass (*Distichlis spicata*), widgeongrass (*Ruppia maritima*) or a species of *Salicornia* may be planted. Sago pondweed (*Potamogeton pectinatus*), *Zanichellia*, *Triglochin*, alkali grass (*Puccinellia*), *Eleocharis parvula*, *Eleocharis rostellata*, *Juncus acutus*, saltbush (*Atriplex* spp.), *Sueda* spp., and some forms of bulrush (previously *Scirpus acutus* and *S. fluviatilis*) are also tolerant of moderately to strongly saline conditions.

Saline waters usually have moderate to high pH levels; and the species suggested above are generally appropriate. Low pH waters (bogs) impose an entirely different set of constraints, although pH is often less directly influential than are the associated parameters. Bogs also tend to have fairly stable water levels (little disturbance), low nutrient contents, and high organic contents, though most materials are only partially decomposed, and dark or stained waters that restrict penetration of sunlight. At extremes, low pH may directly affect plants, though some seem to adapt to ranges as low as 3.5 S. U.; for example, certain colonies of cattail (*T. latifolia*) and, of course, a number of mosses. However, many acidophilic species have adaptations that overcome nutrient or low light constraints rather than obvious adaptations to high hydrogen ion concentrations. For example, many insectivorous plants occur in bogs, obtaining their nitrogen and other scarce nutrients from insects rather than the impoverished substrate or waters. Water lilies (*Nymphaea*), spatterdock (*Nuphar*), and lotus (*Nelumbo*) have large root stocks capable of storing and supplying energy to sustain growth to reach the surface where floating leaves can obtain abundant light energy for photosynthesis.

Typical bog trees include black spruce (*Picea mariana*), tamarack (*Larix laricina*), and balsam (*Abies*), while acidophilic shrubs include representatives from *Spirea, Vaccinium, Kalmia, Chamaedaphne, Magnolia, Ledum, Persea, Alnus, Ilex, Leucothoe,* and *Rhododendron*. Grasses often found in bogs include *Calamagrostis, Deschampsia, Muhlenbergia, Eriophorum,* and *Glyceria*. Emergents in bogs are dominated by mosses, and a few species of *Juncus, Eleocharis, Cyperus, Dulichium arundinaceum, Rhynchospora*, many sedges (*Carex*), *Peltandra*, and species of herbs from the genera of *Lachnocaulon, Helonias, Cypripedium, Habenaria, Pogonia, Spiranthes, Caltha, Sarracenia, Drosera, Parnassia, Polygala, Pinguicula, Calla, Drosera, Darlingtonia, Hypericum,* and *Viola*. Submergents consist largely of *Utricularia* and *Menyanthes*, while *Nuphar* occurs in the floating-leaved niche.

Pocosins and bays often have pond pine (*Pinus serotina*) perhaps sharing the canopy with loblolly bay and a shrubby understory comprised of titi, zenobia, fetterbush, sweet bay, red bay, wax myrtle and various *Ilex* species. Others include red maple, black gum, pond cypress and sweet pepperbush. Ground cover includes greenbrier, Virginia chain-fern, sedges, grasses, blueberry, pitcher plants (*Sarracenia*), and sundews (*Drosera*).

SUBSTRATE

Most common substrates are suitable for most wetland plant establishment, but fertile loam soils are best for wetlands as for other types of plant growth. Sandy loam soils are soft and friable, allowing for easy rhizome and root penetration while heavy clay soils may restrict root and rhizome penetration. Inadequate or excessive nutrient content may limit growth and development, depending on the nutrient content of inflowing waters. If topsoil is replaced or a layer of good soil is placed above the impermeable liner, soil amendments are usually not necessary because wetland plants thrive in a broad range of soil types. Occasionally, fertilizers or liming may be beneficial if the substrate is poor in nutrients or acidic. Nutrients in natural wetland soils often become immobilized under reducing conditions in the substrate. Dewatering the substrate and oxidizing enclosed nutrients makes them readily available to plants, causing vigorous growth and development in the plant community. Macronutrients are rarely limiting in wetland soils, but obtaining samples for laboratory analysis of soil constituents in a created system is always a good practice prior to selecting species for planting.

Sandy loam and clay loam soils normally have adequate nutrients, provide good water and gas circulation, and have moderate texture to support the new plants and to permit root or rhizome penetration. Peaty soils are generally nutrient deficient and their soft, loose texture will not anchor new plantings and many float out or fall over. However, peaty soils are needed if the project will develop a bog since the organic acids will depress soil pH and mats of *Sphagnum* or other mosses may be laid on top of the peat and anchored to the parent material below. Spent mushroom compost and various manures have been layered in newly constructed wetlands by developers attempting to imitate the peaty organic soils formed by natural wetlands, but the benefits are doubtful.

Clays and gravels may be so dense or hard that they inhibit root penetration, they may lack nutrients found in topsoil, or they may be impermeable to water needed by roots. If a clay or bentonite liner is used with topsoil above be careful of digging through the topsoil and planting the roots in the clay or bentonite layer. At one system with only a thin layer of topsoil over essentially native bentonite, plantings died because the roots were located in powder-dry bentonite even though free water was only 6 to 8 cm above. Plants placed in heavy clay soils may never spread because their roots are unable to extend beyond the original planting hole or their rhizomes cannot penetrate the heavy clays and initiate new shoots. Sands and gravels dry rapidly and if the water level is lowered below the level of the roots, the plants may die from dessication. In either case, adding organic material will improve growing conditions and some workers have used municipal leaf compost to increase the organic content of poor soils.

Most wetland plants have broad tolerances for normal levels of most soil nutrients, but a few are restricted to acid soils (typical bog species), some thrive in alkaline, calcium rich soils (*Chara* spp. and *Potamogeton pectinatus*), and others tolerate high calcium levels (*Scirpus validus, Pontederia cordata, Ruppia maritima*, and *Salicornia* spp.). Most species that occur in brackish coastal waters or waters of the Intermountain West tolerate relatively high salt levels, but many that are found in natural wetlands of the East are intolerant of salts in the substrate or water column.

Developers often use fertilizers at the initiation of a new system, with little basis from soil analysis or from plant requirements. Since fertilizer and lime are frequently used in terrestrial planting situations, many workers presume that adding nutrients and/or modifying the pH will benefit new wetland systems. In a few instances, soil amendments doubtless have improved growing conditions for new plantings; however, in many cases, the nutrients were quickly taken up by algae and little was available for macrophytes because typical formulations of water-soluble fertilizers coupled with broadcast application were used. If fertilization is deemed necessary, side dressing with a poorly soluble, time-release fertilizer such as "Osmocote" or "Mag-amp" will obtain the best results. Both are granular, slow-release formulations that should be applied to the substrate prior to flooding. Various tablet formulations may be used if the area cannot be dewatered or if fertilizer is to be applied after planting.

SPECIES CHARACTERISTICS

Since the plants in wetland systems provide the basis for animal life as well as conducting important hydrologic buffering and water purification functions, selection of species appropriate to project goals is important. Generally, marsh emergents (*Typha, Scirpus, Phragmites*, etc.) produce limited foods for vertebrates, though most support abundant insect populations. Muskrats and beaver feed on tubers and rhizomes and a few waterfowl use leaves, stems, or tubers of *Sagittaria, Alisma, Acorus, Pontederia, Sparganium, Cyperus*, and some other types. However, emergents provide cover and protection from the weather and predators that is critical to many wildlife species (Figure 12-3). In many cases, cover is simply a dense stand of vegetation to swim or run into to avoid predators or to get out of a cold wind, but muskrats use emergents for constructing shelters (muskrat houses) and many birds construct floating or suspended nests of leaves and stems. Though generally marsh emergents are needed for cover, some species of knotweeds, especially *Polygonum lapathifolium, P. pensylvanicum*, and *P. amphibium*, produce abundant seed crops, as well as providing important cover for wetland and terrestrial wildlife (Figure 12-4). Other *Polygonum* species either produce lower quantities of seed or production is irregular, but could be important in some regions.

Marsh submergents and floating types tend to have larger seeds and smaller tubers that are usable by many types of wildlife, especially waterfowl. Notable are the various species of *Potamogeton* (Figure 12-5), some of which occur in most freshwater wetlands and almost all of which produce abundant seeds

Figure 12-3. This trumpeter swan brood will find relative safety from predators once it reaches the cover of the dense stand of emergents.

Figure 12-4. Marsh vegetation provides important shelter for deer, fox, pheasants, grouse, and other upland species during harsh winter conditions.

Figure 12-5. A wide spread submergent, Sago pondweed, normally produces abundant seeds or nutlets and energy rich tubers favored by many waterfowl.

or tubers. Sago pondweed (*P. pectinatus*) is perhaps most well known, but most species produce abundant supplies of seed and tubers. Even the less obvious types may be important in some areas. In late summer and fall, the quiet back waters of New England rivers are often covered with uprooted leaves of the diminutive *P. subterminalis* after puddle ducks have dug out and consumed the tubers. Succulent leaves and stems of *Myriophyllum, Cerato-phyllum, Ruppia,* and *Elodea* support populations of more vegetarian species (gadwall, widgeon, swans, and coots) in many regions, whereas *Vallisneria* is used by muskrats, most ducks, and many geese wherever it occurs. Virtually all portions of *Nuphar* are sustenance to moose as long as the surface is ice-free and to muskrat and beaver throughout the year. Even the large seed of *Nelumbo* is used by some ducks, and it provides excellent brood habitat for young wood ducks in swamps, lakes, and reservoirs throughout the South.

Submergents also provide cover and feeding habitat for fish fry and dense beds in marshes, lakes, and rivers are often important spawning habitats. Protection from predators is important but the substrates for micro- and macro-invertebrates make these beds critical to survival of larval and early immature stages of important freshwater fishes, as well as larval and adult stages of myriads of amphibians. Although emergents and floating- leaved species also provide cover and substrate for invertebrate foods, dense stands of filiform-leaved submergents create abundant, optimal environments for invertebrates and immature fish and amphibians.

Figure 12-6. American pondweed, primarily floating-leaved, often found in similar habitats as Sago pondweed, is also a prolific producer of nutlets and tubers.

Floating species tend to be small and succulent, and entire plants along with vegetative propagules are food for birds, mammals, and fish. In addition, the microhabitats in floating mats of *Lemna* or *Spirodela* support abundant invertebrate populations that are heavily used by many vertebrates (Figure 12-6).

If a marsh has herbacious plants and a swamp has woody plants, then a bog is a combination of marsh and swamp since most bogs have both herbacious and woody vegetation. Not surprising, bog genera and in some cases a few species, are common to the marsh and swamp. Of course, bogs have many unique species — *Sphagnum* moss, the ericaceous (heath) family, and some orchids — but the sedges and grasses are often found in nearby marshes and many of the woody species are represented in neighborhood swamps. The preponderance of heath plants, many of which have well-developed water conservation mechanisms, might seem unusual for wetland plants. However, it's not uncommon for major portions of northern bog regions to have annual precipitation values similar to and even less than the arid plains and deserts of the Southwest. The difference, of course, is temperatures and subsequent evapo-transpiration rates. Even though the peat layers below and the bog mat supporting the living plants are saturated, anaerobic, acidic, and low nutrient conditions in peat discourage root penetration and living plants are largely dependent on meager precipitation intercepted by the upper most layers of moss. Fortunately, *Sphagnum* moss is morphologically adapted to capture, wick up, and hold many times its own weight in water.

Though most commonly identified with relatively flat lands and far northern climates, bogs also occur on mountain tops and high valleys in the Appalachians, along the Coastal Plain (pocosins), and on north and northwest facing slopes in the Eifel region of Belgium and Germany. These unique hanging bogs are supported on moderate slopes by emerging groundwater and a northern maritime climate. Of course, the greatest area of bog wetlands is certainly in Manitoba, Ontario, the maritime provinces, upper New England and its counterpart, the taiga, in northern Europe and Siberia, and the heathlands in the British Isles.

Bogs are almost synonymous with *Sphagnum* moss and many people would agree that "If that's *Sphagnum*, it must be a bog." But species selected for a bog should include *Sphagnum* moss because of its role in establishing and maintaining the acidic environment as well as producing the substrate for most other plants. Beyond that, species selection is limited to those capable of surviving the harsh conditions found in most bogs. Herbacious species often include sedges and a few grasses, cottongrass, calla lily, lady-slipper orchids, bogbean gentian, bog rosemary, pitcher plants, bladderworts and sundews. In open water sections, large floating leaved species, water lilies, lotus and spatterdock, dominate because the dark stained waters severely reduce light penetration. Shrubs include cranberry, blueberry, heather, leatherleaf, Labrador tea, bays, fetterbush and titi and spruce, pines, tamarack, cedar, pond cypress, red maple and black gum are common tree species.

In wooded wetlands, submergents are limited or lacking, and trees and shrubs provide food and cover for wetland as well as many terrestrial wildlife species. A few scattered, permanent ponds with dense populations of submergents, floating-leaved, and floating herbaceous species may be inordinately important during the dry portion of the year to a variety of wildlife. The overall significance in terms of life support in wooded wetlands is the hard and soft mast and cover produced by shrubs and trees. Some species produce substantial quantities of soft mast or berries (blueberries, blackberries, cherries, hackberries, persimmon, grapes) that may be critical in years of hard mast failure. The food source that is significant to many important wildlife species is hard mast or nuts from oaks, hickories, and cypress. Though upland oaks produce large quantities of acorns in good years, depending upon the species, production may be limited to alternate or every 3 years. Production in wet-growing oaks is more frequent, often almost every year, and wet oaks seem much less susceptible to damaging spring frosts that frequently devastate acorn production of upland species. Doubtless, adaptation to life in low areas that often harbor pockets of cold air includes delayed flowering avoiding damage by late spring frosts. The result is that swamp oaks rarely fail to bear bumper crops that not only support wetland wildlife, but frequently sustain terrestrial species, especially when upland oak production fails (Figure 12-7).

Similar species are important timber producers with many oaks, pecan, green ash, sweetgum, bald cypress, black walnut, white cedar, spruce, birch, and hemlock having significant commercial value. Though bottomland soils can

Figure 12-7. Acorns of willow oak often carpet the forest floor providing food for many kinds of birds and mammals.

range from well-drained loams to poorly drained clays, moisture conditions are rarely limiting, nutrients are replenished by annual flooding, and tree growth is often superior to neighboring upland sites. Fortunately, wildlife also benefit from many commercially valuable tree species so that these species support two important components of the life support function.

Care is needed if fast-growing vines and brambles such as blackberry are used since brushy types tend to dominate and retard shrub and tree establishment and vines often smother young trees. Whether planted or not, poison ivy quickly becomes established in any moist woods and swamps support luxuriant growths.

Small spring-borne seeds of buttonbush (*Cephalanthus occidentalis*) are food for many vertebrates, while its trunks and branches shelter fish, amphibians, reptiles, and birds, as do willows and birches.

Shrubs and trees not only provide shelter, perches, and open nest sites, but also create more cavity roosts and nests than their terrestrial cousins. Fast-growing but short-lived birches break off, decay, and die, producing cavity nests for chickadees, nuthatches, and prothonotary warblers; sycamores provide comparable shelters for woodpeckers and wood ducks. Similarly, snags and windfallen oaks and hickories shelter deer, bear, and raccoon; and the base and lower branches of all harbor fish, amphibians, and reptiles, especially when water levels are high and flooding over the buttressed trunks of many aquatic trees or the knees of cypress.

Species selection for flood control wetlands should include a mix of marsh and swamp types in appropriate locations. Shrubs and marsh emergents in the upper reaches and potential channels will provide roughness and filtering action to slow water velocity reducing erosion in shrub and tree stands in the remainder. The upstream planting might consist of marsh emergents — cattail, bulrush, rush — interspersed with fast-growing willows, birches, and ashes. If the area will be flooded to 0.5 m or greater for more than 3 weeks during the growing season, *Pontederia* and *S. validus* will do better than *Typha*, *Juncus*, or the other *Scirpus* spp. In the remainder and interspersed among the shrub-marsh section, species representing later stages (oaks, hickories, gums, spruce, or tamarack) are put in place. Since the objective is to obstruct and slow flood waters, species with multiple stems or branches and dense growth forms are preferred (*Salix, Populus, Alnus*). However, planners should carefully examine the anticipated flood depths, duration, and time of year in selecting candidate species from Tables 12-3 through 12-6 or other sources. Almost all species listed will endure extended inundation during the nongrowing season, but many are intolerant of deep waters or flooding for more than 15 to 30 days during the growing season.

Only a few species have been widely used in wetlands designed for water purification: *Typha latifolia, T. angustifolia, Phragmites australis, Scirpus validus, S. cyperinus*, and a few token attempts with *Iris, Pontederia, Acorus, Cladium jamaicense*, and various grasses (mostly *Phalaris arundinacea*). Cattail tolerates a wide range of water chemistries, including waters with very poor quality, and may be the most commonly used in North America, though common reed has been planted in most European systems. The choice of cattail and/or bulrush in North America and common reed in Europe simply reflects the most common species found in natural wetlands on those continents. Despite considerable discussion and limited research, a clear choice for the one best species is not apparent and may never be. Typically, only one or two emergent species are planted in most wastewater treatment wetlands, but many other species soon invade and it is not unusual to find 80 to 90 different species in a single system after only 4 to 5 years. Nor is it unusual to have 1 to 2 species dominate one year and 1 to 2 completely different species dominating in subsequent years.

Water depth controls zonation of wetland plants; secondarily, water quality (especially turbidity and dissolved oxygen content) impact species distributions. Selection of species should be primarily based on planned operating depths, with chemistry considerations included if poor quality water is anticipated. Defining and mapping planting zones based on water depths in the new system facilitates species selection as well as determining quantity of planting materials needed. These zones are defined below.

1. Transitional: areas which are seasonally flooded but not inundated for extended periods during the growing season. These are commonly referred to as "wet meadows" or if wooded, "bottomland hardwoods." They are rarely flooded during the growing season, but water depths may be 0.5 to 1.0 m at other times of the year. As the name implies, this zone is a gradient between terrestrial well-drained soil conditions and saturated soil conditions in the next deeper zone.

2. Shallow: areas which are frequently or continuously flooded during the growing season with depths up to 10 cm. This zone occasionally dries during annual drought periods or may be dry for more than 1 year during extended droughts.
3. Mid: the zone that is continuously inundated with 10 to 50 cm. Generally, it only dries during prolonged droughts.
4. Deep: continuously flooded with 50 to 100 cm or more. Rarely dries.

Classifications of herbaceous wetlands plants in Table 12-1 represent the most frequently observed water quality conditions for robust growth of each species. Those in Table 12-2 reflect the range of water depths that species is found in but planners should design for the low to mid water depths. For example, most species in the mid category thrive at 15- to 25-cm water depths though many can survive in deeper waters. However, some species of *Polygonum, Carex*, and *Phragmites australis* readily cross category boundaries, thriving in dry environments as well as in the transitional, shallow, and mid zones; many other species can survive in the near reaches of the neighboring zone. Latitude and altitude differences occasionally, and extremely wet or dry climates often, compensate for water level requirements, confusing attempts to categorize growing requirements of wetland plants. However, under the appropriate water conditions, the majority of the plants listed in Tables 12-1 through 12-6 will survive throughout the mid-latitudes of the northern hemisphere and many occur from the treeline in the north to the equator in the south. In fact, some species have worldwide distributions (i.e., *Phragmites australis*) and many genera include very similar species present on every continent except Antarctica.

The broad adaptability and aggressive colonization ability of a few species may cause dominance of the entire system by only one of these weedy types. Unless the project is designed to provide treatment for wastewater or will receive poor quality water from some other source, developers should exercise caution in planting *Phragmites australis, Typha latifolia, T. angustifolia*, or any of the *Salix* spp. All are aggressive, weedy plants that easily dominate the entire wetlands if suitable growing conditions are present and all are difficult to control. In addition, in some regions, other species can easily become pests (e.g., *Paspalum*), as have many introduced wetland plants. Obviously, exotic species should not be planted in the new system, regardless of project objectives (Figure 12-8). Salt cedar, water hyacinth, *Meluleuca*, purple loosestrife, and many species of submergents (*Myriophyllum*) have severely damaged many natural wetlands.

As is the case with other plants, perceptions of native undesirables (weeds) vary with desires, goals, and regions. Cattail is considered a weed due to its aggressive colonizing ability in the mid-Atlantic states and portions of New England. But it provides valuable cover and nesting habitat in much of the Midwest and West although stands often become so dense that control is implemented. Common reed is frowned upon by many naturalists since it

Figure 12-8. A native of South America, water hyacinth has been introduced almost world-wide, dominating wetland systems and clogging the waterways of this village in northern Thailand.

produces little food value for wildlife but it can provide excellent wave protection and shoreline stabilization. Reed canary grass dominates many marshes in the Northwest causing much concern, but it is valued in riparian situations in more arid regions of the West and widely used for wet meadows and erosion control in the East. Planners are advised to discuss their plant selections with local wetland scientists to avoid any misunderstandings.

Woody species shown in Tables 12-3 through 12-6 are grouped by tolerance to inundation during the growing season. Once again, latitudinal, altitudinal, and climatic differences blur category boundaries and planners should remember that each category is defined as the ability to tolerate flooding for a certain time period during the growing season. Most species will do much better under less severe flooding conditions and new plantings will require more modest inundation for survival.

Other factors that need consideration include aesthetics, wind breaks, shoreline erosion protection, or other relative capabilities of one species versus another. In the end, plant species selection is reduced to re-examining the original objectives, eliminating those species that will not thrive because of climatic, hydrologic, or substrate conditions, and then selecting species that will provide the life support, flood buffering, or water purification functions that are required to achieve project objectives (Figure 12-9). As in site selection and design, compromises must be made without losing sight of the principal objectives of the new wetland. However, water depth, and/or flooding duration, is the most important controlling factor for species survival, and selections should be based primarily on this criterion.

Figure 12-9. Old growth trees furnish shelter, perching and nest sites for wading birds, eagles, and osprey.

Since the objective is to create or restore a viable, self-maintaining wetland ecosystem and the plant community is the basis for all other life in the system, species selection and planting provides an important opportunity to influence the diversity and complexity of the new system. Planners should attempt to create as diverse (in terms of number of species) and complex a system as feasible, since the self-maintenance attribute of natural ecosystems is directly correlated with the degree of diversity and complexity.

As any farmer knows, maintaining a single-species system (a crop field) requires substantial input of time, energy, and chemical additives, and system failure is not unusual despite all his efforts. Simple systems lack resiliency that would prevent minor disturbances from causing major alterations to the system. For example, insect outbreaks in a single-plant species system have devastated that one plant population with drastic effects on wetland system performance. However, if the system has 20 to 30 species of plants, severe impacts to or the loss of one species will have much less impact on the functioning of the total system.

Consequently, planners should select and plant as many species within each zone as is feasible, recognizing of course, that numerous species will naturally colonize and some of the planted species may be lost over time. Since most functional values of wetlands are dependent upon the interactions of many types of plants and animals with the abiotic components, establishing more rather than fewer types of plants in the initial stages will advance system development and subsequent performance of desired functions. At a minimum, 10 to 15 different species should be planted in the transition and shallow zones, and 5 to 10 species in the mid and deep zones. If natural sources for planting materials are used, core and/or soils associated with dug plants normally contain propagules of the plants occurring in the natural system and many different species will be planted. However, if commercial sources are used, a number of species should be obtained for each zone present in the new system instead of the common tendency to choose one or two familiar, showy or inexpensive species.

SOURCES OF PLANT MATERIALS

Planting materials may be purchased from nurseries, dug from the wild, or grown by the developers of a project. Advantages and disadvantages of these options influence quality of planting stock, general and seasonal availability of materials, and costs for acquisition and planting. Generally, plants collected from similar wetlands in the vicinity will be more suitable for most projects than plants obtained from other sources. Wild-dug materials will inevitably contain a considerable amount of seed and other propagules in the attached soil materials that will enhance establishment of a diverse, complex community of plants in the new system.

WILD COLLECTIONS

Plants collected from the wild are more closely adapted to local environmental conditions than those obtainable from any other source. Wild collections have acclimated to local soils, typical hydrologic regimes and, of course, regional weather patterns. Consequently, wild plants will initiate new growth more quickly and develop more robust growth habits at earlier stages than will plants secured from nurseries as seeds or potted plants, especially if the nursery is distant from the project. This is especially true in wetlands constructed for wastewater treatment or other harsh environments. For example, plants used in acid mine drainage treatment systems are generally collected from similar acidic environments because plants from normal habitats suffer considerable stress and need a lengthy acclimatization period before vigorous growth begins.

If collecting and planting are carefully scheduled, local collections will require limited storage capacity since the newly dug materials will be planted in the new system within a day or two. Digging is scheduled for one day, with overnight storage in damp, cool conditions followed by planting the following day. Depending on air temperatures and humidity, overnight storage may merely consist of a pickup truck with planting materials layered into the box, thoroughly moistened (but not so heavily watered as to wash soil from root hairs), and covered with burlap or synthetic tarps.

Before digging, developers must obtain a thorough biological survey of potential collecting sites to insure that unique species or systems will not be damaged or that unwanted species will not be obtained. Many states prohibit or require collecting permits for digging any native plant and regulations should be researched early in the planning stages. Though creating wetlands is a lofty goal seriously impacting a natural wetland could negate any benefits. Nor will harming a natural wetland endear the project to neighbors and potential supporters. Obviously landowner permission is required as with any similar activity.

Digging plants is typically more expensive than obtaining them from some suppliers, but may be less costly than nursery-raised stock. Other disadvantages include availability concurrent with planting schedules because digging conditions may be difficult early or late in the growing season, undesirable species may be included in the seed bank, local supplies may be limited or difficult to obtain because of regulations or land ownership, and finally, careless collecting could impact natural wetlands or populations of rare or endangered species. But if these factors are not disadvantages, proper collecting techniques can minimize harm to the natural wetland. During collections, small plots (0.2 to 0.3 m) should be taken with a buffer zone of 0.5 to 0.7 m between collecting plots so that the undisturbed vegetation will colonize the collection plots within the same growing season. Be careful not to trample or damage plants between the collecting plots. Properly spaced, careful digging during spring is often very difficult to detect by the end of summer.

COMMERCIAL SOURCES

Ten years ago, only a few commercial nurseries specialized in collecting or growing herbacious wetland plants but a recent compilation of vendors by the NRCS and WES has over 500 listings. Many are forestry nurseries that produce some wetland shrubs and trees even though most of their production consists of commercial tree species. But many more vendors with an impressive selection of primarily wetland plants have recently begun operation. Florida still seems to have the largest number, but California, Minnesota, Maryland, New York, Oregon, and Wisconsin also have many and developers should not have difficulty locating a nearby source in most parts of the U.S. A selection of vendors is included in Appendix C. Other sources include commercial tree and shrub nurseries, seed and grass companies, state nurseries, and suppliers of native flower and landscaping plants. The district offices of the Natural Resource Conservation Service (NRCS) and university extension services have useful regional lists of plant dealers for conservation plantings, orchard and garden plants, and agricultural plants. Some of these include firms that offer plants suitable for wetland projects. The Plant Materials Centers of the NRCS occasionally have planting materials available for testing and demonstration, and are often familiar with sources for specific types or varieties.

Nursery-grown plants are supplied as potted materials, bare-root seedlings, containerized seedlings, or bagged root balls with a short sapling in the case of many shrubs and trees. Potted materials are expensive, often $1.00 to $3.00 per plant, but are easily planted and usually suffer less transport and planting shock. For comparison, digging costs with union scale labor often fall in the $0.30 to $1.00 range, depending on substrate and equipment availability. Bare-root seedlings are typically $0.25 or less per seedling, but are much more susceptible to transport and planting shock. Freshly collected plants from managed natural marshes available from a few sources vary from $0.15 to $1.00 per plant, and are not as susceptible to shock as bare-root seedlings. Do not expose bare-root seedlings to the air for even a few minutes in humid climates or survival will decline drastically. Containerized seedlings may be enclosed in small plastic trays, in paper sacks, molded peat or wood fiber, and larger plastic containers. Avoid black containers if the summer is already warm and planting materials will be exposed to the sun. Cost varies with the type of container and size (age) of the plant, but most are several times more expensive than bare root seedlings. However, added costs may be justified because containerized seedlings will survive in sites that are too harsh for bare-root seedlings. Many species of wetland shrubs and trees may be obtained at little or no cost from state forestry department nurseries, depending on the state and the objectives of the project. Prices at commercial nurseries vary with the species and the size of planting stock, but typically range from $3.00 to $50.00 per seedling with most of the variation related to size.

Though costs may be higher, especially considering the need for express shipment, using commercially supplied plants has strong advantages in availability of large quantities and delivery on site at scheduled planting times.

However, planners must contact potential suppliers well in advance (6 to 12 months) if they require large quantities (more than 500 plants) and to coordinate delivery dates and methods. Most suppliers can readily supply smaller quantities, but large orders must be started as seeds or collecting sites and methods determined well in advance of delivery dates. Large quantities should be shipped as several partial shipments to reduce on-site storage requirements and limit plant mortality. In addition, plants may or may not be available at the desired planting time depending on whether stock is grown outdoors or in greenhouses and local weather conditions. For example, planting conditions may be excellent in southern areas during March and April, but wild collections are difficult in northern latitudes until after ice-out, which may be as late as May. Greenhouse seedlings are usually available very early and late in the year, but generally are not available all through the winter, and seedlings from outdoor beds will often only be available during the first half of the growing season.

One significant disadvantage of using commercial suppliers is a tendency to plan wetland plantings based on the species available from those suppliers, and only a few species are widely available. Intermixing wetland plants and soils from a regional site with the nursery-grown plantings will add diversity and complexity to the system without significantly increasing costs. Another disadvantage lies with the genetic and physiological adaptations of plants to their growing site that may inhibit their ability to survive and grow at other locations with different soil, hydrologic, and climatic characteristics. Developers should try to select a nearby nursery or, next best, a nursery with similar soils and climate conditions and should avoid large latitudinal distances between plant source and destination; longitudinal variation is more acceptable.

Planting materials from commercial sources should always be shipped by express package service, delivered by the supplier, or picked up by project staff. Although all of these will add cost, shipping delays of 1 to 2 days and/or storage under unfavorable conditions can cause substantial mortality with subsequent loss of labor and expenses for planting as well as replacement costs for planting materials.

Plants can be grown specifically for a particular project, but this often requires a year or more of lead time and elaborate facilities such as a greenhouse or outdoor nursery. Few developers, unless they are planning a very large project or working on many projects over a number of years, can justify or sustain the additional effort and cost. In addition, propagation of wetland plants is currently much more of an art than a science, with many failures to be expected during start-up of a new operation. Existing nurseries have overcome those hurdles and developed successful methods to propagate some species and can provide them at reasonable costs.

If costs, commitment, or interest favor establishing a facility to propagate wetland plants, information on seed scarification and seedling nutrient and water requirements for some species is available in the references on wetland plants in Appendix A. Additional sources include a number of scientific journals devoted to plants (botanical journals), the Journal of the Society of Wetland Scientists, newsletters and journals of specialist groups for wastewater treatment with constructed wetlands and aquatic weed control, and a few regional journals covering specific areas with significant wetlands (i.e., the Chesapeake Bay or San Francisco Bay). After reviewing published information, developers should review the old annual reports from nearby U.S. Fish and Wildlife Service National Wildlife Refuges, and contact the Plant Materials Centers of the NRCS and request advice and assistance from established nurseries, bearing in mind that the latter may be less than enthusiastic over helping a new competitor.

SUMMARY

The principle goal of identifying candidate plant species and planting or fostering their establishment is to create as complex and diverse a plant community as practical in the minimum time interval. Complexity and diversity in the animal community is a direct function of a complex and diverse plant community, and the degree of diversity in the biological components governs the stability and self-maintaining capability of the total system. Deliberately establishing a diverse plant community will facilitate early system development and performance of wetlands functions, as well as reduce operating and maintenance needs.

CHAPTER 13

PLANTING VEGETATION

INTRODUCTION

Since marshes are successional stages to bogs and swamps, efforts to establish either bogs or swamps might begin by establishing a marsh appropriate to the region. In addition, much more experience has been obtained in creating marshes than any other type of wetland. Consequently, this discussion of vegetation establishment techniques concentrates on methods used in creating marshes. Planners developing a swamp on a wet site might first establish marsh vegetation and intersperse tree seedlings within the marsh plants, taking care to not use aggressive, tall, or heavy foliage, or emergent marsh species (Figure 13-1). Bog developers will want to distribute clumps of *Sphagnum* and other mosses and acidophilic herbs, as well as shrubs and trees, among the marsh plantings.

Many sites suitable for bottomland hardwoods will not be flooded during the growing season and recent wetland protection and restoration programs have encouraged landowners to restore wetlands on thousands of acres of cropfields. In these cases, directly planting wetland shrubs and trees will significantly advance the development of the new swamp. Old crop fields are especially amenable to planting since site preparation and planting can often be carried out by mechanical means.

Most emergent wetland plants are perennials that reproduce and spread largely by vegetative means, their seed scarification requirements vary and annual seed germination rates are quite low. However, many compensate for varying conditions and low rates by producing large numbers of seeds (e.g., *Typha*, *Phragmites*, *Scirpus*, and *Cyperus*). Conversely, many submergents reproduce prolifically by seed and a few have astonishingly long seed viability periods — in tens or even hundreds of years (e.g., *Potamogeton, Nuphar, and Nelumbo*). Others that may be rooted submergents or rooted with floating leaves or free floating produce vegetative structures, geminules, that survive harsh conditions on the bottom and establish new plants the following spring. Consequently, choice of planting techniques is strongly influenced by the type of wetland and the plant species selected.

NATURAL METHODS

In general, the least expensive method, if a natural seed source is nearby or the substrate contains a seed bank, is to create suitable conditions for natural

Figure 13-1. Wetland diversity is increased by planting as many different species as feasible and carefully manipulating water levels to enhance germination of seeds introduced with transplanted wetland soils.

invasion and establishment. For most species, including many woody shrubs and trees, that consists of moist, almost soupy muds maintained by holding the water level at or immediately below the surface or by periodic shallow flooding and dewatering. Best results for typical wetland plants will be obtained by careful water level manipulation at the onset of warm weather in the spring.

Few emergents' seeds will germinate while completely submerged, preferring instead the warm, moist mudbars exposed after shallow waters recede leaving bare soil conditions. In fact, only 1 cm of water has been shown to filter out enough light to prevent germination in *Typha*. A few hardy terrestrial species will also establish during early phases, but as the wetland species gain height, shallow flooding will inhibit or kill the terrestrial species, removing competition for wetland emergents. At this point, deep flooding will also inhibit the growth and vigor of wetland species and overtopping individual plants will likely cause mortality. Water levels must be carefully managed to maintain adequate depths to preclude terrestrial species, yet not impair the growth of the wetland types. As growth continues and plants reach 0.3 to 0.5 m, water levels should be raised to optimal elevations for the desired species — usually 2 to 4 cm for *Typha, Scirpus, Phragmites, Cyperus, Sagittaria, Alisma, Pontederia*, and *Peltandra*.

Although the seeds of some submergents will germinate under shallow water, most require similar wet soil conditions for optimal germination and growth. Geminules of submergents and floating species are adapted to initiating growth while submerged, as are the large root stocks of the water lilies, *Nuphar, Nelumbo*, etc.

Woody species rarely germinate underwater; most require moist soil conditions or a moist leaf and litter layer similar to terrestrial plants. Seeds of many species float and eventually lodge on hummocks and shorelines where proper moisture conditions are present. It is not unusual to find dense growths of seedlings representing an obvious strand line. Nor is it uncommon to discover that the vast majority of individual trees on lake shores and other relatively permanent water bodies are the same age. During normal conditions, water depths were too great for seed germination but a drought at some point lowered water levels creating the moist soil conditions conducive to germination and seedling growth.

Assuming adequate seed sources, manipulating water levels to foster germination and growth of desired marsh species and retard establishment of terrestrial types is substantially less expensive than planting, but may require 2 to 5 years to achieve full coverage. The time will depend on the type and size of a native seedbank, distance from a nearby seed source and seed dispersal mechanisms, soil fertility, and water management precision. In unusual cases, a dense stand of *Typha, Phragmites, Scirpus, Eleocharis*, or *Carex* may appear the first spring because a large quantity of seed was provided optimal germinating and growing conditions. More likely, stands will be initially sparse, gradually increasing through the growing season and during the second year, but not reaching high densities until the third or fourth years. Woody species will require much longer and if the potential seed source for hard mast producers is more than 200 m, natural establishment may be very slow.

Though less costly, using natural planting methods by manipulating water levels will generally take much longer to establish good coverage than would deliberate planting. In addition, planners may have limited influence or control over what species become established or their distribution within the new wetland. To some extent, both types and distributions can be affected by water management related to elevation differences within each pool (Figure 13-2). However, these techniques are imprecise and results will vary. Conversely, the new system may be invaded by many species originating from seeds in the bottom soils or outside sources, even though most of the system is deliberately planted. Survival and spread of either planted or invading species will depend on which species are involved, soil fertility, water chemistry, and most importantly, prevailing water depths.

Keep in mind that most emergent wetland plants find their best growing conditions in moist soils but they have adaptations to grow in saturated or flooded soils where other plants are unable to survive. If the soils are kept too dry, terrestrial species will survive and compete with the emergents, reducing growth rates and inhibiting spreading by the wet types. The objective is to eliminate terrestrial species by flooding, but not cause stress on the wetland

Figure 13-2. Many different submergent, emergent, floating-leaved and woody species support comparable diversity of fish, amphibians, reptiles, mammals, and birds in this complex wetland.

species from deep flooding. Consequently, careful and precise water level management is important to successful establishment of vigorous, spreading stands of emergent wetland species. Inability to flood uniformly and shallow and dewater quickly because of bed slopes or inadequate water control structures will impede establishment of emergent plant communities. This is one important reason why bed slopes should be flat (i.e., 0.0% slope) and why fairly large controls with high flow capacity are needed.

Continual inundation into early or mid-summer, followed by rapid dewatering without periodic reflooding will foster germination and growth of a completely different group of wetland plants — the moist soil species — a group of largely annuals with prolific seed production.

Similar flooding and drying can be used to establish woody species but the cycle (hydroperiod) is necessarily different. During the growing season only very short term (2 to 5 days) and shallow innudation is employed to stress or retard the most susceptible terrestrial species. Longer (15 to 30 days) but not deeper flooding in winter will have a significant impact on potential terrestrial competitors but not over stress wetland species. Even in winter water depths should be less than 20 cm whereas summer depths should not exceed 10 cm for the first few years to avoid stressing young woody wetland species.

PLANTING VEGETATION

Deliberate planting times and methods vary with the species selected and type of planting materials; that is, seeds, tubers, rootstocks, or whole plants.

In temperate regions, planting herbaceous vegetation is generally most successful in early spring, although the planting period extends from after the onset of dormancy in the fall to mid-summer depending on species. Though not recommended, cattail and bulrush have been planted later — in September and October in southern regions of the U.S. — as long as adequate time remains before killing frost to permit the new plantings to add new shoots. Root development is concurrent and continues after frost has killed the aboveground portions of the plants.

Tubers and root stocks are best planted in the fall after dormancy, but care must be observed with water levels and water quality. If the water quality is poor (i.e., low dissolved oxygen and/or high organic loading), submerged plantings will die from inadequate oxygen during the winter. Conversely, shallow water levels or simply wet substrates may freeze hard enough to kill tubers and root stocks if winter temperatures are extreme. Root stocks with 20 to 30 cm of stalk protruding above the water's surface allow higher water levels, even if the water is low quality, because the cut stalk provides a pathway for oxygen from the atmosphere to the roots. With good quality water, tubers of the most common emergent wetland plants, except cattail, are best planted after dormancy in the fall. Cattail and, to a lesser extent, *Phragmites*, and most of the grasses and sedges develop and spread faster if planted immediately after dormancy is broken in the spring.

Broadcast seeding of tree and shrub seeds has had highly variable success; hand or mechanically planting acorns and seeds of some other species is generally more successful. Pelletized seeds with coatings of fertilizer, fungicides, and stimulants are available for common species. Carefully handled bare-root seedlings, containerized seedlings, saplings, or cuttings are even more reliable, though more expensive to purchase and more laborious to plant.

The planting window for woody species varies with planting method. Seeds are planted throughout the year except for mid and late summer. With seedlings, fall and winter plantings, after the onset of dormancy, have been the most successful. Basically, seedlings are planted when soils are moist and seedlings are dormant. Some shrubs and trees can survive if planted in early summer, but the percentage lost is likely to be considerably higher unless soil moisture conditions remain high throughout the summer. Since projects using wet-growing trees or shrubs are often in floodplains or other areas frequently flooded during the winter, choice of planting materials and water level management is crucial to seedling survival.

Flooding tolerance specifications are largely developed from observations on mature trees and shrubs, and seedling tolerance is frequently quite different. Not only do mature plants have growth form and adaptations (knees, buttresses, etc.) that protect them from flood damage, they also have height to avoid being overtopped by deep flooding. Seedlings overtopped for more than a few days during the first winter will probably not survive. In addition, moderate to high velocities (flow rates) are often associated with deep flooding that may extract or physically damage seedlings which lack a securing root structure. Consequently, if deep or prolonged flooding cannot be prevented, the extra investment in older, taller seedlings (possibly coupled with anchors) is good insurance.

Those that have observed a dense stand of cattail or other wetland plant suddenly spring up in a roadside ditch will probably question recommendations against seeding wetland plants. However, collecting seed and then distributing it over a suitable substrate has not been very successful with most marsh and bog species, probably because many species have differential germination rates and some have stratification requirements that are not easily satisfied. Many species are capable of surviving extended periods of dormancy because of thick, hard, tight seed coatings that must be ruptured, dissolved, or decay before the embryo is freed. In some species, drying followed by immersion promotes germination; in others, drying reduces viability. Some require cold temperatures, others seem oblivious to temperature or storage conditions, surviving, for example, at room temperatures in dry museums for over 100 years! Others can survive after being frozen in ice or mud for much of the winter. Even within the same species and the same seed crop in natural habitats, germination tends to be erratic, with some seeds germinating after one year and others remaining dormant and not germinating until 4 to 5 years later. Germination of the large tough seeds of *Nelumbo* has been initiated by experimental treatment with acid to increase the permeability of the seed coat. Finally, as with most seeds, the proper conditions of moisture, temperature, oxygen, and often a light stimulus (perhaps UV radiation) are required for germination. With all the variables and contingencies, attempting to establish a new wetland plant community by seeding is at best unreliable and at worst may simply fail.

Delayed seed germination is partially the basis for a fairly simple planting method that quickly generates a diverse plant community in the new system. Because of the large quantities produced and prolonged viability of wetland plant seeds, the substrates of most natural wetlands contain an abundance of seed of many species, commonly called the seedbank. Cores (10 to 12 cm diameter and 15 to 25 cm long) of wetland soil and included propagules (seeds, roots, tubers, and rhizomes) collected in existing marshes have been successfully transplanted to new wetlands in many regions. However, attempts to simply dig out wetland soils and spread the materials across the bottom of new systems have not been as successful. Core planting is laborious and costly due to the need for collecting, transporting intact, and installing the heavy cores. Also, the cores as well as surrounding substrate must be kept moist, as in any other planting method. But success with core plantings frequently develops a complex, diverse community in a very short time period. Additional advantages include: plants developed from propagules in good soil conditions do not suffer from shock and die-back commonly associated with whole plants, and the reservoir of seeds and other propagules provides a source for new plant development not only through the first growing season, but in some cases for years later.

PLANTING METHODS

Many marsh and bog creators will have fairly clean, weed-free areas for planting after construction equipment has finished grading the site. But in cases where the project merely installed a dike(s) with a water control structure(s) in the lower end of a drainage without shaping the new wetland, some site preparation may be necessary. New wetland protection and restoration regulations have also discouraged landowners from farming prior converted wetlands and encouraged many to restore these wetlands. Many are undergoing old field succession and may be densely covered with annuals and weedy species that will compete with the new wetland plants.

If the terrain and water supplies are suitable, deep flooding (0.5 to 1.0 m) for 2 to 3 weeks in late spring and early summer will eliminate most herbacious and woody terrestrial forms but not undesired wetland species. Obviously simple flooding is the least expensive method of removing weedy species and it also produces good soil moisture conditions for planting after standing water is removed. But if a new bottomland hardwood site will be planted with oaks, pecan, cypress, or gums and the site is largely covered with water tolerant buttonbush, willows, or cottonwood, more intensive measures may be needed to insure survival and early dominance of the planted seedlings. In addition, mechanical site preparation will often loosen compacted soils enhancing conditions for seedling root growth.

Hand clearing, grubbing, small garden tractors and limited herbicide applications may be adequate if the proposed wetland is small. Large areas will require site preparation with construction or agricultural equipment depending on the terrain. Relatively flat sites such as old fields, can be prepared with rubber-tired farm tractors pulling a disc or similar farm implement. Crawler tractors can be used on rolling or steeper sites as long as the site is vegetated to reduce erosion shortly after discing. In any case, site preparation should be accomplished within a few weeks of planting during the growing season or at most a couple of months during the non-growing season. The objective is to remove undesirable weeds so the farm implement is adjusted to turn over the soil to 10- to 20-cm depths similar to crop field management.

Planting densities vary with target operating dates and species used. Most projects use 0.5- to 1.5-m spacing for herbaceous vegetation and 3- to 6-m spacing for trees and shrubs. Roughly, 0.5-m spacing will require approximately 40,000 seeds or seedlings/ha, 1-m spacing requires 10,000, 2-m spacing requires 2500, and 5-m spacing only requires 400 seeds or plants/ha. Closer spacing decreases the time required for spreading and complete coverage, whereas distant spacing requires longer periods for closure. Aggressive weedy species — cattail and giant reed — spread much more rapidly than rushes, sedges, bulrushes, arrowheads, and their relatives. Willows, alders, and other woody species that form colonies or clones spread rapidly through vegetative reproduction while oaks, hickories, gums, etc. that are dependent upon seed production and germination will not increase until initial plantings mature, perhaps 15 to 20 years.

While mechanical methods can seed large areas at relatively low cost, density of hand seeding or planting quickly becomes a factor in project costs. Low-density plantings require much less seed or planting material but often require much longer development periods to achieve high levels of coverage. Conversely, high-density plantings for many marsh and a few shrub species may establish almost complete coverage by the end of the first growing season.

Most emergents in new wetlands are started with hand planting of whole plants or dormant tubers and rhizomes. The root collar, evidenced by a dark line or a line dividing two color zones, provides a handy reference point for installing whole plants at the proper depth since it marks the previous soil line. It may seem obvious, but supervisors should emphasize that whole plants should be installed with the root down. Plants inserted upside down will develop slowly, if at all. Insuring that the roots are spread out and hang straight down and not clumped together, horizontal or even curled back upward greatly improves survival and growth. Tubers are commonly forced into soft substrates deep enough to prevent them from floating out, while rhizomes are angled slightly upward in shallow slits or trenches and then covered over. In either case, the substrate should be well moistened, if not saturated, and must not be allowed to dry after planting.

Most often, developers use portions of or whole herbacious plants with shoots, roots, and rhizomes. If complete plants are used, the materials should be small, new growth transplanted in early spring. With later plantings when natural vegetation is taller, stalks should be cut to 20- to 25-cm lengths to prevent wind throw until the roots develop secure anchorage in the substrate. In either case, tops and roots of many of the grasses will die back and new growth is initiated from buds at the plant base. More hydrophytic species begin new growth from buds on the shoots and the pause is foreshortened. Transplanting shock is likely to be evident in both types, even though fairly large masses of roots and substrates are transplanted, and maintaining optimal water levels or soil moisture contents is critical.

Since wetland plants are infrequently exposed to drought conditions, most lack waxy surfaces, small leaves or thick coverings, and other water conservation measures that many terrestrial species possess. Consequently, disturbance and damage to fine root structures (root hairs) during transplanting frequently causes considerable stress because the roots are unable to take in enough water to offset the high rate of loss from the stems and leaves. Once new root development has occurred and water uptake ability is restored, wetland plants will thrive in moist soils; complete saturation or flooding is unnecessary. However, developers should anticipate seeing tops and, in some cases, portions of the root systems, die back in a high proportion of new plantings. New growth will later begin from buds or root primordia on roots and rhizomes or shoots.

Hand planting tubers, root stocks, or whole plants is labor intensive and costly. Although often recommended, weighting tubers and dropping them into the water is not overly successful. Placing tubers in weighted cotton mesh bags has been satisfactory in some cases with some submergents. Generally, a tree planting bar (dibble) or tile spade is used to make a slit in the substrate, the root stock is inserted, and the slit is sealed. Plants should be placed so that the previous soil line (discoloration line on the stalk) is level with the new soil line, but deep enough to prevent floating out when the area is flooded. Power augers are used to create holes for core plantings in dry soil, and spades are used if the planting is under water or in wet soil.

Though seeding most emergents and submergents is generally not successful, seeds of many wet-growing grasses have been collected from the wild or purchased from nursery sources and sown with good results. But considerable variability has been noted in different regions. For example, reed canary grass (*Phalaris arundinacea*) is easily established by seeding in northern latitudes and, in fact, spreads prolifically in the Northwest, but is much less successful in the South, where most developers use plugs. Broadcast, even aerial seeding of some species can be very successful if water levels are held at the surface of the soil so that seeds sink into the "soupy" substrates. Lowering the water level a few centimeters below the surface a few days later creates excellent germination conditions.

On dryer sites, the substrate is tilled or scarified, as for terrestrial plantings, and the seed is planted with a seed drill similar to any grassland planting. If the seed is broadcast, the substrate should be tilled, raked, or dragged before and after seeding and before flooding. Most wetland grasses (exceptions include a few species of *Glyceria, Paspalum,* and *Panicum*) thrive on moist to saturated soils, without standing water, more typical of wet meadows than marshes. Consequently, the substrate should be shallow flooded and dewatered repeatedly, or pool levels should be maintained at or immediately below the surface to foster germination and growth. The exceptions, some *Glyceria, Paspalum,* and *Panicum* species, grow in standing water or as floating mats and are generally better planted with root stocks or rhizomes.

Acorns and other seeds may be collected from natural sources or purchased from many commercial nurseries. In either case, seeds should be float-tested prior to storage or planting. Merely pour the acorns into a tub or other large water container, stir a bit and remove any floating (unsound) acorns. Viable acorns and other tree seeds can be kept in heavy plastic bags in cold storage (2 to 3°C) for 1 to 3 years; some large seeded herbacious species remain viable for 10 to 20 years. Alternatively during winter, seeds may be stratified — buried in bags or loose below the frost line in moist soil or fine sand.

Most tree species are seeded at 5 to 7 cm or deeper if dry soil or rodents could be a problem. Spacing for direct seeding should be closer than for seedlings since only 25–50% of the seeds will produce a viable tree.

Hand planting shrub and tree seedlings is quite simple but often done incorrectly. Nursery-grown shrub and tree seedlings are generally larger (minimally 40- to 50-cm above-ground lengths) to enhance the seedlings ability to compete with other vegetation. Seedlings should be planted with their root collars (previous soil line) just below the surface of the ground and roots must be straight down and not curved. If using a dibble, jam the dibble into the ground at a 30-degree angle, move the handle away from you to form a V-shaped trench, withdraw the dibble, insert the seedling below the root collar and then lift it up to bring the root collar in line with the soil surface to straighten out the roots. While holding the seedling, jam the dibble into the ground 10 to 15 cm behind the trench and push it forward again partially filling the trench. Use your shoe to tamp the loose soil around the seedling.

Some shrub and weedy species, *Salix, Populus, Cephalanthus occidentalis, Liquidambar styraciflua,* and *Fraxinus pennsylvanica* can be started with cuttings. Cuttings (0.5-m long and less than 2-cm diameter twigs and including at least one vigorous bud) should be collected in the spring just prior to the end of dormancy. They may be stored in well-moistened sand overnight and then forced into saturated soils. *Salix, Populus and Alnus* are sometimes used as nurse crops to reduce competition from annual or perennial herbaceous species. These grow rapidly and shade out many herbaceous weeds but are eventually overtopped by typical bottomland hardwood species. Cottonwood has been used as a nurse crop protecting oak plantings and harvesting at 10 to 15 years for pulp wood provides an interim cash crop.

Planting tree seedlings should follow standard tree planting methods and seasons, but water levels will need to be high enough to ensure well-moistened soils but not so deep as to endanger the seedlings. Mesic or moist soils rather than inundated, facilitate hand or mechanical planters and provide the best growing conditions for the new seedlings. Start-up management may require periodic flooding and drying during the first few growing seasons and shallow flooding during fall and winter. Remember, growth requirements for wet growing trees and shrubs are similar to herbacious wetland plants. They have the ability to grow in much wetter environments than terrestrial species, but optimal growing conditions are moist to wet but not necessarily flooded soils. Conversely, maintaining dry soils will favor terrestrial species that may outcompete the planted wetland types.

Herbaceous species of forested wetlands may be introduced by obtaining soil cores from an existing system and placing at similar elevations in the new system. Frequently, the understory consists of species that exhibit maximum growth in early spring before canopy closure and many of these are available from wildflower nurseries.

Hand planting tools, available from forestry and agricultural suppliers, include planting bags to protect seedling roots from drying, tree planting bars (dibbles), hoedads (maddox), dibbles modified for container plantings, spades, power augers, seeding bars for acorns and other large seeds, seedling protector tubes, sun screens, stakes, anchors, and labels (Appendix C).

To reduce costs, many developers have used mechanical devices to expedite the planting process; simplest is perhaps using a trencher to cut a shallow ditch or row into which the planting material (seed or seedlings) is hand placed and tamped by foot (Figure 13-3). Depending on substrate composition and

Figure 13-3. Ditching machines and planters greatly increase planting rates but the new rows should run perpendicular to the long axis of the unit to avoid short-circuiting, especially in a constructed wetland cell.

stability, this method can substantially expedite planting. An additional advantage is that the bottom of the ditch is lower than the surface of the substrate and may be within the pool water level or at least will tend to collect and retain water longer than the top of the substrate, thus providing better moisture conditions for the new plantings. Farm equipment, such as tree planters, tobacco and tomato planters, modified to handle the species to be planted, also form a trench, place the root stalk or whole seedling in the trench, and partially cover it. If the terrain is suitable for farm equipment, the new substrate will not be damaged by equipment, and planners have access to any of these, planting times can be substantially shortened and costs reduced.

Seeding and planting rates vary considerably with desired densities but some general rules have been developed. For direct seeding, one planter can sow 2.5 to 3 ha per day on a level, prepared site at typical spacings. Aerial seeding varies so much with type of seed, seed mixtures, equipment, distance from landing field and terrain that developers should check with a local applicator. Initially helicopters seem more expensive but shorter turn-around times coupled with much better coverage in confined situations often balance the higher costs per hour. Two or three people with an agricultural planter modified for the chosen seed can sow 15 to 20 ha per day in good conditions. One planter can hand-plant 600 to 800 seedlings per day in good conditions whereas two or three people with a machine planter can plant 5,000 to 10,000 seedlings per day. Rough terrain or dense vegetation can reduce these rates by half.

Depending on the principal function of the wetland and planned water flow patterns, planting rows should generally traverse the narrow axis of the pond rather than the long axis. This is especially important for wastewater treatment systems because rows running the length of the cell will result in water flowing down each row, thus avoiding the retarding, filtering action of the planted vegetation. In some instances, flows with high volume and velocity have even prevented subsequent colonization of between-row areas. Unless the project will create a plantation, nonlinear or random planting rows will enhance the esthetics and natural appearance.

A few projects have successfully created small, backyard scale bogs in previously terrestrial environments. After the basin was excavated, a liner was installed and then much of the basin was filled with peat. Hummocks were deliberately formed in the topmost surface of the peat mimicking the typical hummocks and hollows in a natural bog. Live, wet layers of *Sphagnum* moss, salvaged from utility or road excavations, were layered across the uneven peat surface and herbacious and shrub species planted. Sprouting seeds within the living moss layers (and perhaps the peat) contributed additional species.

But *Sphagnum* moss is relatively slow growing, shoots only extend 5 to 7 cm a year, and creating an extensive bog could require a considerable length of time unless a very large amount of living *Sphagnum* could be obtained from a salvage source or peat mine. In addition, poor growing conditions will not support vigorous growth or expansion of herbacious, shrub, or tree species. But it may be worth experimenting with the uppermost layers from a peat mine if one is nearby.

Plant anchoring methods may be needed if the substrate is soft, planting materials are buoyant, or wave action or erosion will disturb the new plantings. At one new system, grading left soil substrates very soft and flooding created "soupy" conditions deep within the substrate. Planters merely forced cattail seedlings into the substrate without forming holes for each. But the new plantings were sparse at best and completely absent in many areas. Upon inspection, seedlings were discovered in windrows along the shore of the downwind dikes. The soft substrate was inadequate to anchor the seedlings against wave erosion and in a windy region, continual wave action plucked the seedlings out and washed them against the shores.

Various erosion blankets (straw, coconut fiber, or synthetics) will protect substrates, hold plantings in place, and trap sediments to help stabilize the site. Barriers placed outward from the plantings to protect against wave erosion can be constructed of piled rock, drift fencing, or floating booms consisting of poles or logs anchored with chain or cable to the substrate. Netting pegged down at intervals will hold plantings in place and, depending on mesh size and placement, may provide protection against wildlife depredation. However, mesh may entrap wildlife, and synthetic materials are resistant to degradation and can easily injure wildlife.

Mature cattail, bulrush, and reed plants receive little use by waterfowl, but tender new shoots with high protein content may be heavily grazed. Muskrats, nutria, and beaver feed on roots and tubers and may need to be excluded from new plantings with woven wire fencing. Depending on adjacent wetland or terrestrial habitats and animal populations, it may be necessary to protect shrub and tree seedlings from mouse, rabbit, or deer damage by placing wire screening or seedling protectors around the new plantings.

POST-PLANTING MANAGEMENT

Needless to say, plantings should be allowed to become well established and overcome planting shock before other stresses are introduced. Depending on the objectives, expecting significant flood water retention, wildlife food production or wastewater treatment within the first growing season after planting is unrealistic and could cause plant mortality or system failure. If possible, stresses from flooding, wastewater, etc. should be gradually introduced into the system so the plants and other components have an opportunity to adapt to the new environmental conditions. Adequate plant establishment could take 2 to 5 full growing seasons for herbacious vegetation and 10 to 20 years for woody species. During the startup period, the vegetation should be monitored frequently and if dead or unhealthy patches are found, they should be replanted and/or the introduced stress factor(s) reduced. Additional planting material may be obtained from healthy regions within the new system, from the original source, or from another source if poor growth or mortality is widespread. In the latter case, water level operations and other factors should be evaluated to identify causes prior to seeking more planting materials. Most wetland plants are very adaptable and quickly become established in new environments if proper water and nutrient conditions are available. If substantial plant loss occurs, poor planting times or techniques, inappropriate water levels, or inadequate nutrients are more likely to be the cause than poor planting stock.

Proper water depth and its careful regulation are the most critical factors for plant survival during the first year after planting. Many plantings have failed because of mistaken concepts that wetland plants need or can survive in deep water. Small, new plants lack extensive root, stem, and leaf systems with aerenchyma channels to transport oxygen to the roots. Consequently, flooding often causes more problems for wetland plants during the first growing season than too little water, especially if the water has low dissolved oxygen content. In fact, optimal conditions for transitional species, emergents, and woody species consist of saturated, but not inundated soils. Submergents and floating species require actual flooding soon after germination or planting because most depend on buoyant structures and water pressure for physical support to achieve an upright growth form. The objective of water level management is to create unfavorable conditions for terrestrial species by shallow flooding or saturating the soil, but not to stress wetland species by deep or prolonged inundation. Shallow flooding (1 to 2 cm) can limit invasion of weedy or terrestrial species once the wetlands plants have stems higher than 5 to 6 cm. But it is absolutely essential that stems and leaves of desirable species project well above the water surface to avoid drowning new or even older established plants.

This is an important reason for designing systems with little or no slope on the substrate and easily maintained water control structures that precisely regulate elevations. Water level management must create very similar water depths and/or duration of flooding throughout the newly planted area, and must precisely maintain that level despite fluctuating inflows. Many new plantings have been lost because stormwater inflows could not be discharged rapidly enough to avoid overtopping and drowning small plants. Similarly, inability to drain the entire area on schedule due to undersized control structures or uneven substrates will cause spotty areas of poor growth or even mortality.

Current evidence suggests that gas transport from the atmosphere to root structures in saturated soils is influenced by a number of factors, including differences in the respective gas proportions (each individual gas moves towards the region with a lower concentration), similar gradients in relative humidity, and temperature differences between various parts of the same plant. In addition, some evidence suggests that wind blowing across the surface of leaves and stems creates a Venturi effect, pulling intra-plant gases out into the atmosphere. Replacement due to atmospheric pressure forces air into other sections of the plant and the combined effect establishes a flow pattern through the plant that is maintained until the wind dies.

Doubtless a combination of these factors operates in many species though one may be more important than another in certain environments. Regardless of which is paramount, access to free air above the water surface by portions of the plant is essential for any to operate, and managing that contact is the key to managing many wetland plants. Stems, leaves, "knees," and swollen trunks exposed to free air are analogous to breathing tubes; blocking those tubes will stress and eventually kill the plant. Since water level manipulation is the simplest method for exposing or blocking these breathing tubes, careful management is critical to survival of new plantings as well as older stands and, conversely, a method for limiting establishment and growth of undesirable species or restricting colonization into unplanned areas.

Wet meadow species that grow in the transitional or shallow zones should be watered during the first year by shallow (< 1 cm) flooding with intermittent drying periods, depending on the species. Generally, transitional species can tolerate reasonably long dry periods, whereas shallow zone species should not be allowed to completely dry during the first growing season. Some species, such as various spikerushes (*Eleocharis*), many grasses (*Spartina, Festuca, Poa, Agrostis,* and *Alopecurus*), most ferns, and many herbs will tolerate extended dry periods. Others such as *Leersia, Glyceria, Cladium, Eriophorum,* and *Carex* are not as tolerant and should not become completely dry the first growing season. A few, for example some *Juncus, Polygonum, Cyperus, Paspalum,* and *Scirpus* species, can withstand 2 to 3 cm of inundation and complete drying of the surface as long as adequate soil moisture is present.

Figure 13-4. Annual high waters eliminate most woody species on this sandbar in the Little Missouri Badlands of North Dakota but willows and smartweeds proliferate and re-cover it each summer.

Water levels for woody species should be managed similar to those for transitional/shallow zone plants with the principle goal of shallow, intermittent flooding to restrict invasion or retard growth of terrestrial species but not impact wetland types (Figure 13-4). Surface drying will not be detrimental if the soil has adequate moisture, but extended or deep flooding will retard the establishment of desired species. Bear in mind that the categories in Tables 12-1 to 12-6 are based on field observations of mature individuals of each species. Small, young shrubs or trees are not nearly as tolerant because they lack extensive root, stem, and leaf structures to supply adequate oxygen to their roots.

Emergent species (shallow to mid depth zones) should be planted in saturated but not flooded soils and allowed to grow stems with leaves that project above planned flooding levels the first season. After stems reach 5 to 10-cm, water levels can be raised 2 to 3 cm above the substrate and proportionately increased as plant height increases until desired elevations are reached. Unless otherwise dictated by project goals, best results will be obtained by maintaining 2- to 3-cm water depths throughout the new stand of emergents during the first growing season.

As soon as submergent and floating leaved plants show new and vigorous growth, water levels should be slowly and gradually increased to support erect, upright growth forms. Shallow overtopping is not detrimental, but exposure and drying should not be allowed to occur. Maintaining stable water levels and keeping the plant continuously submerged is critical for these species.

However, submergents require oxygen, light, and nutrients similar to other types, but they must obtain oxygen from surrounding waters and receive adequate solar radiation to carry on photosynthesis even though completely submersed. Flooding the new plantings with turbid waters or waters with low dissolved oxygen will stress and perhaps cause mortality of submergents. Floating species or those with floating leaves can survive in poor waters so long as other conditions are suitable. Growth forms in many submergents vary with water quality conditions and, in poor water, they often have long, virtually bare stems with a profusion of leaves near the water surface. In good conditions, leaves are regularly distributed throughout the length of the stem unless water depths are too great.

Just as inappropriate water levels can inhibit establishment and growth of desirable wetland plants, unsuitable levels can be used to control prolific growth and spread of weedy species. Flooding may retard invasion by terrestrial opportunists and deeper flooding may retard undesired colonization of additional areas by planted species. For example, expansion of cattail stands may be controlled by maintaining depths of 40 to 60 cm especially if dissolved oxygen levels are low. Mowing or cutting the stems of many emergents, including *Typha* and *Phragmites*, followed by flooding well above the cut stems for several weeks during the growing season will thin or kill stands of these species. Basically, any method of interrupting the required oxygen or solar radiation supplies will stress and eventually kill most species.

If low temperatures and thick ice cover are anticipated, water levels should be slowly raised to accommodate ice thickness and reduce frost penetration into the substrate. In spring, for emergent species, water levels should be lowered and kept at or just below the surface of the substrate until new growth has reached 15 to 20 cm when levels may be raised to 2 to 5 cm above the substrate and gradually higher during the summer. For submergents, pool levels should be lowered in spring to 2 to 5 cm until new growth is evident, and then raised to normal operating levels.

CHAPTER 14

ATTRACTING AND STOCKING WILDLIFE

We seem to have a natural tendency to think that since this is a new system and we have planted vegetation, we also need to introduce the kinds of wildlife and fish that we wish to have in the system. However, introduction should only be seriously considered after careful evaluation of factors affecting natural methods, the most important of which are type, distance, and connections with nearby natural wetlands.

Without a substantial commitment of time and effort, most introductions of native vertebrates will not succeed. Furthermore, most birds, mammals, fish, and reptiles, the showy species that everyone wishes to have immediately, are highly mobile, and if suitable habitats are present in the new system, these groups will naturally invade. If natural wetlands are nearby and birds, mammals, and turtles do not begin using the new wetland, some essential requirement is lacking and no amount of introduction effort is likely to succeed. Fish will move in if the system is connected to other surface waters at almost any time of year, but especially in spring. Animals released in a new system with inadequate habitat will simply move to nearby wetlands that supply their needs or die.

Ecologists broadly refer to an animal's requirements with three terms that define important concepts: trophic level, niche, and habitat. An animal's habitat is its address, its home, its dwelling place, the physical location that provides cover or shelter, and opportunities to find food, water, and mates. Each animal is adapted to and requires a specific type of habitat and that is the address at which it can be located. Habitats might include grasslands, marshes, forests, deserts, and subdivisions of each. Its profession within that habitat, the manner in which it makes a living or how it obtains energy and nutrients (food) to survive and grow, is its niche.

Niches include one type of lichen growing on rocks, another type of lichen growing on tree bark, antelope grazing on forbs, deer browsing on twigs, bison grazing on grass, fox feeding on mice, coyotes feeding on rabbits, herons feeding on fish, raccoons feeding on carrion, fungi decomposing plant litter, and bacteria decomposing fecal material. Niches may be very narrow and specialized — snail kites (*Rostrhamus sociabilis*) living almost solely on apple snails (*Pomacea*) and night-blooming cacti pollinated by a single species of moth — or broad and generalized — snapping turtles (*Chelydra serpentina*) preying on invertebrates, fish, amphibians, birds, and mammals, foraging on aquatic plants and scavenging on dead plants and animals.

Broad categories of niches are grouped into trophic levels: plants are producers, herbivores are primary consumers, carnivores are secondary consumers. Omnivores feed on plants (primary producers), herbivores (primary consumers), and perhaps carnivores (secondary consumers). Decomposers (invertebrates, bacteria, and fungi) degrade organic material and, finally, others (parasites, scavengers, and saprophytes) obtain nutrients and energy from all levels. Some organisms fit nicely into one trophic level; that is, most plants are producers since they create organic materials from inorganic substances, but even some plants are saprophytes, obtaining nourishment from dead plant material. Many microbes are decomposers, but some bacteria are chemosynthetic, capable of deriving their energy from chemical transformations and some may be decomposers or producers, depending on environmental conditions. Most animals blur the lines of distinction; that is, many animals (raccoons, coyotes, bears, and pigs) scavenge, feed on plants, and feed on other animals.

Producers, microscopic and macroscopic plants, provide the basis for most ecosystems, trapping and using solar energy to manufacture complex organic substances from simple inorganic materials. Though many plants would naturally invade a created wetland, the last chapter addressed methods to facilitate establishing this important group to expedite operation of the system. Our purpose was to "jump-start" the new wetland — to decrease the time period for establishing a diverse plant community. Natural methods of plant dispersal and invasion would have accomplished the same result, but they may have taken a bit longer. Since at some time of the year, most animal populations are larger than that population's habitat will support, individuals are forced to search for new homes and they will discover the new system during their wanderings. Therefore, if suitable, unoccupied homes are available, most types of animals will discover the system within the first year of operation.

As plants begin producing potential foods for animals and transforming raw environments into suitable homes, animals from nearby systems will quickly discover the new habitats. Consequently, deliberate introductions are rarely necessary or warranted. Attempts to establish populations of animals are obviously doomed to failure until appropriate habitats and niches are available. If desired species occur in wetlands in the vicinity, but fail to appear in the new system, re-evaluate potential habitats and available foods, modify system parameters as needed, and allow time for pioneering before considering introductions.

In the worst case condition, the new system is remote (tens or hundreds of kilometers) from any similar natural wetland or does not have any surface water connections with adjacent streams, lakes, or wetlands. Developers should then review the literature on similar systems, consult with local authorities, and select species for introduction that occur in natural wetlands, shallow lakes, or streams in the region.

Trapping or collecting animals, including insects and other invertebrates in some states, and many plants is prohibited or regulated by state or federal laws. Consult with the nearest office of the state natural resource agency on collecting locales, collecting methods, and permits required. Unless the project goal is to provide habitat for a unique (rare, threatened, or endangered) species, only commonly occurring, abundant animals should be collected from locations that will not jeopardize a unique species or population.

If planting materials were obtained from natural wetlands, most microscopic and many macroscopic organisms will have been included in soil cores and soil associated with plant roots. Assuming that soil cores were kept moist and cool, virtually all microscopic organisms and the eggs or immature stages of many invertebrates will have been introduced, along with a myriad of plant propagules, during the planting process. If not, obtaining and introducing soil cores is a practical means of bringing in these organisms.

Before thinking of introducing microorganisms, remember that they are ubiquitous, occurring and/or spreading rapidly into almost any environment that contains a source of energy and nutrients. In some cases, the environment may seem to be hostile; that is, acid mine drainage, but only the most extremely acidic waters (pH < 2) lack abundant microbial populations deriving energy by transforming metallic ions and nutrients from other inorganic sources. With the exception of a few very complex, large molecular weight compounds, some type of microorganism is capable of breaking down almost any substance known to occur in wet environments. A few naturally occurring and some anthropogenic polycyclic compounds are resistant but not immune to biodegradation, taking very long periods for decomposition and, in some cases, requiring initiation via radiation exposure. However, given time (remember, generation times may be only 20 to 40 hours in some microbes), a population of microbes will develop that is adapted to degrading virtually any compound that contains a potential energy source. Consequently, regardless of the objectives for the new wetland, inoculation of microorganisms is rarely warranted.

As with some plant propagules, eggs and small life forms of invertebrates are often carried in the feathers, fur, or on the feet of birds and mammals. Even nonflying insects and mollusks may be injected into the system by a bird or mammal passing through. Many flying insects migrate over long distances and it would be very unusual if most did not appear almost overnight. Entomologists have long known that many northern insect populations are restored after hard winters by migrants carried north by spring winds. Occasionally transported at high altitudes, flying insects will quickly find and populate even the most remote system. A probable bumblebee (at least, it was yellow and black) smeared my aircraft windshield at 3353 m above sea level over Kentucky one spring afternoon. Many nonflying types hitch rides with migratory birds and wandering mammals.

If fish are desired, they are the most likely to need to be introduced. Mosquitofish (*Gambusia affinis*) are the most desirable because of their control of potential pest mosquito populations. Topminnows (*Fundulus*), shad (*Dorosoma*), shiners (*Notropis*), killifish (*Fundulus*), and other forage species are needed to provide food for game species. Pike (*Esox*) occur in many natural marshes and bogs, as do bullheads and catfish (*Ictalurus*), sunfish (*Lepomis*), bass (*Micropterus*), crappie (*Pomoxis*), and perch (*Perca*). Gar (*Lepisosteus*), bowfin (*Amia*), crappie, bullheads, and catfish are often found in southern swamps during periods of high water. Suckers (*Catostomus*), carp (*Cyperinus carpio*), catfish and bullheads or other bottom-feeders should not be used because their foraging uproots submergents and suspends fine sediments that block radiation for submerged plants. A few species may be obtained from hatcheries, but most will need to be trapped or netted, with appropriate permits from the state natural resources agency, in local water bodies and released in the new wetland.

Amphibians (frogs, toads, and salamanders) and reptiles (turtles, snakes, lizards, and crocodiles) are archtypical of wetlands (Figure 14-1). Amphibians lay their eggs in fresh water and reptiles lay bird-like eggs in moist earth or rotting vegetation. The simplest method of introducing each is to collect, with proper permits, egg masses of amphibians or egg clutches of reptiles in spring or early summer and place them in the new system. Most frogs and toads attach masses of eggs to stems and leaves or other underwater portions of living or dead plants. Simply cut off the stem, place it in a container with water from the wetland, and later set the stem into similar water depths in a quiet backwater of the project. Most will survive a few hours in transit if the container is protected from warming. If longer periods are expected, large shallow containers with a high proportion of water surface for oxygen exchange should be used. Egg-laying times range from the first warm days of spring to midsummer, depending on species.

Figure 14-1.   Unless the new wetland is far removed from any natural system, frogs, toads, and salamanders will quickly take up residence.

Many salamanders place their egg masses in moist layers or piles of organic debris that are difficult to locate without causing considerable disturbance to the natural area. Adults of many are aquatic and can be netted or trapped, similar to small turtles, but a portion of the trap must project into free air or trapped individuals will drown. Others can be hand-captured or caught with drift fencing and pits, transported in moist, cool containers, and released at similar sites in the created system.

Mudpuppies (*Necturus*) and a few relatives, commonly present in many natural wetlands, are unusual amphibians that attain sexual maturity and reproduce without transitioning into an adult air-breathing form. Immatures and adults can be netted or trapped similar to fish.

Freshwater and saltwater turtle eggs have been successfully dug up, transported, and placed in a simulated nest in well-drained but moist soil or sand with little or no vegetation. Incubation is dependent upon solar heating and a shaded area is unlikely to maintain warm enough temperatures (Figure 14-2). Eggs may also be incubated in a pail of moist sand maintained at average soil temperature (6 to 10 cm below the surface) and hatchlings released in shallow waters with abundant vegetation. Swimming ability in hatchling turtles is limited and many are unable to swim to the surface for air. They often need physical support structures to climb above the water's surface. Commercial bird egg incubators are usually not satisfactory. Incubation periods vary from 30 to 150 days, depending on species and temperatures. Be careful in attempting to vary incubation periods with high or low temperatures unless you want only one sex; hatchling sex ratios of many turtles vary with incubation temperatures.

Nests of egg-laying snakes and lizards are typically found in moist accumulations of organic debris — rotting logs, brush piles, and clumps of herbaceous vegetation near or just below ground surface. Collect the eggs and duplicate the natural nest at the project site. Many snakes and lizards retain their eggs until hatching rather than depositing them in nests. Live-bearing snakes and lizards must be hand-captured or trapped (drift fencing with regularly spaced pits) and released in similar habitats.

Song birds, waterfowl, and perhaps wading birds are likely to be the first large animal visitors to the new wetland, depending on time of year (Figure 14-3). Migratory birds tend to return to the area where they learned to fly to nest and raise their young, and many show some degree of fidelity to their migratory stopovers and wintering grounds. If necessary, flightless young of many species and, in some cases, flightless adults can be used to stock new sites. Adult Canada geese (*Branta canadensis*) maintained in captivity for years and later transported to distant release sites for their first flight either became resident in the release area or returned to it after spring migration. Ducks and geese for release can be purchased from private propagators or captured before they reach flight age in natural wetlands. In ducks and geese, females return to their natal area and males accompany their mate.

Figure 14-2. Hatchling turtles emerging from underground nests follow open skylines leading them to the nearest water and most new wetlands will be well supplied after a year or two.

With a few exceptions, most waterfowl will discover and populate the new wetland without assistance. Wood ducks (*Aix sponsa*), and probably other cavity-nesting species, have such strong fidelity to natal areas that they often fail to pioneer into nearby suitable habitats (Figure 14-4). In fact, one female may use the same box for a number of years and female wood duck ducklings banded in a nest box have been recaptured as nesting adults in that same box in succeeding years. Wood ducks can be transferred by capturing the female and hatching brood in a pole-mounted box, transporting box and contents to the new site, and placing the box on another pole before gently uncovering the opening. Some females will desert their broods and return to the trapping site, but others have a stronger attachment to the brood than to their original nest site. If the female deserts, developers will be faced with capturing and raising the brood before release. In either case, ducklings attaining flight at the new site will return to it for nesting (Figure 14-5).

Many private propagators raise a variety of ducks, geese, and swans, including many exotics. Developers should not be tempted to release exotic species in created or restored wetlands, regardless of how appealing. This cannot be over emphasized with regard to mute swans (*Cygnus olor*), the common swan that graces ponds and lakes in parks and zoos around the world. Though considered aesthetic by many, its aggressive territoriality often prevents any other species of waterfowl from using the same habitat, and it is not adverse to attacking pets and humans.

Figure 14-3. Great blue herons and other wading birds will locate a new wetland but are unlikely to establish nesting colonies until shrubs or trees have matured to provide suitable nesting substrates.

Figure 14-4. Extensive programs to erect and maintain nesting boxes have replaced tree cavities lost to logging and restored wood ducks to much of their former range.

Figure 14-5. Wood duck nest boxes (far left) substitute for natural cavities increasing wood duck numbers and simple board ramps (foreground) provide basking sites for alligators and turtles. Unless located in deep water, nest boxes should have metal predator guards.

Many ducks are omnivores, feeding heavily on invertebrates in some seasons and using seeds or vegetation during others. A few, widgeons (*Mareca americana*) and gadwall (*Anas strepera*), are largely vegetarian feeding on submergent plants as are many diving ducks though some divers concentrate on invertebrates. Swans make extensive use of submergents, but also graze on tender young emergents and grasses. Geese tend to graze marsh or meadow vegetation, strip seed heads from grasses and sedges, and readily adapt to cereal grains, as do many dabbling ducks and swans (Figure 14-6). Canada geese thrive on pasture and golf course grasses so long as a farm pond or water trap is nearby.

Canada geese and mallards (*Anas platyrhynchos*) introduced or naturalized in an area closed to hunting will likely become pests. Both adapt to a variety of foods and living conditions and quickly lose their fear of man. Visitors exacerbate the situation by feeding them and soon their droppings litter neighborhood yards, golf courses, and swimming pools; neither should be introduced in urban environments, and casual visitors should not be encouraged to remain. If your project is near a golf course, it may be well to discourage use by Canada geese since highly fertilized greens are choice grazing areas and their droppings won't endear you to course managers or golfers.

Wading bird (herons, egrets, ibises, and bitterns) populations have been restored to a few areas by raising and releasing young or, in one instance, by holding adults in large, semi-natural cages for a number of years. Most are colonial nesters and are unlikely to take up residence in a new system unless existing colonies are nearby, even though migrants and dispersing young may visit the site for many years. The same principle, releasing young at or slightly before flight age, appears applicable to some and should be tested on other species. Cormorants (*Phalacrocorax*) have been attracted to new sites by placing nesting platforms in snag trees in one area, and flightless young brown pelicans (*Pelecanus occidentalis*) were used to restore an island nesting population at another. In each case, restoration efforts have entailed substantial commitments of time and resources. Wetland raptors, osprey (*Pandion haliaetus*), and bald eagles (*Haliaeetus leucocephalus*), have also been restored by raising and releasing young birds, but these projects are expensive, time consuming, and must be continued for many years.

If necessary, wetland mammals can be live-trapped and released in similar habitats in the created system. Large numbers of inexperienced young make trapping and restocking easier in late summer and early fall than in other seasons. In addition, many mustelids — mink (*Mustela vision*), weasel (*Mustela* spp.), and otter (*Lutra canadensis*) — have delayed implantation, and females caught in the fall are likely to be pregnant and will give birth at the new site next spring. Even skunks (*Mephitis* or *Spilogale*) may be live- trapped, the trap gently covered with burlap, and the trap slowly and smoothly lifted for transport in a pickup bed to the release site. Mice, voles, and shrews are readily captured in small live traps, but shrews must have abundant food available during captivity to supply their high metabolic needs or they often starve overnight in the trap (Figure 14-7).

Figure 14-6.   Mallards and pintails feed largely on aquatic plants and invertebrates but in fall and winter its not unusual for their twice daily feeding flights to be 20-40 kilometers to find waste grain in cropfields.

Figure 14-7.   Mice and other small mammals follow bands of dense riparian vegetation leading them to new habitats. Many are important foods for larger mammals and birds as well as adding further diversity and interest for recreational users.

This brief outline of introduction methods should not be interpreted as encouraging deliberate efforts to stock animals in created or restored wetlands. It is included merely to demonstrate that under worst-case conditions, many wetland animals can be stocked using existing techniques described in the references or in recent journals. Most animals that can survive in the new system will naturally colonize it; if they do not invade naturally, introductions are likely to fail because suitable habitat is not available.

Microorganisms will be introduced along with transplanted vegetation and soil and by birds and mammals, as will many larger invertebrates. If the system has surface water connection with existing streams or other water bodies, even if only during high flows in the spring, many fish and aquatic insects will rapidly colonize it. Most reptiles and amphibians will gradually move in from nearby systems; turtles reside in farm ponds many kilometers from the nearest water body. Birds will suddenly show up some morning; many migrate in late evening hours or at night. Wandering young mammals dispersing from home sites will follow stream courses or travel overland to an isolated system. Only with very unusual circumstances will introductions of any type of animal be warranted and, then, only after thorough determination that suitable habitats exist, but seed populations are remote or isolated by an impassable barrier. Fortunately, stocking is rarely necessary, or else creating or restoring a wetland for life support functions would be a very expensive, long-term undertaking.

CHAPTER 15

CONSTRUCTED WETLANDS — SPECIAL CONSIDERATIONS

INTRODUCTION

Constructed wetlands have recently received considerable attention as a low-cost method for cleaning many types of wastewater. Though the concept of deliberately using wetlands for water purification has only developed within the last 20 years, in reality human societies have indirectly used natural wetlands for waste disposal for thousands of years. Man has dumped his wastes into nearby streams or wetland areas since prehistoric times. And as they do for natural ecosystems, wetlands processed these wastes and discharged relatively clean water.*

Constructed wetlands, in contrast to natural wetlands, are man-made systems that are designed, built, and operated to emulate natural wetlands or functions of natural wetlands for human desires and needs. Constructed wetlands as used for wastewater treatment may include swamps — wet regions dominated by trees, shrubs, and other woody vegetation, or bogs — low nutrient, acidic waters dominated by *Sphagnum* or other mosses — but most commonly they are designed to emulate marshes.

Contaminant removal processes in wastewater treatment wetlands are similar to microbial transformations present in conventional package treatment plants, lagoons, or other conventional wastewater treatment systems. Conventional systems require large inputs of energy, complex operating procedures, and subsequent costs to maintain optimal environmental conditions for microbial populations in a small treatment area. The low capital and operating costs, efficiency, and self-maintaining attributes of wetland treatment systems result from a complex of plants, water, and microbial populations in a large enough land area to be self-sustaining.

Wetlands constructed for wastewater treatment, at least initially, are comparatively simple, often single plant species systems (Figure 15-1). A properly designed and constructed cell with adequate treatment area covered in a dense stand of *Typha*, *Scirpus*, or *Phragmites* will efficiently remove target contaminants from influent waters while providing habitat for a few muskrats (*Ondatra*), blackbirds, and some songbirds but little else. If operated at maximum efficiency for wastewater treatment, it will not have adequate capacity

* After Hammer, D.A., Water Quality Improvement Function of Wetlands, *In* Nierenberg, W. A., Ed., *Encyclopedia of Environmental Biology*, Vol. III, Academic Press, Orlando, FL, 1995.

Figure 15-1. Constructed wetlands often only have a few species planted and if the wastewaters are high strength, the plant community may be dominated by one or two species as in this *Typha* stand.

to store flood waters nor can it release substantial quantities to amplify low stream flows in dry conditions. Wastewater treatment has been maximized through optimized design and operating criteria and all other functional values have been subordinated. But the water improvement function is still efficient and enduring even though other wetlands functional values are substantially reduced or nonexistent.

Treatment system longevity is poorly documented since few operating scale systems have been in operation for more than 15 years. Litter/detritus accumulation rates have been measured at 2 to 3 cm/year in municipal systems with no loss of treatment efficiencies (Figure 15-2). Therefore designs should incorporate this accumulation factor in dike height specifications and dikes should have one meter of freeboard for a 30-year operating lifetime or greater for longer operational status. At that point, the system may need to be cleaned out and restarted, and after testing to identify possible toxic substances, accumulated litter may be composted, burned or land applied similar to conventional biosolids.

DESIGN

OVERVIEW
Design types and composition of constructed wetlands vary considerably depending on when the project was designed (later designs include improvements), project objectives, wastewater applications, geographic/climatic location,

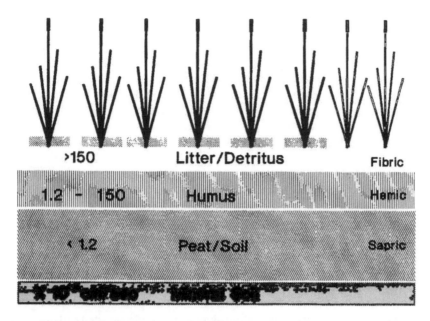

>150   Litter/Detritus   Fibric

1.2 - 150   Humus   Hemic

< 1.2   Peat/Soil   Sapric

Figure 15-2.   After death, marsh vegetation falls to the surface of the substrate, creating the litter/detritus/humus layers that provide enormous quantities of reactive surface area — attachment sites for microbial organisms.

and designer's experience. The original designs attempted to emulate natural wetlands with subsequent "improvements" adding artificial media, "best" plant species, mechanical devices (re-circulating pumps, aerators, greenhouses, etc.) in attempts to improve efficiency. Many improvements have been attempts to achieve theoretical performance efficiencies in gravel systems and many of these improvements resulted in more costly systems to build, operate, and maintain and failed to fulfill the original objectives.

Our objectives in designing constructed wetlands for wastewater treatment include developing a system that is:

1.   capable of providing high-level treatment and discharging relatively clean water;
2.   inexpensive to build;
3.   largely self-maintaining, requiring little or no operation and maintenance time or expense;
4.   manageable by operators with very limited training; and,
5.   capable of providing aesthetic/recreational/educational benefits.

Consequently, some attempts to reduce the size (required treatment area) to the absolute minimum, to install artificial media (gravel), to use vertical flow and/or batch loading, to add recirculation and/or aeration or to cover the wetlands with a structure (greenhouse) fail to meet our objectives. Virtually all of these may reduce the initial construction cost but all will increase operating costs. More importantly, all these modifications will increase operational complexity and costs thereby eliminating the original attributes that made constructed wetlands highly attractive. A simple, slightly larger system may be more expensive to construct but it will be much easier and less costly to operate. Projecting operational costs over a 20- to 30-year plant life time often results in lower total costs for the larger system that was more expensive to initially construct. Furthermore, probability for failure increases almost directly with increasing complexity. Adding pumps, piping, artificial media, greenhouses, fans, etc. lays the basis for failure of a critical component at an inopportune time (midnight Saturday night!) and dramatically increases the likelihood that operator inattention will increase the probability of failure.

MAJOR TYPES

The vast majority of wetlands constructed for wastewater treatment are classified as soil (surface-flow, free-water surface) systems, that is influent waters flow across and largely above the surface of the substrate materials. Substrates are generally native soils (Figure 15-3). This group includes the North American surface-flow designs and the European soil-based designs. In the other class — gravel (subsurface-flow, submerged vegetated bed, rock reed, microbial filter) systems — waters flowing through the system theoretically pass entirely within the substrate and free water is not visible. Substrates in gravel systems are typically various sizes of gravel and crushed rock. Relatively few gravel systems treating municipal waste are operating in North America but most of the European municipal systems were designed as subsurface-flow, soil based systems. In fact, subsurface flow in the European designs is through the litter/detritus layer, not through the underlying soil substrate, making them similar to North American surface-flow designs. The previous terminology was imprecise since most of the gravel-based systems fail to maintain subsurface flow patterns and in fact commonly have as much or more flow above the surface of the substrate. Also a significant proportion of the flow in soil or surface-flow designs is in fact within the substrate — the litter/detritus/humus layers. Furthermore, current loading rates and performance in European soil designs are similar to U.S. surface-flow designs. Since surface and sub-surface flows occur with both major design classes, it is more appropriate to classify designs on the basis of the substrate materials, i.e., soil-based systems for the U.S. surface-flow and the European soil types and gravel-based for the U.S. subsurface flow designs. A soil versus gravel distinction also delineates the semi-natural state of U.S. surface flow and European soil types and the artificiality of the gravel substrate, subsurface flow designs. The latter are in fact large particle-size sand filters and performance is not substantially different with or without vegetation so long as the flow is actually subsurface. However, gravel clogging generally causes surface flow and then vegetative materials become as important as in the soil types.

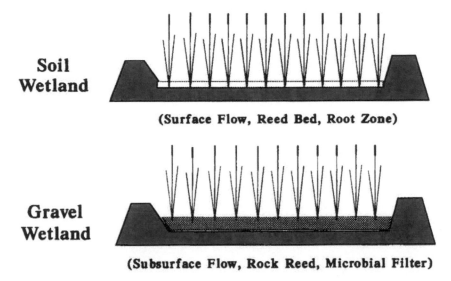

Soil
Wetland

(Surface Flow, Reed Bed, Root Zone)

Gravel
Wetland

(Subsurface Flow, Rock Reed, Microbial Filter)

Figure 15-3. The two major classes of constructed wetlands are differentiated by substrate media. Soil wetlands have loam soils whereas gravel or crushed rock is used in gravel wetlands.

With a few exceptions, only soil systems have been used for mine drainage, agricultural waste, urban stormwater, industrial wastewaters or other applications to date. Because many of the operating gravel systems have experienced serious clogging problems, only soil systems can be recommended for anything more than tertiary polishing of effluents with low concentrations of nutrients in small scale applications such as individual home septic tank systems. In addition, gravel systems have not been shown to reliably accomplish nutrient (nitrogen and phosphorus) removal.

SITING

Wetland site selection is generally controlled by the desire to provide gravity flow for wastewater to the system, between system components, and within each component of the system to eliminate costs and maintenance of pumping wastewaters (Figures 15-4 and 15-5). With a completely new treatment system, the only constraint may be the location of the main sewer line but the majority of constructed wetlands have been add-on systems to provide higher levels of treatment than existing package plants or lagoons. Site evaluation is equally important in either case but the add-on projects are generally constrained to inspecting adjacent or nearby lands.

SIZE AND CONFIGURATION

Though proponents of mass estimates or hydraulic loading lay claims for the merits of one versus the other method, current recommendations using either method tend to estimate similar wetland areas for specific influent-effluent

Figure 15-4.   A 30-ha constructed wetland designed into an area bounded by treatment lagoons on the right and an encircling railroad bed at Weyerhaueser's paper mill near Columbus, Mississippi.

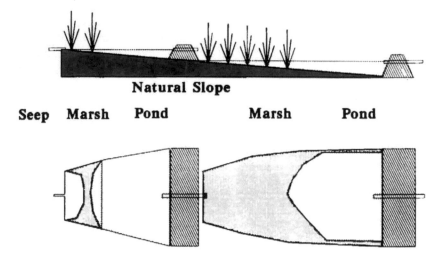

Figure 15-5.   Mine drainage wetlands often have a slight slope due to the nature of the terrain in which seeps are located. Initially, treatment occurs in the upper shallow portion of each cell but as masses of iron are deposited, the shallow zone moves farther and farther down the cell.

concentrations. That is not surprising, since both methods are applied conservatively at present because of our limited understanding of processes and mechanisms and because either is to some extent a manifestation of the same concept. However, the mass-treatment area method attempts to relate the quantity of contaminants to the surface area for attachment of microbial populations that decompose or alter those substances. The retention-hydraulic loading method assumes that microbial populations need a certain period of time to modify pollutants and estimates the storage volume required to constitute that minimum turn-over period. The latter was developed in conventional sewage treatment because, in any one region, municipal wastewater flows have similar pollutant concentrations.

However, comparisons between regions, for example, between the U.S. and Europe, become confusing since water usage, and subsequently, pollutant concentrations vary substantially. Average U.S. water use is approximately 400 L/person/day whereas the average in England is 200 to 250 L/person/day; and in central Europe, the average may be as low as 100 to 120 L/person/day. Since the total amount of organic waste (mass) generated by one person is similar in each instance, the concentration of organic waste (proportional amount in a given quantity of water) is quite different between the U.S. and central Europe. Comparing hydraulic loading rates (retention times) is unrealistic, but comparing organic mass loading provides a common basis for evaluating loading rates and system performance. Similar loading comparison problems arise because of widely varying concentrations in industrial, agricultural, or acid drainage wastewaters.

Since most wastewater treatment wetlands have been designed for minimum size and cost to provide the required level of pollutant removal, maximizing effective treatment area and reducing short-circuiting or unused treatment area results in rectangular shapes. Generally, inlet distribution and outlet collection piping is located completely across the upper or lower end of the cell and a rectangular shape theoretically enhances broad sheet flow across the width of the cell. Circular or elliptical shapes would likely have unused portions in each cell, but a V shape pointing downstream might provide an alternate design without creating short-circuiting. It may not be too effective for nitrogen removal, but should be tested for biochemical oxygen demand (BOD) and suspended solids (TSS) removal. A few projects have used linear systems contoured into the hillside because little level land was available and these tend to blend into the surroundings better than the standard rectangle. Generally, wastewater treatment wetlands should have a 3-4:1 length-to-width ratio and rectangular shape if minimal treatment area is available (Figure 15-6). However, other more aesthetic configurations would be practical if the wetland area was increased and ancillary benefits are also likely to increase with larger size.

Larger size has an added benefit in that contaminant loading will be less, loading fluctuations more readily accommodated, and greater diversity of plant and animal species possible, hence improving the water purification function as well as increasing other beneficial functions. Although planners tend to design the smallest wetlands with the fewest components, that is, reduce the treatment system to the bare essentials, for most wastewater treatment projects, treatment efficiency and system resiliency increase with added size and biological complexity. Small simple systems are vulnerable to upset from fluctuating

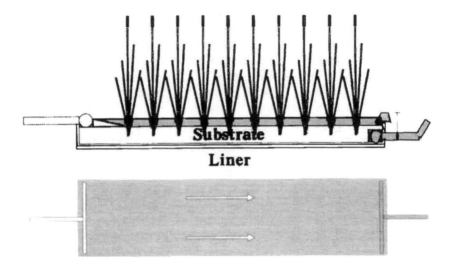

Figure 15-6. A typical rectangular wetlands cell with inlet and outlet distribution piping and an impermeable liner. Inlet and outlet piping should extend across the width of the cell to enhance sheet flow across the entire treatment area.

loading rates or pest outbreaks in important vegetation components. Too small and too simple will likely cause higher operating and maintenance costs and hamper system performance.

If only BOD$_5$, TSS and moderate pathogen removal is required, square cells or even aspect ratios (length:width) of <1 may be used. But if high removal efficiency is needed for BOD$_5$, TSS, pathogens and/or nutrients and metals, cell aspect ratios approximate 4:1, i.e., cell length is four times greater than cell width. Wide, short cells perform fairly well for BOD$_5$ and TSS removal but poorly for ammonia removal. Higher aspect ratios (>5:1) substantially increase construction costs without commensurate increase in removal efficiency. In addition, cells with high aspect ratios tend to be grossly overloaded in the upper portions, with adverse impacts on the vegetation and cell aesthetics, appropriately loaded through the middle portion and under-loaded in the lower section.

Each component of the wetland system is basically a shallow pond or lagoon. Design and construction techniques described in earlier chapters are appropriate for general features such as dikes, berms and typical flashboard/stoplog water control devices with two important exceptions. Dike freeboard must accommodate an organic matter (peat) accumulation rate of 2 to 3 cm/yr in the marshes and extend 30 cm above normal water level elevation for each ten years of projected operation. Adequate freeboard and water level control is also necessary to provide capacity for flow beneath expected thickness of ice cover in colder climates. If nutrient removal is a factor in the north, lagoon storage in winter and wetland operation in summer is often necessary since nitrification rates are very low below 4° C.

Bottom slopes for treatment marshes and ponds are essentially flat. Width slope for the marshes must be flat to ensure equal flow distribution of wastewaters. Length slope may not be >0.05 % in the marshes. Some earlier designs incorporated slopes of 0.5% and rarely as high as 4% primarily in gravel systems, in the belief that increasing the elevation difference between the upper and lower ends of the cell would provide the hydraulic head to force waters through persistently clogging gravel substrates. Most systems with >0.05% slope and an aspect ratio >1 have experienced severe problems in maintaining wetland vegetation. Not surprisingly since as little as 0.1% slope on a 300-m long cell creates a water level difference of 30 cm between the upper and lower ends and operators find it impossible to maintain the desired 4 to 10-cm water depths needed by most emergent wetland plants throughout the length of the cell. Attempts to establish 6 to 10 cm at the mid point mean that the upper end dries out and the lower end is too deep with wetland plant mortality occurring above and below the mid point.

SYSTEM COMPONENTS

Some of the most successful designs (marsh-pond-marsh) for municipal wastewater treatment consist of a number of distinct components sequentially located in each cell as follows:

Emergent Marsh: The first and third compartments are shallow basins with densely growing marsh vegetation — typically cattail (*Typha*), bulrush (*Scirpus validus* or *cyperinus*), reed (*Phragmites*) or rushes (*Juncus, Eleocharis*) — in 10 to 20 cm of water (Figure 15-7). The first marsh functions in BOD_5, suspended solids (TSS), metals, pathogens and complex organics removal as well as in ammonification. Initial operating water depth is 8 to 10 cm above the soil surface gradually increasing during the next 15 to 20 years to 15 to 20 cm as peat accumulates in the marsh.

Pond: The second compartment is a constructed pond with 0.75- to 1.5-m water depths similar to an aerobic lagoon/oxidation pond. Duckweed (*Lemna*) grows on the surface of the pond and various algae within the water column. Submerged pondweeds with linear, filiform leaves (*Potamogeton, Ceratophyllum, Elodea, Vallisneria*) are planted in shallow portions of the pond to increase microbial attachment surface area. The pond functions in further reduction of BOD_5 and most significantly for nitrification and phosphate removal. Operating depth is typically 0.9 to 1.3 m throughout the years of operation.

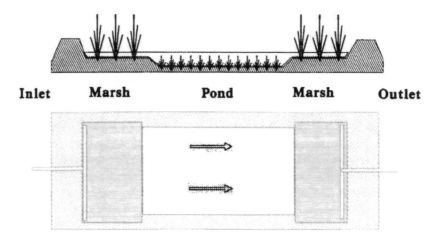

Inlet   Marsh   Pond   Marsh   Outlet

Figure 15-7. The marsh-pond-marsh design concept to enhance ammonia removal in constructed wetland treatment systems. Alternating zones of shallow water/emergent plants and deeper water/submergent plants provides the combination of environments required for nitrification/denitrification.

Emergent Marsh: The second marsh is physically and operationally the same as the first. The second marsh functions in BOD_5, suspended solids (primarily algae and *Lemna*), metals, pathogens and complex organics removal as well as in denitrification.

Comparatively the marshes function most efficiently for BOD_5, TSS and pathogen removal but the pond, because of the greater amount of oxidized environment, is more efficient at transforming ammonia to nitrate and precipitating phosphorus.

FLOW MANAGEMENT

Water flow control in constructed wetlands is critical to proper operation since achieving design performance is dependent on exposing incoming wastewaters to the maximum amount of effective treatment area within each cell. The objective is to thinly spread influent waters across all of the surface area, i.e., create sheet flow conditions throughout each cell. Consequently, design of suitable yet simple control devices is essential to proper operation and performance.

Influent distribution between cells is normally accomplished by splitter devices that may be as simple as a small concrete well with one pipe coming from the primary treatment system (perhaps a lagoon) and two or more pipes exiting the well (Figure 15-8). Simple "V"-notched weir plates installed at the exit orifice provide equal flow distribution to downstream cells but planners must insure that weir plates are installed at precisely the same elevations. If

Figure 15-8. The main inlet is on the left (upwelling) and outlets with "V" notched weir plates distribute flows to three wetlands. Note that flow is equally divided between two outlets with the third (right foreground) closed off.

desired, "V" notches with different angles — 30°, 45°, or 60° — allow differential flow application to individual cells and placing a solid plate (no "V" notch) over the outlet pipe will shut off the flow to any cell. Measuring flow depth in the notch and consulting tables in a hydraulics text provide means to estimate flow volumes to each cell. To avoid forming-on-site construction costs, small preformed, concrete septic tanks have been used for inexpensive wells.

Exit pipes from the splitter well join an inlet distribution pipe located across the width and in the upper most portion of each wetland cell. This distribution pipe functions to produce equal discharges along the length of the pipe and subsequently across the width of the cell. Although many spreader devices — level-lip spreaders, toothed weir plates, gated irrigation pipe — have been used, these are susceptible to problems from inaccurate placement and substrate settling. A contractor's elevation error of only 2 to 3 cm in placing the pipe and/or an equal amount of settling will significantly influence future flow distribution. And correction may be difficult if not impossible. Consequently a simple PVC pipe equipped with "T" discharge fixtures is most commonly used (Figure 15-9). In practice, little training or expertise is needed to visually evaluate different flows from individual discharge points and adjustment merely requires inserting a wooden board into the "T" and rotating the "T" fitting up or down to decrease or increase the discharge from each "T" fitting. The inlet distribution pipe with "T" fittings is placed upwards on the inside of the dike or strapped to support blocks to elevate it above the substrate and

Figure 15-9. Checking and if necessary, adjusting individual flows is facilitated with "T" fittings on the inlet distribution pipe to insure equal influent flows to each portion of the cell.

accommodate future accumulations of peat within the cell. It should be at least 25 to 40 cm above the surface of the substrate. In either case, a rock splash pad (6- to 10-cm diameter) is needed below each "T" fitting to prevent substrate erosion. Concrete should not be used since it precludes plant growth. Plant growth and litter/detritus accumulation at the discharge points is important so that waste-water flows immediately into the litter avoiding potential odor problems.

The effluent collector pipe consists of a slotted or perforated PVC pipe placed in a ditch with the top elevation of the pipe some 6 to 10 cm below grade, i.e., below the surface of the soil substrate and surrounded by crushed rock or gravel at the outlet end of each cell. Again the function of the collector is to encourage flow throughout the width of the cell so the collector should traverse most of the width. Generally, a discharge pipe connects to the collector near the midpoint but that location is not so important as long as the same elevation is maintained throughout the length of the collector pipe. The dis-charge pipe connects to a stoplog or "elbow" water control structure that provides for precise water level elevation control within the cell (Figure 15-10). Contractors occasionally place filter fabric around the collector pipe for pro-tection against soil and other materials during construction but this must be removed before placement of the rock since microbial growth will clog the fabric. Concrete pads or troughs should not be used since they prevent plant growth above and around the collector pipe. Plant growth and the rock sur-rounding the collector pipe provide a final filtering step to remove algae and other suspended solids.

Figure 15-10. The flashboard or stoplog control structure in the background provides reliable water level regulation despite considerable fluctuation in inflows and discharge flow rates are measured in the Parshall flume in the foreground.

PLANT MATERIALS

Wetland vegetation substantially increases the amount of habitats available for microbial populations in the water column, in the litter/humus layer and in the rhizosphere. Attributes of preferred plant species include:

1. adaptation to local climate and soils (native species);
2. tolerance to pollutants in the wastewater;
3. high biomass production;
4. perennial species;
5. rapid growth and colonization;
6. non-weedy, aesthetic habit; and,
7. values for wildlife habitat.

Commonly used species include:

Cattail	*Typha* spp.
Bulrush	*Scirpus* spp.
Rush	*Juncus, Cyperus, Fimbristylis, Eleocharis* spp.
Arrowhead	*Sagittaria* spp.
Iris	*I. versicolor, pseudacorus*
Plantain	*Alisma* spp.

Figure 15-11.   Pest outbreaks can devastate simple wetlands with only a few plant species.

Giant Reed	*Phragmites australis*
Submergents	*Potamogeton, Ceratophyllum, Najas*
Water Lilies	*Nymphaea, Nymphoides, Nelumbo*, spp.

Planners should include the broadest feasible mixture of plant species to maximize plant diversity and enhance stability of the constructed wetland. Single species systems are vulnerable to pest outbreaks and other perturbations and, depending on the species, could degrade performance (Figure 15-11). For example, broad-leaved plants (*Sagittaria, Alisma*) tend to drop their leaves simultaneously in the fall causing a substantial increase in organic loading and occasionally poorer overall performance (Figure 15-12). Stems and leaves of *Typha, Scirpus, Phragmites* and the rushes are shed throughout the growing season and persist after dieback in the fall. In fact, fallen stems frequently endure for 1-5 years undergoing slow, gradual decomposition that creates and maintains the porous litter, detritus layers important in treatment processes (Figure 15-13).

Planting sources and methods are the same as discussed in Chapters 12 and 13. It is especially important to gradually introduce wastewaters and allow the biotic community to gradually adapt to poor quality water. This is not only true for emergents but flooding the new plantings with turbid waters or waters with high organic loading and/or low dissolved oxygen will stress and perhaps cause mortality of submergents. Floating species or those with floating leaves can survive in poor waters so long as other conditions are suitable.

Figure 15-12. Broad-leaved species such as arrowhead add to diversity and aesthetics but in the fall, they tend to drop all their leaves simultaneously adding measurably to the organic loading on a constructed wetland.

Figure 15-13. *Phragmites* is an aggressive, weedy species that dominates most wet environments in Europe and consequently, is the most commonly used species in European constructed wetlands.

CONSTRUCTION

Construction is similar to any other wetland but following contract specification as closely as possible becomes more critical. Grading must meet the
specifications in the plan within described tolerances to achieve proper functioning of the system. Out-of-tolerance lateral bed slopes may not only cause
ponding but are likely to cause channeling or short-circuiting that reduces the
effective treatment area in the cell and depresses performance. Similarly,
improper grading along the cell length could make it impossible to set and
maintain proper water depths as well as causing channeling and short-circuiting. Obviously inability to manage water depths would severely retard establishing or managing the desired plant community and retard functioning of
the constructed wetlands.

Meeting permeability specifications is important in all wetlands but even
more so in constructed wetlands since groundwater protection is much more
important. Be careful that contractors do not excavate deeper than planned
and penetrate an impermeable layer into a permeable layer. Permeability
testing should be agreed upon prior to construction and frequent testing is
necessary during construction. Compacting *in situ* or fill material must be done
with proper equipment and only when moisture conditions are satisfactory.
Bentonite or soda ash blankets or synthetic liners must be carefully installed
following specifications in the designs to insure proper functioning.

OPERATION

Operate the wetland with clean water or very low-strength wastewater for
the first month after planting. During the fifth week, initiate operation with
one-half strength wastewater or with one-half the design flows and continue
for three months. After the end of the fourth month, begin operation with
full-strength wastewater or with full design flows. Check proper operation of
all piping, pumps, and water control structures and monitor vegetation throughout the start-up period.

Normal operating water depth should be 10 to 20 cm in each marsh component and 0.9 to 1.2 m in the pond. Proper water depth and careful regulation
is the most critical factor for plant survival during the first year after planting.
It is absolutely essential that stems and leaves of desirable species project well
above the water's surface with poor quality water to avoid drowning new or
even older established plants.

This is an important reason for designing systems with little or no slope
on the substrate and easily maintained water control structures that precisely
regulate elevations. Water level management must create very similar water
depths and/or duration of flooding throughout the newly planted area and must
precisely maintain that level despite fluctuating inflows.

After plantings have become well established, operation and maintenance typically consists of driving around the dikes at least once each week to check for any erosion or seepage or animal damage, mowing the dikes to improve visitor access and aesthetics and collecting water quality samples as needed. Routine weekly inspections are necessary to ensure appropriate flows through the inlet distributor and outlet collector piping. Vegetation health and vigor, water levels in each component and all piping should be checked and adjusted as necessary. *Scirpus* and some other species may fail to die-back the first fall and tops may even remain green throughout the first winter in constructed wetlands. But they then resume normal fall die-back the second season. Neither is cause for concern. Apparently the warmer wastewaters confuse the new plantings the first year but they are able to adjust and return to normal phenologies the second year.

Dikes and flow control structures should be inspected for leaks and corrective action implemented. Flow distribution within cells should be occasionally inspected to detect channel formation and short-circuiting and corrected by planting vegetation or adding soil in any channels. Check on the grass and wetlands vegetation at least once a week. Gardeners can identify any visible signs of stress or disease on the grass — yellowing, chlorosis, leaf damage, etc. — the same signs would indicate that the cattail or lotus are having problems. Maintenance workers should also check the pumps, valves, "T" fittings, etc., at least one each week to ensure that pumps and all piping are operating properly, i.e., check for clogging and make sure that the flow coming out of each "T" fitting is the same in each cell.

MONITORING

GENERAL

Determining the performance of a constructed wetlands is dependent upon accurate determinations of the effluent volume of wastewater and the concentrations of the various pollutants and comparing those values with influent parameters and/or regulatory permit limits. To accurately monitor performance of the treatment system, operators must be able to measure inflows and outflows and contaminant concentrations in each. Since system performance is evaluated on the basis of removal/transformation of pollutants, influent and effluent monitoring provides the basic data for comparison.

VOLUMES

Flows can be measured with simple "V" notched weirs or Parshall flumes that are read by someone frequently or with automated devices that are visited weekly or monthly. A number of standard flow measuring devices are available and should be employed to insure accurate flow measurements. In the worst case, designers may be able to use a per capita average for that region or economic conditions for wastewater produced or a per capita average for that region or economic situation for potable water use. However, neither of these will include numerous variables and/or sources (stormwater runoff, industrial wastewaters, small business wastewaters, etc.) that may contribute considerable quantities of water to the total volume requiring treatment.

WATER QUALITY PARAMETERS

Water quality parameters (basically the pollutants) that need to be sampled and analyzed for many situations are similar since we likely are interested in treating or removing the BOD, TSS, bacteria and other pathogens, and perhaps nutrients (N & P) in most wastewaters deriving from organic sources. This includes domestic wastes, food processing wastes, municipal waste, and most agricultural waste. Industrial wastewaters will require the same parameters plus analyses for complex organics, salts, metals, etc. depending upon the manufacturing process or products. In certain situations, true and apparent color measurements may be needed, i.e., pulp/paper mills, distilleries, molasses plants, etc.

One of the most reliable automated sampling devices is produced by ISCO in Nebraska (Appendix C). A complete unit capable of operating in remote regions via solar power, ranges from $3500 to $4000 in the U.S. and the equipment is very simple to use, i.e., user friendly, durable, portable, and reliable. With a solar panel and battery, it can operate at remote locations with the only service or maintenance consisting of someone picking up the water samples on the pre-programmed sampling frequency. Automated samplers are water sample collection devices — not water quality analyzing. They only collect the samples at specified intervals or flow levels, then someone comes by and picks up the samples collected over the last programmed sampling period and takes the samples to a lab for analysis. An important advantage is that the automated sampler will collect a series of samples over a specific period, for example, one sample per hour for a 24-hour period which gives us a composite sample for the entire 24-hour period, or one sample each day at a specific hour of the day for a week or month, etc.

Sampling frequency will also vary for each type of wastewater or source but should be designed to be compatible with flow/volume samples. Normally a minimum would be collecting a composite 24-hour sample on a week day once or twice each month and a grab sample each time the composites are picked up.

BIOTA

In addition to water quality parameters, monitoring biological components can provide valuable information on the health and vigor and consequently, successful operation of the constructed wetland. Monitoring major biological components only requires weekly walking or driving around all dikes in the system inspecting for stress signs or damage to the wetland vegetation —

chlorosis, poor growth, stunted, off-color, insect damage, etc. Concurrently, operators should casually inventory bird, frog and other large animal populations to detect any significant changes that could suggest problems in the system.

A CASE HISTORY

The Minot, North Dakota constructed wetland system provides polishing for wastewaters from some 43,000 residents. Minot previously used a 5-cell facultative lagoon system (278 ha) but odor problems and the fact that Minot's discharge represents most of the flow in the Souris River during dry periods, led to incorporation of low (1 mg/l) NH_3 parameters in the discharge permit. System upgrading included sludge removal and installing aerators in the first lagoons and construction of a 4-cell marsh-pond-marsh wetland with 51.2 ha of effective treatment area (wetted surface). I patterned the wetland design after an earlier livestock wastewater treatment wetland we built in northern Mississippi. The lagoons were dredged out, aerators installed in two lagoons and the wetland was built in late fall of 1990 at a cost of $1.6 million or $0.23 per gallon per day of capacity. Operation began in the summer of 1991.

Minot's northerly location (48° 15'N) causes subfreezing air temperatures from November through March and ice cover on water bodies often approaches thicknesses of 1 m. Consequently, system design included 180 days storage in existing lagoons with flows to the wetland of 22,000 m³/day during the 180-day growing season. Design considerations included anticipated influent levels of 20 mg/l for BOD_5, 20 mg/l for TSS, 7 mg/l for NH_3, 30 mg/l for TKN, minimum influent levels of dissolved oxygen > 3 mg/l. Desired discharge level for NH_3 was < 1.0 mg/l. NH_3 was the critical design parameter and loading rates were established at 3 kg/ha/day NH_3 though occasional peak loadings of BOD_5 and TSS needed to be accommodated.

Highest removal efficiencies for NH_3, BOD_5, and TSS were anticipated with alternating zones of dense emergent vegetation (i.e. *Typha* and *Scirpus*) in shallow water and submergent vegetation (*Potamogeton, Vallisneria*) in the deep sections. Creating wildlife habitat to provide opportunities for outdoor recreation, environmental education, and improve public acceptance of the wastewater treatment system was also desirable.

The design included four cells 177-m wide and 747-m long (length to width ratios of 4.2:1) with marsh-pond-marsh-pond-marsh zones sequentially within each cell (Figure 15-14). Marsh zone A was designed to function for BOD_5 and TSS removal and ammonification. Design operating depth was 15 cm in the 100-m long zone and it was planted in *Typha latifolia*. Pond zone B functioned for nitrification, had an operating depth of 60 cm in the 221-m long zone and was planted with *Potamogeton pectinatus* and *Vallisneria americana*. Marsh zone C functioned for nitrification/denitrification and nutrient removal with an operating depth of 30 cm in the 122-m long segment and was

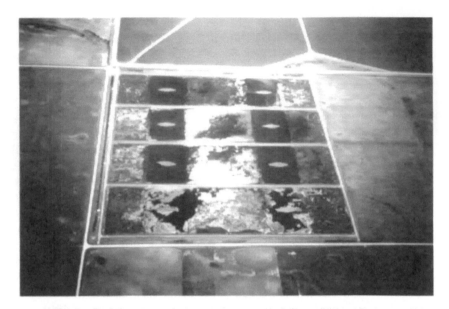

Figure 15-14. The Minot constructed wetlands system includes multiple shallow water/emergent plants and deeper water/submergent plants to provide high removal rates for NH, and fecal coliform bacteria.

planted with *Scirpus validus* and *Lemna*. Pond zone D functioned for nitrification/denitrification, nutrient removal and to increase dissolved oxygen with operating depths of 60 cm in the 221-m long zone planted with *Potamogeton pectinatus* and *Vallisneria americana*. The final segment, marsh zone E, functioned for denitrification, TSS (algae, *Lemna*) and fecal coliform removal in the 100-m long segment with an operating depth of 15 cm planted in *Typha latifolia*. Small islands were included in pond zones B and D for spoil disposal and wildlife nesting/loafing sites. Transition zones between zones A and B and zones D and E have 6:1 slopes with planted *Sagittaria latifolia*.

Monthly flows to the system have averaged 15,655 m³/day but ranged from 1896 to 28,762 m³/day with the flow equally distributed to all four cells in 1991, all flow applied to only Cell number 4 to foster vegetation growth in the other cells from April to June 1992, applied to Cells 1 and 4 in April and May 1993 and subsequently through all four cells through the end of 1994.

Average influent levels for BOD, have been close to expected (12.8 mg/l), but TSS levels have been higher (36.8 mg/l) and NH, levels have been lower (2.1 mg/l). Discharge values have been excellent: BOD, — 7.9 mg/l, TSS — 17.1 mg/l, NH, — 1.0 mg/l — (but 0.48 mg/l excluding November and December values when water temperatures were 0.5–1.0°C) and coliforms 23.6 CFU/100 ml <u>without chlorination</u>. Highest discharge NH, levels (5.33 mg/l) occurred during cold weather when water temperatures approached freezing or when all flow was applied to only one cell and the loading rate increased to 8.2 kg/ha/d NH, (Figure 15-15). Average loading of 1.6 kg/ha/d

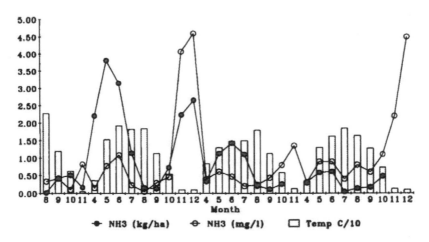

Figure 15-15. Ammonia removals have been excellent during normal operating periods but nitrification rates are very low in cold waters and discharge levels increased during the coldest months.

NH_3 has been well within design limits. Effluent NH_3 concentrations were not related to hydraulic, BOD_5, or TSS loading rates but closely followed NH_3 loadings.

The Minot wetlands system has performed very well for normal wastewater discharge parameters. It has also provided excellent wetland wildlife habitat and cooperative efforts by the U.S. Fish & Wildlife Service installed nesting platforms and other habitat amenities. Alternating depths and different vegetation types that optimize nitrogen removal also provide excellent combinations of shelter and feeding areas for many different kinds of wetland and upland wildlife. The design has also accommodated additional wildlife inputs of fecal coliforms without chlorination.

SUMMARY AND COMMENTS

Much can be learned from both natural wetlands receiving wastewaters and from constructed systems with a few years of operating history. Despite some attempts to reduce wetland treatment systems to minimal components and treatment areas with seemingly most efficient combinations of substrate, vegetation and loading rates, most successful systems are indistinguishable on casual examination from natural marshes. In fact, poorly performing systems that I have visited did not appear to be viable marsh ecosystems. Generally the absence of an important component, attribute or characteristic was obvious to anyone with experience in natural marsh ecosystems. Conversely, successful systems are often quite similar to a natural marsh and it's beginning to appear that the basis for design of wetlands constructed for wastewater treatment should be to simulate the structure and functions of a natural marsh ecosystem.

Because constructed wetlands are open, outdoor systems they receive inputs of animal and plant life from adjacent areas and from distant sites and over time are likely to become more and more similar to natural wetlands in a region. Though we may design and build a system with a specific substrate and only one or two plant species currently thought to be highly efficient, over time, many types of plants and animals will take up residence. Consequently, a constructed wetlands is likely to become more similar to a natural wetlands as the system matures and ages. To prevent these invasions and attempt to maintain a monoculture would be difficult and costly and may be self-defeating. Living organisms that become established in an operating system are not likely to detrimentally impact treatment efficiency and may very well improve system operation.

If we examine the history of the development of constructed wetlands for wastewater treatment, it is obvious that astute observations and careful monitoring of natural wetlands receiving polluted waters was the basis for early proposals for the use of wetlands in wastewater treatment. Subsequently, Max Small (marsh-pond-meadow) and others designed constructed wetlands patterned closely after natural systems. However, as always, we thought we could improve upon Mother Nature, and some designs began to appear that were dependent upon a highly conductive substrate, the engineered soil or gravel bed types. But as we discovered clogging problems with horizontal flow, gravel based systems, modifications included applying the wastewater to the upper surface in a vertical flow design with batch loading. Later, recirculating was added and some have even included forced aeration with or without a greenhouse enclosing the entire system. Unfortunately, most "advances" since the early marsh-pond-meadow design have failed to achieve the original objective of developing a simple, low-cost, low-maintenance, and effective wastewater treatment system. The latest designs (marsh-pond-marsh) are quite simply an attempt to return to the first developments that patterned constructed wetlands after natural wetlands. In fact, one might observe that we have come full circle in developing this technology (Figure 15-16). We started out copying natural wetlands, increased complexity in construction and operation but failed to attain significantly better efficiencies, and recently returned to the original concept that attempted to emulate natural wetlands. The "newest" but in fact "old" concepts have proven to be the least costly to build, have higher removal efficiencies for a wider variety of pollutants, are less costly and complex to operate, and also provide substantial ancillary benefits.

In retrospect, "improvements" from the earliest designs that emulated natural wetlands, deviated from the original objectives which were to provide a wastewater treatment system with:

1. low construction cost (capital costs);
2. self maintaining attributes (little time, expense or operating skill); and,
3. high pollutant transformation capabilities.

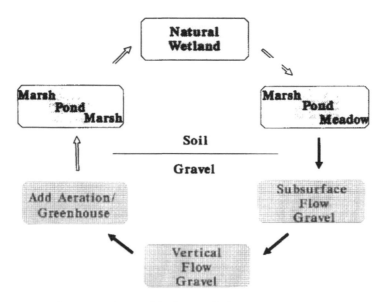

Figure 15-16.   The circular history of development in the constructed wetland wastewater treatment technology progressing from systems that emulate natural wetlands through complex, artificial designs and eventually returning to designs that simulate natural wetlands.

With increasing design complexity to improve treatment efficiency (and perhaps reduce required treatment areas), our designs added substantial construction and operating costs and complexity such that some recent types with recirculation, vertical flow in a batch loading mode and in a few cases, complete enclosures, have become as costly to build and operate as conventional treatment systems. Operational complexity has increased concurrently and many of these complex systems are conceptually and financially similar to trickling filters or other conventional wastewater treatment systems. The original objectives of low-cost, efficient, self-maintaining systems have been lost in attempts to improve removal efficiencies. In addition, simple designs are most amenable to applications in the developing nations and simple constructed wetlands treatment systems built and operated with local skills and materials may provide the means whereby many developing nations can reasonably address their water pollution problems.

CHAPTER 16

OPERATION AND MAINTENANCE

INTRODUCTION

Natural ecosystems, including wetlands, are largely self-maintaining; that is, the complex of animals, plants, soil, water, and air that make up the total system perpetuates itself through time with only minor if any changes occurring as the result of disturbances or perturbations. The old-growth forest continues as an old-growth forest for hundreds and even thousands of years as do prairies, deserts, tundras, and many other natural systems. Over time, the animal and plant components have adapted to and modified the abiotic components (parent soil materials, hydrology, climate, atmosphere, and radiation) slowly establishing a complex of interacting components that is resistant to change. If a major disturbance impacts the system (e.g., a hot fire), thereby eliminating major structural components, the remnants begin the process anew and, through a series of successional stages, restore a very similar if not identical system.

This process of plants and animals adapting to and modifying the abiotic components and, in so doing, establishing increasingly complex systems with the more complex stage gradually replacing its less complex predecessor, is known as ecological succession. The final, self-perpetuating combination of living and nonliving components is referred to as the climax stage or condition, and each interim state is known as a successional stage. Increasing complexity within each stage is generated by increasing diversity; that is, greater numbers of different species and fewer individual members of each species, resulting in rapidly increasing numbers of connections or pathways between species. A direct result of increasing interactions through more and different connections is the stability of the overall system or the inherent ability to resist modification or change despite impacts from outside factors.

Most types of wetlands are not climax stages but, in fact, represent successional states. Largely because of our relatively short life span, we casually observe a cattail marsh over 50 to 60 years and tend to think of it as a stable, self-perpetuating system — the marsh is still a marsh. However, depending on the region, closer examination would reveal subtle yet irreversible changes on-going in the marsh all the while (Figure 16-1). In the absence of major impacts or disturbance, species of plants and animals and numbers of individuals in each species change from year to year, and the overall pattern of change is predetermined and predictable. Basically, the depression gradually fills,

Figure 16-1. Wetland plants are capable of surviving relatively dry conditions without undue stress although a prolonged dry spell would allow invasion of terrestrial species.

water depth decreases, and plant and animal communities change from those adapted to deep water through those of shallow water and eventually to those that thrive in dry, terrestrial environments. In the pothole region of the Dakotas and adjacent provinces, the terrestrial plants are grasses and forbs and animals adapted to living on these plants — bison, elk, antelope, prairie dogs, ground squirrels, and mice. The muskrats, ducks, fish, and wading birds that lived directly or indirectly on marsh plants are gone. The marsh has become a prairie. The new stage is, in this case, the climax under existing climatic conditions and the prairie will perpetuate itself until major climate changes occur. In more humid climates, the marsh may gradually become a bog, perhaps a swamp, and eventually a forest. Most wetland systems of interest to us are merely an ephemeral, transitional state if viewed in the total context of ecological succession.

Viewed from the perspective of geological time, the overall process of gradually succeeding stages culminating in the climax form in either case is predictable and irreversible; but examined from a short-term human viewpoint, the marsh or swamp persists for many years and appears to be a self-perpetuating stable system. Without an understanding of the changes that have gone on before and the changes that will inevitably occur in the future, we perceive the marsh (or any other wetland system) as a permanent feature of the landscape, an enduring combination of living and nonliving elements interacting to perpetuate the whole. Managing a new wetland system with the misconception that it is an unchanging, stable, or permanent system will at least cause needless concern for the managers and, at worst, lead to inappropriate management changes and excessive costs. Even within the short human time frame, considerable change will and should occur. The astute manager understands the inevitability of change and the need for appropriate disturbance and interrupts, and reverses or influences the manner of change to accomplish project goals and the functional benefits sought from the new wetland.

To this point we have caused major changes in the hydrology of our site to create conditions suitable for wetland plants and animals, and we have introduced selected members of those communities. The new wetland is presently in a very early stage of succession and now our management objective is to foster those changes that will increase the diversity and complexity of the wetland, yet prevent or reverse changes that will cause replacement of the wetland by a more advanced successional stage. We hope to shape the new system so that it develops the structure and function of a mature, stable, self-maintaining system, within our perspective, that we visualized in formulating our earliest plans. Doubtless, those plans were rooted in observations of a nearby wetland or perhaps engendered by impressions from literature, films, or talks with others; but the model system had adapted and changed through many years until reaching the state that we now view. Certainly, we should not expect the newly created or restored system to immediately duplicate the structure and function of model wetlands. Unfortunately, many do. Many are disappointed that the new wetland is not identical to the reference wetland in the first year or two! Others have concluded that the project failed since the 4- to 5-year-old, created wetland only has 70% or 80% of the plant species or animal species in a reference natural wetland. What will it have after 10 years? Or 20 years? Most likely the created and natural will become indistinguishable. Many projects have "failed" because the site was graded, flooded, and planted, and then left to its own resources, and the diversity and complexity of a well-developed system were still absent 5 or 6 years later.

In the first place, it is grossly unrealistic to expect to create even the simplest type of natural wetland system in that short time; secondly, deserting a new wetland and expecting it to suddenly become a fully developed system reveals a limited understanding of the basic factors that create and maintain natural wetlands. If the hydrology is correct, many "failures" may well become example wetlands, but development could require tens or hundreds of years. Had the developers actively manipulated the single most important factor (hydrology) they could have substantially increased system development and made good progress towards creating a complex, diverse, and relatively stable system in only 2 or 3 years. Active management of the new system is needed to expedite progressive development to the desired successional stage and then arrest further development through simple but critical manipulations of driving factors, principally hydrology.

FIRST YEAR

Directional management in the form of water level manipulations is crucial during the first growing season, irrespective of the type of wetland. After planting is completed in the spring, the emergent vegetation area of the pool should be inundated with 2 to 3 cm of water for 5 to 7 days to inhibit germination and/or retard growth of opportunistic terrestrial species. This level will saturate but not inundate the soils in much of the transition zone, and depths in deeper areas (the submergent/floating leaved zone) will range from 5 to 150 cm, depending on topography. Water levels are then lowered to at or just below the surface of the emergent zone substrate and maintained at that elevation for 15 to 20 days. By this time, evidence of renewed growth of planted materials should be apparent throughout the transitional, shallow, mid, and deep zones in the system; emergents can be expected to have produced shoot lengths of 5 to 10 cm. If soils from other wetlands were introduced, seeds of many additional species will germinate, creating a green carpet throughout the zone. The water level is then raised to flood the emergent zone slightly deeper, (3 to 5 cm for another 5 to 7 days) and then lowered for 15 to 20 days as before. If significant mortality of planted material has occurred, leaving large unvegetated regions, levels should be lowered to the substrate elevation or just below it and new plants installed as needed during this period (Figure 16-2).

Figure 16-2. Turtle tracks highlight the drought impacting this wetland. Although many wildlife species will be temporarily displaced, most will profit from the rejuvenated productivity following reflooding.

By midsummer, most emergents should have stems in excess of 30 cm, seed bank species should be above 15 cm, and submergents/floating-leaved species proliferating across the deeper zone. However, the transitional zone likely contains a substantial number of and perhaps high proportion of terrestrial species that are beginning to outcompete the desired wetland plants. In addition, the emergent zone may now contain a significant proportion of transitional or moist site species that became established during the second drawdown period. Depending on stem height of wetland plants that came in with soil materials, water levels should be raised to 10 cm or higher (15 cm if it will not overtop desired plants) above the emergent's substrate for 3 to 5 days, increased to 15 cm for 3 to 5 days, and eventually to 25 cm. During this gradual raising, care must be taken to insure that the water surface does not overtop desired species. The objective is to gradually raise the level, following increases in stem height of desirable species without overtopping the latter.

If possible, gradually increase the depth to between 30 and 50 cm in the emergent zone and maintain that level for 10 to 20 days. Regardless of the final elevation, carefully monitor plants in the shallow to mid zones (basically, the emergents) for signs of stress, indicated by reduced growth rate or in severe cases, yellowing or chlorosis. If stress becomes apparent, immediately reduce water levels to 2 to 3 cm over the emergent substrates. Any additional plantings that were added during the drawdown will be more susceptible to stress and should be carefully watched during these flooding periods. The goal in raising levels is to inundate as much of the transition zone as possible, flood and kill terrestrial species in the transition and emergent zones, yet not overstress and retard growth in the emergent zone. Submerged and floating leaved species also need to be carefully monitored, though most will have little difficulty with deeper waters unless high turbidity or low dissolved oxygen are potential factors.

After 20 days, reduce levels to 5 to 7 cm over the emergent substrates and maintain that elevation until late summer. Water levels should be gradually increased to 10 to 15 cm over emergent substrates, well before the onset of cold weather and higher if thick ice cover is expected. During winter, water depths should be managed to maintain free water below the ice cover or at least avoid freezing the substrate and damaging roots and rhizomes. However, the water or ice surface should not overtop stems of emergent species or oxygen supplies will be cut off and substantial mortality may occur. Do not dewater the system at anytime during freezing temperatures unless you are attempting to reduce the vegetative cover or eliminate weedy, emergent species.

For wooded wetlands, water level manipulations should follow a similar pattern, though depths and duration of flooding must be less. During late spring and early summer, flood the area to depths of 2 to 3 cm for 2 to 3 days, followed by 10 to 15 days of drawdown, and repeat the cycle. In late summer, flood the new wetland to 3 to 5 cm for 5 to 6 days and then dewater for the remainder

Figure 16-3. Most forested wetlands thrive on winter flooding but will only tolerate short periods of inundation during the growing season.

of the summer. After leaf fall, gradually raise water levels to 2 to 3 cm for 5 days, dewater for 10 days, followed by 3 to 5 cm for 5 days, dewater for 10 days, and then maintain 10-cm depths for 5 days, alternating with 20-day drawdowns for the remainder of the winter. Be careful of lengthy drawdowns during freezing temperatures or the shallow root system of newly planted shrubs and trees may be damaged. Again, the goal is to create unfavorable conditions for terrestrial species without overstressing wetland plants by using the principle that flooding conditions are not optimal for wetland trees or shrubs but they can tolerate inundation better than terrestrial types. Depending on the region of the country, normal winter rains may well create the winter flooding/drying conditions that favor wetland species and simplify or confound management efforts (Figure 16-3).

SECOND YEAR

During the first warm spells in the spring of the second year, raise the water level to flood most of the transition zone and maintain that level until warm weather is firmly in place and new growth started (may require close examination) on last year's plantings; then lower water levels to at or immediately below the surface of emergent substrates for 5 to 10 days, raise levels to 1 to 2 cm over the substrate for 2 to 3 days, and lower them again to the surface for 5 to 10 days. The goal is to create warm, moist mud flats to enhance germination and growth of any wetland species brought in with soil materials during planting. As new growth of these plants becomes evident, slowly raise water levels to 1 to 2 cm over the substrate and then gradually higher until depths in emergent zones are 8 to 10 cm. Do not overtop stems of any desirable new growth during this period. With one exception, maintain 8 to 10 cm depths for the rest of the summer. In late summer when emergents have adequate stem height and growth is slowing, increase water levels in the pool to flood at least the lower half of the transition zone to depths of 2 to 4 cm and maintain that level for 5 to 7 days, and then return to normal operating elevations.

SUBSEQUENT YEARS

Fall and winter operation is similar to the previous year. During the third summer, manipulate water levels as in the second year. In the fourth spring, flood the transition zone early, lower water levels to 2 to 4 cm above emergent substrates for 10 to 20 days, and then raise levels to normal operating levels (8 to 10 cm) for the remainder of the summer. Fall and winter operation is the same.

Procedures for the fourth year become normal operations for subsequent years if the project is destined to establish a marsh. Sequentially raising and lowering levels during the first few years fosters invasion and/or growth of additional plants introduced with soils or carried in by wind, birds, or other seed dispersal agents. If natural wetland soils were incorporated or if natural wetlands are nearby, the new system should have 40 to 60 species of wetland plants by the end of the third growing season. Do not be alarmed if one species proliferates rapidly one year and then fails to appear the next. This commonly happens with the smaller *Eleocharis* spp., many floating types, and of course, rooted or nonrooted algae. Wild rice (*Zizania aquatica*) is notoriously fickle, but in fact seems to vary with those all-important warm, moist, mudflat germinating conditions in early spring. Algae blooms wax and wane with dissolved nutrient conditions. If they become a problem, dewater during summer to flush the system, maintain moist but not flooded soils, and reflood in the fall. After deep flooding in early spring, dewater as quickly as possible to flush out redissolved nutrients and limit future nutrient inputs.

More commonly, weedy species (*Typha, Phragmites, Salix,* or *Populus*) will proliferate and soon dominate the system, crowding out other species and reducing system diversity. If extensive areas need control, cut or mow stems of *Typha* and *Phragmites* at the soil surface during a drawdown in late summer and then flood as deeply as practical over the winter. In cold regions, fall/winter drainage followed by mowing or burning and deep reflooding through the spring months may accelerate depletion of the carbohydrate reserves. Reserves are lowest during the fruiting period and mowing followed by deep flooding for the remainder of the year will reduce the plant's ability to send new shoots through the water column to air. Complete control may require treatment over two or more years. As an alternative, consider blasting potholes in dense emergent stands with the ammonium nitrate-fuel oil mixture used by waterfowl managers for years. This inexpensive mixture is now commercially available,

Figure 16-4. Two herbicide applicators commonly used by cranberry farmers for treating individual plants (weeds).

but it is a high explosive and should only be used by competent personnel in safe areas (i.e., not in an environmental education wetland surrounded by housing developments or commercial businesses!). Stems or trunks of *Salix*, *Populus*, *Melaleuca*, or *Cephalanthus* need to be cut and the stumps painted with an herbicide labeled for near or in water use (Rodeo); again, more than one application may be necessary.

Purple loosestrife (*Lythrum salicaria*) is a native of Europe that was introduced during early settlement and has spread into most wetland habitats north of the 35th parallel. It is a highly aggressive, weedy species able to colonize and dominate the transitional and shallow to mid zones in virtually any marsh and even open-crowned wooded wetlands (Figure 16-4). Moderate control is possible with flooding to overtop the stems for 5 weeks or longer during the growing season, but complete control will likely require multiple applications of a broad-leaved herbicide (glyphosate or Rodeo). Recent experimental releases of a root-mining weevil (*Hylobius transversovittatus*) and leaf-feeding beetles (*Galerucella* spp.) from Europe have shown promise as biological control methods. Positive experimental results have also been obtained with Triclopyr (triethylamine). Drawdown and mudflat exposure during the first half of the growing season in an infested area will produce an explosive growth of purple loosestrife. Seed collections from some commercial sources contained purple loosestrife only a few years ago; so if commercial seed is used, specifications to exclude purple loosestrife need to be included with the order and shipments should be carefully screened before planting.

In more southerly regions, water hyacinth (*Eichhornia crassipes*), Brazilian elodea (*Egeria densa*), Parrot's-feather (*Myriophyllum brasiliense*), Eurasian watermilfoil (*Myriophyllum spicatum*), hydrilla (*Hydrilla verticillata*), Australian pine (*Casuarina equisetifolia*), salt cedar (*Tamarix* spp.), *Melaleuca* and other introduced exotics have become serious pests in natural wetlands. Complete control of most is only possible with repeated, proper application of recommended herbicides by weed control specialists but experiments are underway using a weevil (*Oxyops vitiosa*) for biological control of *Melaleuca*.

Nutria are widely known for damaging, in some cases severly, marsh vegetation and muskrats occasionally cause similar problems. Other potential depredation problems include beaver, deer, rabbit and mouse damage to tree seedlings such that protective covers (commercially available) may be necessary. Above ground, course growths of older cattail plants provide little food value to any vertebrates but tender new shoots with high protein contents have been devastated by Canada goose flocks.

Depending on regional soil fertility and nutrients carried in by inflows, simply manipulating water levels at the appropriate times may sustain a complex, diverse and productive marsh for many years. However, in most circumstances, available nutrients become bound in reduced conditions in the substrates and basic productivity declines. First indications are typically the proliferation of water crowfoot (*Ranunculus* spp.) and relatively clear water, followed by the appearance of bladderwort (*Utricularia* spp.) as nutrient supplies become limiting. The marsh should be dewatered and dried out as much as possible during one growing season. If purple loosestrife is a concern, delay the drawdown until late summer. If not, and project goals are life support, recreation, or education, do not despair over the lost season. Dewatering in early summer will cause an explosive growth of transitional species (*Polygonum, Bidens, Eleocharis,* and *Echinochloa*) in the dewatered regions that produce large quantities of seed used by many types of wildlife. Invertebrate populations also respond to nutrients from decomposing vegetation and those released by oxidizing the substrates, and the combination will attract spectacular populations of wildlife when the system is reflooded in the fall. Nutrients released from plant and animal decomposition and from substrate "sinks" are available for use by all segments of the system the following spring, supporting dramatic rebounds in productivity.

BOGS AND FENS

Created or restored bogs and fens must be managed quite differently, avoiding extreme hydrologic disturbances that benefit marshes or seasonal flooding needed in swamps. In either a bog or fen — the determining factor appears to be the pH of the water — water levels should be similarly fluctuated during the first and second growing season, except depths must be much less. Mosses are just as readily stressed or killed by prolonged flooding during the growing season, as are emergent plants; since most mosses have heights of less than 10 cm, flooding depths should be correspondingly lower. In addition, prolonged flooding during the growing season will be detrimental to introduced shrub and tree species. Finally, low water levels must not expose moss root structures or desiccation will cause stress and mortality.

Similarly manipulating water levels during the first two years will foster establishment of a variety of species, some of which will later die out but some will remain, enhancing the overall diversity of the system. After the second winter, water levels should be lowered to and maintained at normal operating elevations for the duration of the project. In contrast to marshes, bogs require low nutrient, acidic but stable, constant-elevation waters. Fluctuating water levels, increased nutrient contents, and/or decreased hydrogen ion concentrations will favor marsh species, shifting the bog towards a marsh. For example, a wastewater treatment project that discharged highly-treated but nutrient-rich waters into Houghton Bog in Michigan has caused predictable vegetational changes with extensive stands of cattail now present. Opposite conditions in these parameters reverses the trend and a marsh will gradually take on the characteristics of a bog. With adequate water control structures, maintaining stable water levels is not difficult, but limiting nutrient inputs from runoff may require pretreatment in a small marsh upstream from the bog.

Fens occur in similar habitats as, and occasionally surround bogs. They are outwardly more similar to marshes, but waters in most fens are very low in nutrients and relatively stable. Many fens appear to be located in groundwater discharge areas; hence, the low nutrient and hydrogen ion concentrations. They share other characteristics with marshes in that dominant vegetation is an emergent form (typically *Carex* spp.) and pH of the water is neutral to slightly alkaline. Low (2 to 5 cm), stable water levels with low nutrient content and moderate alkalinity will enhance growing conditions for fen vegetation. Limiting nutrient input from runoff is an important management principle and adding lime to decrease acidity may be required, depending upon pH of influent waters and precipitation. Since both could become difficult to implement over a long period, attempts to establish fens at locations other than groundwater discharge areas should be cautiously undertaken.

FORESTED WETLANDS

Long-term management of wooded wetlands consists of simulating normal winter flooding and summer drying with an occasional flooding (less than 5 days) during the active growing season; typically, bud eruption to mid summer, but slightly later for some shrubs. Water level fluctuations during the first few years will increase diversity by permitting transitional and some terrestrial species to become established in minor high spots, microhabitats, within the swamp. If terrestrial types proliferate, increase the frequency but not the duration or depth of flooding during the growing season until after desirable species are approaching maturity. Conversely, if wetlands species do poorly, investigate soil moisture conditions during dry periods and/or decrease flooding frequency, duration, and depths during the growing season (Figure 16-5). Unless goals are to foster the most flood-tolerant species, complete inundation

Figure 16-5. Excessive depths and duration of flooding during the growing season caused substantial mortality in this well-established swamp allowing buttonbush, river birch, and other weedy species to invade much of the area.

in the winter should not exceed 15 days, with drying periods of equal or greater length. If only highly tolerant species are desired, increase flood frequency and duration during the growing season after 8 to 10 years of growth and carefully monitor vegetation for signs of stress. In most cases, even the moderately tolerant species can withstand considerable flooding for one season, though stress will be evident (leaf color, leaf fall, and bark cracking). But two or more flooded growing seasons will cause substantial mortality.

OTHER CONSIDERATIONS

Shallow wetlands in regions with prolonged hot, dry periods may lose considerable water with gradually receding levels falling from shallow shorelines. Summer storms that restore levels may appear to bring a reprieve, but managers must be cautious, due to the ever-present danger of botulism under these conditions. Warm, moist mudflats with high organic contents and perhaps other poorly understood factors seem to create suitable environmental conditions for bacteria-producing botulinum toxins. Shallow reflooding of previously exposed mudflats apparently dissolves and disseminates these extremely toxic compounds, affecting a wide variety of vertebrates. To avoid catastrophic losses, wetland managers may need to either flood potentially dangerous mudflats with deep water, immediately reduce levels below pre-rainfall levels, or implement scaring/hazing methods to prevent wildlife use.

Controlled burning can be a useful tool to open up dense stands of vegetation or shape composition of plant communities; for example, to foster grasses at the expense of forbs, reduce accumulations of organic matter, and restore hydraulic capacity and influence insect and mammal populations. In older systems, hot fires can reverse successional changes by reducing peat deposits and resetting the system back to its earliest state.

Obviously, these recommendations are perfectly applicable to the ideal system in which each zone — transition, shallow, mid, and deep — is flat-bottomed and the only variations in substrate elevations are between zones. Few natural wetlands exhibit this pattern and few if any projects will have the natural site relief or the funding to create these conditions. Water level changes will not and need not be instantaneous because of limits on size of water control structures and maximum inflow/discharge rates. However, this lag results in slow increases and decreases in water depth, and coupled with the ranges for flooding depths and duration in the above, creates extended, repeated periods with warm, moist but not inundated soil conditions that are optimal for germination and growth of wetland plants. Once these plants are firmly established, water levels are gradually raised and maintained at normal operating elevations for the rest of the summer to inhibit invasion of terrestrial species.

MOSQUITOS

Mosquitos inhabit almost all wetlands, but population size and inherent problems vary substantially with type of wetland and region of the world. Largely because of previous human experience, a few mosquitos may cause serious problems for wetland managers in the South and West, whereas clouds of biting pests are accepted as normal living conditions in the Midwest and Northeast. Black, moose, deer, and horse flies add to swirling hordes of mosquitos and black flies have even caused significant gosling mortality at Seney Refuge in the Upper Pennisula of Michigan. These northern areas have extensive natural wetlands producing high populations of blood-sucking insects, and human populations have learned to accept the inevitable. After dark, outdoor activities are simply not possible; but for example, in Tennessee, my neighbors will complain to the mayor that evening, not next morning, if bitten by a single mosquito while in the yard after dark!

Common misconceptions aside, the largest populations and most aggressive types seem to occur in cold, high latitudes and not in the tropics. I have experienced only minor mosquito problems in the Carribean, in the Brazilian rain forest towards the end of the rainy season, and in various parts of Southeast Asia. However, encountering unending clouds of mosquitos on the first warm, calm, day of the summer in Churchill, Manitoba brought a quick understanding of shorebirds perching on any elevated structure and gulls nesting high in trees!

Potential mosquito problems vary strongly with distance to human residences and activities since foraging range varies significantly between different species of mosquito. In addition, females tend to avoid shaded waters for egg-laying and prefer calm rather than moving waters; typical egg to adult cycles range from 5 to 9 days and highest populations occur in organically rich waters. Control methods include deep flooding to strand flotsam and debris in the spring, repeated dewatering to strand mosquito larvae before they metamorphose into adult stages (i.e., on 5-day intervals), system design and vegetation management to preclude stagnant backwaters with no or limited connections to the main pool, shading the water surface, introduction of mosquitofish (*Gambusia affinis*) and insuring they have access to the entire water column, careful monitoring and, if needed, removal or dispersal of floating mats of *Lemna, Spirodella*, or other floating species and introduction of bacterial control (*Bacillus thuringiensis, B. sphaericus*). If all else fails, proper application of appropriately labeled insecticides may be necessary, but caution is advised. Overzealous insecticide application has stressed emergent vegetation in at least one system.

Suitable system designs, minor vegetation or water level management, and introduction of *Gambusia* are inexpensive preventive measures that should be incorporated in any system where minor mosquito production will cause adverse reactions by society. However, before initiating more involved procedures, carefully inspect or enlist the assistance of a good vector control specialist to inspect every aspect of the wetland system. This includes such minor components as an old soda pop can, discarded tires, undrainable depressions in wooded areas, hollow stumps, water control structures, open piping, and anywhere else that you can possibly imagine that standing water might occur. Quite likely, problem mosquito production originates from some small but frequently overlooked component of the system or ancillary artifact. Identify and correct these sources prior to initiating an extensive control program and, if the latter is chosen, enlist the services of specialists.

SYSTEM PERTURBATIONS

Although many wetlands, especially marshes, thrive on disturbance or change and astute management will deliberately introduce perturbations as needed, additional variations in "normal" operating conditions will occur more often than predicted or preferred. These perturbations can be categorized as:

1. periodically occurring and predictable; and,
2. infrequently or rarely occurring and hence unpredictable but probable.

Predictable disturbances can be anticipated and were hopefully incorporated in earliest planning. If not, modifications may be possible during system startup. In contrast, unpredictable disturbances confront managers with serious and potentially catastrophic situations. In a sense, all disturbances that were not included in system planning and design can be considered unpredictable since good planning has considered all potential problems and accommodated those that are practical. However, designing for all possible events, for example a 1000-year storm event, would needlessly increase costs, but one could theoretically occur at any time.

Predictable disturbances include seasonal changes in precipitation, runoff, nutrient and sediment inputs, groundwater levels, stream or river flows, evapotranspiration rates, air and water temperatures, vegetation growth patterns and animal activities, and in some areas inflow changes because of upstream use, or increased discharge to sustain aquatic life in receiving waters or comply with water rights regulations. Most should have been included in the design, but one or more was likely overlooked and will need to be accommodated during start-up. In contrast, unpredictable perturbations include record storm events, vegetation damage, failure of a dike, water control or other structural component, significant changes in adjacent land use or water rights, design flaws, and pest or disease outbreaks.

Predictable disturbances are likely to be first encountered during start-up. In addition, system start-up periods provide opportunities to test and debug mechanical, electrical, and hydraulic systems and assure adequate performance. Soil amendments or chemicals to modify water chemistry (such as lime or fertilizers) may be needed to stimulate initial growth of macrophyte and microbial populations. High precipitation events test capacity of water flow control devices and water recording instruments, but may cause channeling that should be corrected with vegetative or mechanical baffles or weirs. Conversely, droughts test water retention capacity (including the liner) and endurance of plant and animal communities. Many predictable disturbances are seasonal perturbations that are specific to local climate, soils, waters, and vegetation and operating procedures can generally be modified during start-up to incorporate those that were overlooked during design.

Unpredictable disturbances may be serious problems since only limited capacity has been included to accommodate them. Record storms, earthquakes, or tornados, sudden appearance of a sinkhole in the pool, or any other event that damages physical structure such that the entire system is jeopardized are all improbable but could create the need for extensive and costly repairs. A gradual increase in residential housing may not be predictable in the design stage, but should be identified as a potential factor influencing future operations. Construction activities might also cause episodic and excessive inputs of sediments that may smother the system. Heavy foraging in early spring during a one-time use by a large population of migrating geese could seriously retard annual growth, and if it occurs in the earliest years, may necessitate replanting. If the same flock used the wetland for loafing between feeding forays to nearby fields, the nutrient additions could initiate or accentuate algal blooms later that summer. Unusual insect outbreaks have devastated wetland vegetation and were only brought under control with insecticides. Wild fires could destroy significant proportions of the living biotic component and decomposing materials, causing substantial changes in the hydrology and biology of the system. Extreme and/or prolonged drought may cause operational changes or even require additions from other sources such as groundwater. However, do not be tempted to use irrigation return water regardless of the severity of the drought. Deposition of irrigation water contaminants in wetland soils could have long-term, far-reaching detrimental impacts.

If the project objectives are to establish a flood buffering or wastewater treatment wetland, initial introductions of flood water or wastewater could cause considerable disturbance. Planners should gradually phase-in low to moderate levels of flood waters and dilute concentrations of wastewaters during the first years of operation lest the abrupt introduction create a serious perturbation that hinders full system development or detrimentally impacts new plant communities.

This discussion of disturbances is not meant to be an all inclusive list of minor deviations from normal operating conditions or probable catastrophes. It is included to stimulate planners to identify as many potential disturbances as possible, categorize each as predictable or unpredictable, and to choose whether or not to incorporate provisions for each in the system design and operating procedures. Most predictable events are recurrent and need to be accommodated, whereas the rare, improbable, and unpredictable events will need to be dealt with as they occur.

At this stage in the technology, every created wetland is an experiment, with the exception of classic waterfowl marshes; but as each new disturbance arises and is dealt with and as each new idea is applied and the results monitored and recorded, the accumulated information base will support derivation of predictive relationships and wetland creation will slowly become more of a science than an art. Consequently, a long-term monitoring program and detailed records on operations, disturbances, results, and modifications is the responsibility of every wetland developer and manager.

ROUTINE MAINENANCE

Since system hydrology is the most important and most easily manipulated factor, maintaining control and monitoring system inflows, outflows, and water levels is essential to managing the new wetland. Insuring integrity of dikes, berms, spillways, and water control structures should be a regularly scheduled activity. Each of these components should be inspected at least weekly and immediately following any unusual storm event. Any damage, erosion, or blockage should be corrected as soon as possible to prevent catastrophic failure and expensive repairs.

Vegetative cover on dikes and spillways should be maintained by mowing and fertilizing as needed. Frequent mowing encourages grasses to develop good ground cover and extensive root systems that resist erosion and prevents establishment of shrubs and trees. Roots of the latter could create channels, with subsequent leakage or even dike failure. Muskrats and other burrowing animals can damage dikes and spillways and unrepaired burrows may lead to dike failure. If wire screening was not installed in the dikes, a thick layer of gravel or rock over trouble spots may inhibit burrowing. However, if damage continues, trapping and shooting may be needed for temporary relief until wire screen can be installed. Burrows are most easily repaired by setting an explosive charge from the top of the dike to collapse the network of tunnels and then filling the subsequent crater with compacted clay. Fences, pathways, roadways, and visitor facilities should be inspected concurrent with the weekly dike inspections and repaired as necessary. Pesticides or other chemicals should not be used unless extreme circumstances warrant use, in which case care is necessary to avoid damaging the system. This also applies to insecticides, since heavy applications of insecticides have damaged emergent vegetation in some wetlands.

Livestock grazing may cause serious damage to wetland vegetation, especially during the early years when plants are becoming established. Rubbing and loafing beneath shrubs and trees often damages new growth and accelerates soil erosion, exposing roots and causing mortality. Perimeter fencing may be required if livestock are anticipated to be a problem.

MONITORING

Wetlands are complex, highly productive systems with diverse and abundant populations of animals and plants interrelated in myriad pathways between biotic and abiotic components. To attempt to measure and understand every component and each pathway for energy and nutrient flow is far beyond the scope of normal operating guidelines. Yet because our knowledge of the components and processes is incomplete, we lack universal, readily identified indicators of development and successional changes, system robustness and viability or, conversely, danger signals. Continuous change (flux) also confounds our understanding and complicates the design of a monitoring plan that is appropriate throughout the lifetime of the project. For example, simple photographs may be adequate to document plant establishment and growth in early stages (marsh) but are much less useful as the wetland progresses to the later shrub and tree stages. Waterfowl and shorebirds can be visually counted in the marsh but songbirds must be inventoried by their calls and squirrels, deer and turkey require indirect call counts, track, scat or trapping methods. Measures of other developing, steady-state or failing functions similarly vary and early symptoms are difficult to recognize, the most appropriate adjustments are not well understood, and the results of alterations may not be evident for long time periods. Consequently a long-term monitoring plan is essential to develop an information base for continuous comparisons of functional status and biological integrity of the system. The monitoring program serves two principle functions:

1. To document system development and progress toward achieving project goals and objectives;
2. To quantify functional values; and,
3. To identify symptoms of problems or failures at an early, remedial stage.

The monitoring plan need not be elaborate or lengthy, but it must provide clear documentation of project and monitoring objectives, organizational and technical responsibilities, specific tasks, methods and basic instructions, quality assurance procedures, schedules, reports, resource requirements, and costs. Since the life of the project may easily span many decades and numerous personnel changes, written documentation is essential to insure that data sets are at least comparable if collection or analysis procedures change, as is likely to happen. A carefully defined monitoring plan should be part of the operations manual so that it is readily available to serve as a benchmark for data collection throughout the life of the project.

Recorded measurements of water surface elevations, and inflows and outflows in each unit of the wetland are basic monitoring parameters to correlate and understand changes in survival and distribution of biological components, especially macrophytes including woody species. Water levels are generally measured with staff gauges located in the deepest part of a pool within easy viewing distance of a dike or on a sidewall of a water control structure. Gauges should be positioned so that readings represent elevations above mean sea level and can be directly related to substrate elevations so that areal extent and volumes can be calculated from the "as built" construction drawings of the system. Inflow and discharge can be similarly measured with staff gauges in simple "box" or V-notched weirs at each location. If considerable variation is anticipated and knowledge of flows is critical to maintaining desired water levels, automated flow measuring and recording devices (hydrographs, etc.) may be needed. Under normal steady-state operating conditions, weekly readings of pool levels, inflows, and discharges are adequate.

During start-up in all systems and during operation of wooded systems, determining and recording flooding frequency, duration, and extent of coverage is important. This is basically the frequency and time period (hydroperiod) that the water surface elevation is above the elevation of the wetland substrate in any part of the system. Surface water elevations, determined from a staff gauge or recording graph, are related to substrate elevations on the "as built" topographic map to calculate flooding depth, frequency, and duration of flooding in each area and the volume of the water in the wetland. The latter is useful in predicting the time needed to flood the system to a certain elevation with a given rate of inflow or to dewater it with available discharge rates.

Meteorological information will be needed to interpret changes in hydrologic conditions and in the biotic components. If an established recording station is not close by, managers should acquire basic instruments measuring air temperatures, precipitation, relative humidities and wind velocity. Again cost and complexity of the project should dictate the degree of instrumentation and data acquisition. Though most of our meteorological database was accumulated from manual observations and recording, the more complex stations currently available record information on a variety of parameters in PC formats for later downloading or for transmittal to a base station. In the long term, the initial cost of more powerful equipment is often more than offset by reduced operating costs and usefulness of the information.

Monitoring should obviously include measurements of both form and function and managers are likely to be most interested in measuring wetland functions for comparisons with their objectives. But many functions are difficult to measure or develop very slowly and managers often measure wetland form as a means of determining system development. Form measurements (plants, animals, etc.) are also often an indirect method of addressing functional measurements. Either or both are very important to understanding wetland development and progress toward achieving project goals and objectives. Since sampling procedures for many of the biotic components almost represent a separate field of study, only a brief overview of some techniques is presented below. References in Appendix A provide detailed descriptions.

Wetland macrophytes, rooted, and floating plants are the primary producers supporting all other life in the system. Consequently, knowledge of changes in the macrophyte component are essential to understanding changes in any other component or process (Figures 16-6 and 16-7). Wetland vegetation is subject to seasonal, annual, and long-term changes in species composition and in distribution; that is, location and extent of coverage within the system. Some species may change little from year to year, while others may flourish for a year or two and then disappear. Monitoring these changes requires information on which species are present at what locations in the system during a specific time interval. A number of methods are used to determine each, but most are basically enumerations of individuals of each species along line transects randomly or regularly bisecting the system or from quadrants distributed throughout the system. In either case, the sample points should be permanently fixed locations sampled with the same procedures at the same times each year. Seasonal and annual photographs made from permanently marked stations are a simple method to document conditions in the plant communities. In addition, color aerial photos and color aerial infra-red (IR) photographs may be used in conjunction with ground sampling to map species distribution, coverage, and changes from year to year. IR photos detect different chlorophyll contents of different species, as well is in diseased or infested individuals, providing species composition of the community along with early indications of stress in many plants. Measurements of standing crop biomass from clipping or harvest sampling and samples of litter/duff material and organic content of the substrate from core samples are useful in understanding biomass production, decomposition, mineralization, and accumulation rates.

Figure 16-6. Monitoring plants is relatively simple yet provides easily interpreted indication of basic changes in the wetlands system.

Figure 16-7. Aerial photography and/or fixed point ground photography records plant community development and coverage during the season and annually.

Figure 16-8. Birds occupy a variety of niches and are easily monitored indicators of system health and well being.

The large vertebrates are probably the most commonly sampled components next to vegetation, primarily because of their prominence and our interest in their populations. Birds are often sampled with direct counting methods that attempt total counts of the entire wetland or selected portions (Figure 16-8). Mammals, reptiles, and amphibians are indirectly sampled through track counts, scent posts, or trap, mark, and recapture methods. Fish are collected with gill nets, trap nets, or electro-shocking devices. In contrast to other vertebrates, fish (especially bottom feeders) are often exposed to contaminants in the sediments and analysis may provide early warning of accumulation and bioconcentration. Unfortunately, very few vertebrates make good indicators and most types are unlikely to reflect minor but perhaps significant long-term changes in the system.

Understanding change in the system largely means understanding fluctuations in the biological communities as a result of wetland development, successional changes, hydrologic changes or some disturbance. Monitoring development, viability, and robustness (i.e., the health and well-being of the system and its individual components) requires a comparative baseline for that system or comparative information from a similar system. Successional changes tend to follow general patterns that can be compared with a reference wetland or literature accounts. Disturbances may originate internally as well as externally and may be biological; that is, invasion of a new species, as well as physical or chemical (i.e., temperature changes or variation in quality of inflows). In addition, changes in one parameter may be manifested by single or multiple, and indirect as well as direct cause and effect alteration in an apparently unrelated component or process. Because of the tremendous number of interactions, the web of pathways for energy and nutrient exchange between species, between species aggregates or communities comprising a trophic or functional unit, and between biotic and abiotic components, monitoring basic parameters such as energy flux or nutrient exchange is impractically complex and expensive. Consequently, monitoring programs attempt to identify and collect information on potential indicators that are expected to rapidly reflect changes and are relatively easily measured. Selected macrophytes, vertebrates, and macroinvertebrates are common components of most wetland monitoring plans.

Besides plants, invertebrates are perhaps the best indicators of progress towards developing complexity as well as on-going health and well-being in a wetland system. Macro-invertebrates are sensitive to changes in environmental parameters, relatively long-lived, and resident in the system occupying nearly all levels in the trophic structure. Taxa diversity and abundance respond rapidly to environmental changes and sampling methods are simple and easily conducted. Commonly used methods range from grab samples to artificial substrates for colonization over a period of time and later collection. Analysis is dependent upon accurate identification since changes are detected by changes in types and numbers of different taxonomic groups. Some groups are readily identified, but others may require considerable training or the assistance of a specialist. Many macro-invertebrates also serve as early indicators of contaminant problems since many are detrital feeders or scavengers.

Managing a wetland system to maintain a healthy and functional community of aquatic plants and animals achieving the project goals requires feedback on changes that occur in components and processes. Monitoring selected communities over time provides a comparative basis for judging system development, normal conditions and response to management changes or disturbances. A well thought out and rigorously implemented monitoring plan is an essential component of the operating guidelines for any wetland system.

CHAPTER 17

WETLAND RESTORATION AND CREATION

The astonishing reversal in public attitudes towards wetlands resulted in a number of pieces of legislation initially removing federal subsidies for wetland destruction and later encouraging and even compensating landowners for protecting and/or restoring wetlands. In terms of acreages affected, the Swampbuster and other provisions of the Food Security Act have probably been the most important though the North American Wetlands Conservation Act and the North American Waterfowl Management Plan have made significant contributions. Swampbuster provisions are punitive similar to those in 404 regulations of the Clean Water Act in that farmers lose all federal subsidies if they convert wetlands to crop production. In contrast, The Partners for Wildlife (USDI, Fish & Wildlife Service), the Conservation Reserve Program (USDA, Agricultural Stabilization and Conservation Service) and the Wetlands Reserve Program (USDA, Natural Resource Conservation Service) provide financial incentives to landowners to protect or restore previously converted wetlands.

But the outcome of these efforts is very encouraging. During the 1970s average annual loss rates approached 300,000 acres the bulk (>80%) of which was attributed to agricultural conversion. By the early 1980s average annual loss rates had dropped to around 150,000 acres with agriculture responsible for less than half the total. Agricultural conversion further declined by the end of the decade contributing only 25% of the average annual total of some 100,000 acres and that trend is believed to have continued through the early 1990s, i.e., agricultural conversion represented less and less of the total loss.

On the positive side, since 1987 the Partners for Wildlife Program has restored over 200,000 acres, the Conservation Reserve Program has restored some 300,000 acres and the Wetland's Reserve Program has restored almost 300,000 acres. In addition to the 800,000 acres of wetlands that have been restored on private land, public agencies restored an estimated 100,000 acres on public lands. If the annual loss rate is now approximately 100,000 acres per year and the restoration rate appears to be similar, we may have at least reached the point of overall "no net loss."

But note that the changes have largely occurred in the agricultural sector; the development loss rate has declined somewhat from a peak of approximately 150,000 acres per year in the late 1970s to 85,000 to 90,000 acres per year or an estimated 75 to 80% of the current loss rate. And most analysts expect that rate and that ratio to continue for the foreseeable future. In contrast, the reversal in agriculture has been so rapid that the same individuals that funded, designed,

Figure 17-1. NRCS employees that designed drainage systems only 10 to 12 years ago are now efficiently designing water control structures and plugging ditches to restore wetlands.

supervised, or conducted drainage are now just as efficiently installing water control structures and ditch plugs to restore the same wetland (Figure 17-1)! The differences are likely due to a number of factors:

1. development sites are limited;
2. many types of development are site specific with little relocation latitude;
3. until recently, Section 404 regulated fill but not drainage;
4. a nationwide permit allowed many actions under 404 regulations;
5. Swampbuster economic penalties are severe; and,
6. economic incentives were provided for agricultural restoration.

In addition, wetlands protected or restored by the agricultural sector tend to be larger, occasionally expansive and relatively intact systems whereas a recent review found that 25% of the 404 permit applications would affect less than 0.25 acres and half would impact roughly one acre. No indication was available of the nature or condition of those minute areas but it seems unlikely that many of them were pristine or intact wetland systems. And few were likely to have been replaced. Significantly, approved wetland losses under 404 are unlikely to decline though increased use of mitigation (replacement) may compensate for those losses. It's important to keep in mind that compensatory mitigation is only applicable within the prescribed wetland regulatory process; at present nonregulatory wetlands (those built for other than mitigation purposes) account for 99% of all man-made wetland restoration and creation.

Figure 17-2. Though reservoirs often drown floodplain wetlands, a few projects have merely shifted the locations. Lateral dikes restrict reservoir waters and upstream waters are pumped into the reservoir during summer but allowed to flood during winter in some 10,000 acres of valuable bottomland hardwoods along Kentucky and Wheeler lakes in Tennessee and Alabama.

In addition, regulation of wetlands in the agricultural community has yet to generate the firestorm of controversy caused by a few cases of 404 applications. Though originally written for large projects with government involvement, recent extensions place minor activities on private lands under 404 jurisdiction. A few highly publicized cases of inflexible application and harsh penalties have contributed more to the general distrust of government than they have to wetlands protection. Unfortunately, expanded interpretations of the Endangered Species Act have similar ramifications and the implications of this combination justifiably strikes fear in the heart of many landowners. Public outcry has forced the issue into the legislative arena where current efforts to redefine wetlands, categorize wetlands, or reduce protection have generated more controversy than solutions though providing compensation could defuse the contentious "takings" issue.

On the other hand, 404 has made wetland considerations an integral part of project planning and instances of successful, cooperative projects that prevented or mitigated potential wetland impacts vastly out number the highly publicized negative cases (Figure 17-2). So much so that financial institutions now often require a wetland survey along with an inspection for underground storage tanks and the more conventional types of information that make up the loan file.

But in a system of government designed to share the costs and benefits among its members, it seems unjust to require an individual to bear the costs of a wetland that benefits all of society. Furthermore, even a casual review of the recent record of wetland gains and losses clearly demonstrates the carrot has been more effective than the stick in restoration though one might argue the carrot would have been much less effective without the stick (Swamp-buster) in protection. Perhaps the key is the combination of carrot and stick and that suggests an approach to resolve the issue.

Location constraints on developers and the inherent nature of Section 404 make it unlikely that 404, as commonly applied, can provide increased pro-tection to wetlands. Will we then be unable to reduce annual wetland losses below 100,000 acres per year? If so, our grandchildren may benefit from wetlands but their grandchildren may be less fortunate. However, the basic reason why many of the small projects permitted under 404 fail to replace impacted wetlands is the impracticality of "in-kind, on-site" replacement mit-igation of "postage-stamp" projects. As that interpretation shifts more towards larger, high value and even off-site replacement concurrent with the establish-ment of mitigation banks and markets, it will be practical for developers impacting small acreages to replace the impacted wetland. Granted that many one-acre permitted projects may need to be involved to support a 50- or 100-acre regional mitigation wetland. But at least replacement would be prac-tical, costs known and included in project plans and the permitting process streamlined. And in many cases, we might have an opportunity to exchange a small, likely degraded, perhaps low value wetland for part of a larger, high value system protected and managed in perpetuity.

Properly located, designed and operated mitigation banks offer opportuni-ties for:

1. increasing the quantity of high quality wetlands through restoration of altered or degraded systems along with acquisition and management;
2. improving wetland scale, size, and location if planners use a landscape approach;
3. demonstrating to landowners that formerly "low-value" lands have the potential to generate income;
4. improving understanding of wetland functions and values to encourage increased use of created/constructed wetlands in development projects;
5. improving the success of restoration and creation projects because larger efforts will likely use wetland expertise and have better financing for long-term management and monitoring;
6. providing time and cost frameworks that developers can incorporate in their planning and financing; and,
7. increasing our wetland resource base — net gain — if developers use appropriate and successful creation techniques.

To be successful, important considerations for mitigation banks include location: in or out of the region, political boundary, watershed or hydrologic unit, type of wetland, wetland form and function and value of wetland

Figure 17-3. Many of our large "natural" wetlands — Lake Agassiz, Lake Mattamuskeet, Okeefenokee Swamp — have been restored following drainage attempts and are currently maintained only through deliberate management efforts. Mitigation banking provides similar opportunities today.

(Figure 17-3). Though universal agreement may not be possible, it is likely that a consensus could be reached among a majority of regional wetland specialists on each of these issues. And developing that consensus is likely to create a regionalized landscape/watershed plan for judiciously located restored and/or created wetlands in the new mitigation banks. This proactive approach could foster replacement of small, impacted wetlands and further reduce current loss rates. And can we create these wetlands? Of course. Many examples were included in previous chapters.

However, bear in mind that most wetlands require management, not simply protection, and management will influence the values society receives from a wetland. In addition, the possibilities for selecting optimal locations of highly valued wetland types creates enduring opportunities for demonstrating wetland values to the public. If we are to maintain or increase our wetland resources, we must go beyond the vague, imprecise "benefits" that currently generate public support for wetland protection. The future of wetlands rests on our ability to demonstrate benefits, functional values, in valid, quantified terms on par with other segments of our economic system. Else public support for wetland protection may vanish as quickly as it appeared. Obviously this has important ramifications for recently successful wetland actions in the agricultural sector as well as for Section 404 in the development sector.

Implementing a wetland categorization system, such as would be required in current legislation, will require a substantial amount of research to develop a methodology to evaluate wetland functional values. Given the myriad of wetland types, composition, sizes, and locations, that effort is likely to entail the efforts of most of the existing and many new wetland scientists for the next decade or more.

Though much of the discussion in the foregoing chapters was addressed to those project planners that are creating wetlands, the same principles and many of the guidelines are every bit as applicable to restoring wetlands. The same level of planning effort, careful development, and arrangements for long-term management and financing should be applied in any restoration project. The major differences are simply that the restoration project has a much higher probability of succeeding since the site previously supported a wetland and the creating project is likely to be much more expensive. Regardless, both substantial restoration and creation will be needed if we hope to achieve any reasonable measure of success in insuring that future societies benefit from similar magnitudes of functional values from wetlands.

APPENDIX A

REFERENCES

Adriano, D. C. and Brisbin, Jr., I. L., Eds., *Environmental Chemistry and Cycling Processes* , U.S. DOE, Savannah River Ecology Laboratory; NTIS, CONF-760529, 1978.

Alexander, M., *Introduction to Soil Microbiology*, John Wiley & Sons, New York, 1967.

Allen, J. A. and Kennedy, Jr., H. E., Bottomland Reforestation in the Lower Mississippi Valley, USDI, Fish and Wildlife Service, National Wetlands Research Center, Slidell, LA, 1989.

Allen, H. H. and Klimas, C. V., Reservoir Shoreline Revegetation Guidelines, Tech. Rep. E-86-13, U.S. Army Engineer Waterways Experiment Station, 1986.

Ambrose, R. E., Hinkle, C. R. and Wenzel, C. R., Practices for Protecting and Enhancing Fish and Wildlife on Coal Surface-Mined Land in the Southcentral U.S., FWS/OBS-83/11, USDI, Fish and Wildlife Service, 1983.

Ann. Ponds — Planning, Design, Construction, Agriculture Handbook No. 590, USDA, Soil Conservation Service, 1982.

Ann. Biological Field and Laboratory Methods for Measuring the Quality of Surface Waters and Effluents, Report-670/4-73-001, Office of Research and Development, U.S. Environmnetal Protection Agency, 1973.

Ann. Wetland Mitigation Banking, IWR Report 94-WMB-6, U.S. Army Corps of Engineers, Institute for Water Resources, 1994.

Ann. Native Plant Material Sources for Wetland Establishment: Freshwater Case Studies, Tech. Rep. WRP-RE-5, U.S. Army Engineer Waterways Experiment Station, 1995.

Ann. Directory of Wetland Plant Vendors, Tech. Rep. TR WRP-SM-1, U.S. Army Engineer Waterways Experiment Station, 1992.

Ann. Wetland Habitat Development with Dredged Material: Engineering and Plant Propagation, Tech. Report DS-78-16, U.S. Army Engineer Waterways Experiment Station, 1978.

Ann. Natural Wetlands and Urban Stormwater: Potential Impacts and Management, 843-R-001, U.S. Environmnetal Protection Agency, 1993.

Ann. Water Measurement Manual, Water Resource Technical Publication, No. 2403-00086, U.S. Government Printing Office, 1974.

Ann. Wetlands: Their Use and Regulation, Office of Technology Assessment OTA-0-206, U.S. Government Printing Office, 1984.

Ann. Impact of Water Level Changes on Woody Riparian and Wetland Communities, Vol. I-VI, USDI, Fish and Wildlife Service, 1977–1978.

Ann. Wetland Restoration, Enhancement, or Creation, Chapter 13, Engineering Field Handbook, USDA, Soil Conservation Service, 1992.

Ann. *Standard Methods for the Examination of Water and Wastewater, 19th ed.*, American Public Health Association, Washington, DC, 1995.

Ann. Proceedings: "Workshop on Structural Marsh Management, New Orleans, LA," U. S. Environmnetal Protection Agency, Dallas, TX 1995.

Ann. Final Environmental Statement — Operation of the National Wildlife Refuge System, USDI, Fish and Wildlife Service, 1976.

Ann. U.S. Army Corps of Engineers Wildlife Resources Management Manual, Environmental Lab., U.S. Army Engineer Waterways Experiment Station, various.

Ann. Riparian Ecosystems: Their Ecology and Status, FWS/OBS-81/17, USDI, Fish and Wildlife Service, 1981.

Arber, A., *Water Plants*, Cambridge University Press, London, 1920.

Bavor, H. J. and Mitchell, D. S., Eds., Wetlands Systems in Water Pollution Control, *J. Water Sci. & Techn.* 29(4), 1994.

Beal, E. O., A Manual of Marsh and Aquatic Vascular Plants of North Carolina, Tech. Bull. No. 247, North Carolina Agric. Exp. Sta., Raleigh, NC, 1977.

Biological Reports: entitled The Ecology of . . . or An Ecological Character-
ization of ......, USDI, Fish and Wildlife Service, various. a comprehensive
series (40+ issued) on community profiles, management and economics of
important wetlands of the U.S. issued as Biol. Rep. xx(x.x) or FWS/OBS —
xx/xx.

Bishop, E. L. and Hollis, M. D., Eds., *Malaria Control on Impounded Water*,
U.S. Government Printing Office, 1947.

Brooks, R. P., Samuel, D. E. and Hill, J. B., Eds., Proceedings of a Conference
on Wetlands and Water Management on Mined Lands, Pennsylvania State
University, University Park, PA, 1985.

Coleman, J. M., Dynamic Changes and Processes in the Mississippi River
Delta, Geol. Soc. of Am. Bull., July 1988, pp. 999-1015.

Cooper, P. C. and Findlater, B. C., Eds., *Constructed Wetlands in Water
Pollution Control.*, Pergamon Press, Oxford, U.K., 1990.

Cox, G. W., *Laboratory Manual of General Ecology*, W. C. Brown Publishers,
Inc., Dubuque, IA, 1970.

Cowardin, L. M., Carter, V., Golet, F. C. and LaRoe, E. T., Classification of
Wetlands and Deepwater Habitats of the United States, FWS/OBS-79/31,
USDI, Fish and Wildlife Service, 1979.

Cross, Diana H. (compiler) Waterfowl Management Handbook, Fish and Wildlife Leaflet 13, USDI, Fish and Wildlife Service, 1988.

Davis, G. J. and Brinson, M. M., Responses of Submersed Vascular Plant
Communities to Environmental Change, FWS/OBS-79/33, USDI, Fish and
Wildlife Service, 1980.

Davis, S. M. and Ogden, J. C., *Everglades: The Ecosystem and Its Restoration*,
St. Lucie Press, Delray Beach, FL, 1994.

Dennis, W. M. and Isom, B. G., Eds., Ecological Assessments of Macrophyton
Collection, Use, and Meaning of Data, ASTM Publication 8453, Philadelphia,
1983.

DuBowy, P. J. and Reaves, R. P., Eds., Constructed Wetlands for Animal Waste
Management, Purdue Research Found., Lafayette, IN, 1994.

Duffy, W. G. and Clark, D., Marsh Management in Coastal Louisiana: Effects and Issues, Biol. Rep. 89(22), USDI, Fish and Wildlife Service, 1989.

Errington, Paul L., *Muskrats and Marsh Management*, The Stackpole Comp., Harrisburg, PA, 1961.

Erwin, K. L., Selected Bibliography: Wetland Creation and Restoration, Ass. of Wetland Managers, Berne, NY.

Ewel, K. C. and Odum, H. T., Eds., *Cypress Swamps*, University Presses of Florida, Gainesville, FL, 1984.

Farb, P., *Face of North America*, Harper & Row Publishers, Inc., New York, 1963.

Flint, R. F., *Glacial and Quaternary Geology*, John Wiley & Sons, Inc., New York, 1971.

Galatowitsch, S. M. and van der Valk, A. B., *Restoring Prairie Wetlands: An Ecological Approach*, ISU Press, Ames, IA, 1994.

Gale, W. F., Botton Fauna of a Segment of Pool 19, Mississippi River, Near Fort Madison, Iowa, 1967-1968, *Iowa State J. Res.* 49:353, 1975.

Garbisch, Jr., E. W., Highways and Wetlands, Compensating Wetland Losses, FHWA-IP-86-22, USDOT, Federal Highway Administration, 1986.

Garlo, A. S., Wetland Creation/Restoration in Gravel Pits in New Hampshire, In Webb, Jr., F. J., Ed., Proc. 19th Ann. Conf. on Wetlands Restoration and Creation, Hillsborough Comm. College, Plant City, FL, 1992.

Gilbert, T., King, T. and Barnett, B., An Assessment of Wetland Habitat Establishment at a Central Florida Phosphate Mine Site, FWS/OBS-81/38, USDI, Fish and Wildlife Service, 1981.

Godfrey, Paul J., Kaynor, E. R. and Pelczarski, S., Eds., *Ecological Considerations in Wetlands Treatment of Municipal Wastewaters*, Van Nostrand Reinhold, New York, 1985.

Good, R. E., Whigham, D. F. and Simpson, R. L., *Freshwater Wetlands*, Academic Press, New York, 1978.

Greeson, P. E., Clark, J. R. and Clark, J. E., Eds., Wetland Functions and Values: The State of Our Understanding, American Water Resources Assoc, Minneapolis, MN, 1978.

Gregory, R. W., Elser, A. A. and Lenhart, T., Utilization of Surface Coal Mine Waste Water for Construction of a Northern Pike Spawning/Rearing Marsh, FWS/OBS-84/03, USDI, Fish & Wildlife Service, 1984.

Hammer, D. A., Ed., *Constructed Wetlands for Wastewater Treatment*, Lewis Publishers, Chelsea, MI, 1989.

Hammer, D. A., *Creating Freshwater Wetlands*, Lewis Publishers, Chelsea, MI, 1992.

Hammer, D. A., et al., Mitigation Banking and Weland Categorization: The Need for a National Policy on Wetlands, Tech. Rev. 94-1, The Wildlife Society, Washington, DC, 1994.

Hammer, D. A., Water Quality Improvement Functions of Wetlands, In Vol III, Nierenberg, W. A., Ed., *Encyclopedia of Environmental Biology*, Academic Press, Orlando, FL, 1995.

Haslam, S. M., *River Plants*, Cambridge University Press, London, 1978.

Haynes, R. J., Allen, J. A. and Pendleton, E. C., Reestablishment of Bottomland Hardwood Forests on Disturbed Sites: An Annotated Bibliography, Biol. Rep. 88(42), USDI, Fish and Wildlife Service, 1988.

Holman, R. E. and Childres, W. S., Wetland Restoration and Creation: Development of a Handbook Covering Six Coastal Wetland Types, UNC-WRRI-95-289, Water Resources Research Institute, Raleigh, NC, 1995.

Hook, D. D. and Lea, R., Eds., The Forested Wetlands of the Southern United States, Gen. Tech. Rep. SE-50, USDA Forest Service, SE Forest Experiment Sta., 1989.

Hook, D. D., et al, Eds., *The Ecology and Management of Wetlands*, Timber Press, Portland, OR, 1988.

Horwitz, E. L., Our Nation's Wetlands: An Interagency Task Force Report, 041-011-00045-9, U.S. Government Printing Office, 1978.

Hutchinson, G. E., *A Treatise on Limnology*, Vol. I, John Wiley & Sons, Inc., New York, NY, 1957.

Hutchinson, G. E. *Limnological Botany*, Academic Press, Inc., New York, NY, 1975.

Josselyn, M., Ed., Wetland Restoration and Enhancement in California, Rep. NO. T-CSGCP-007, Calif. Sea Grant College Program, La Jolla, CA, 1982.

Kadlec, J. A., Effect of a drawdown on a waterfowl impoundment, *Ecology* 43:267, 1962.

Kadlec. J. A. and Wentz, W. A., State-of-the-Art Survey and Evaluation of Marsh Plant Establishment Techniques: Induced and Natural, Vol. I: Report of Research, Technical Report DS-74-9, U.S. Army Engineer Waterways Experiment Station, 1974.

Kent, D. M., Ed., *Applied Wetlands Science and Technology*, Lewis Publ, Boca Raton, FL, 1994.

Kentula, M. E., Brooks, R. P., Gwin, S. E., Holland, C. C., Sherman, A. D. and Sifneos, J. C., *An Approach to Improving Decision-Making in Wetland Restoration and Creation*, Lewis Publ, Boca Raton, FL, 1993.

Kimber, A. and Barko, J. W., A Literature Review of the Effects of Waves on Aquatic Plants, LTRMP 94-S002, USDI, National Biological Survey, 1994.

Kusler, J. A., Our National Wetland Heritage: A Protection Guidebook, Environmental Law Institute, Washington, DC, 1983.

Kusler, J. A. and Brooks, G., Eds., Wetlands Hydrology, Tech. Rep. 6, Assoc. of State Wetland Managers, Berne, NY, 1988.

Kusler, J. A. and Kentula, M. E., Eds., Wetland Creation and Restoration: The Status of the Science: Vol. I and Vol. II, EPA 600/3-89/038a, U.S. Environmental Protection Agency, 1989.

Kusler, J. A., Quammen, M. L. and Brooks, G., Eds., Proc. of the National Wetland Symp.: Mitigation of Impacts and Losses, Assoc of State Wetland Managers, Berne, NY, 1988.

Landin, M. C., Ed., Wetlands: Proceedings of the 13th Annual Conference of the Society of Wetland Scientists, New Orleans, LA, Society of Wetland Scientists, South Central Chapter, Utica, MS, 1993.

Leedy, D. L., Maestro, R. M. and Franklin, T. M., Planning for Wildlife in Cities and Suburbs, USDI, Fish and Wildlife Service, 1978.

Leedy, D. L. and Adams, L. W., A Guide to Urban Wildlife Mangement, USDI, Fish and Wildlife Service, 1984.

Linde, A. F., Techniques for Wetland Management, Res. Rep. 45, Wisconsin Dept. of Nat. Res., Madison, 1969.

Linduska, J. P., Ed., *Waterfowl Tomorrow*, USDI, Fish and Wildlife Service, 1964.

Loftis, D. L. and McGhee, C. E., Eds., Oak Regeneration: Serious Problems, Practical Recommendations, Gen. Tech. Rep. SE-84, USDA, Forest Service, Asheville, NC, 1993.

Lyon, J. G., *Practical Handbook for Wetland Identification and Delineation*, Lewis Publ, Boca Raton, FL, 1993.

Majumdar, S. D., Brooks, R. P., Brenner, F. J. and Tiner, Jr., R. W., Eds., *Wetlands Ecology and Conservation: Emphasis in Pennsylvania*, Penn. Acad. of Sci. Publ., Easton, PA, 1989.

Manci, K. M., Riparian Ecoystem Creation and Restoration: A Literature Summary, USDI Fish & Wildlife Service, Ft. Collins, CO, 1989.

Marble, A. D., *A Guide to Wetland Functional Design*, Lewis Publ, Boca Raton, FL, 1992.

Mason, H. L., *A Flora of the Marshes of California*, University of California Press, Berkeley, CA, 1969.

McKevlin, M. R., Guide to Regeneration of Bottomland Hardwoods, USDA, Forest Service, SE Forest Experiment Station, Asheville, NC, 1992.

Mitsch, W. J., and Gosselink, J. G., *Wetlands — Second Edition*, Van Nostrand Reinhold Company, New York, 1993.

Mosby, H. S., Ed., *Wildlife Investigational Techniques*, The Wildlife Society, Bethasda, MD, 1963.

Moshiri, G. A., Ed., *Constructed Wetlands for Water Quality Improvement*, Lewis Publishers, Chelsea, MI, 1993.

Nelson, R. W., Horak, G. C. and Olson, J. E., Western Reservoir and Stream Habitat Improvements Handbook, FWS/OBS-78/56, USDI, Fish and Wildlife Service, Ft. Collins, CO, 1978.

Odum, E. P., *Fundamentals of Ecology*, W. B. Saunders Company, Philadelphia, 1971.

Patrick, William H., Jr., Ed., Current Topics in Wetland Biogeochemistry, Wetland Institute, LSU, Baton Rouge, LA, 1994.

Payne, N. F., *Techniques for Wildlife Habitat Management of Wetlands*, McGraw-Hill, New York, 1992.

Phillips, R. C., Westerdahl, H. E., Mize, A. L. and Robinson, S. A., Draft Summary of Literature Describing the Functional Ability of Wetlands to Enhance Wastewater Quality, Tech. Rep. WRP-CP-2, U.S. Army Engineer Waterways Experiment Station, 1993.

Poston, H. J., *Wildlife Habitat: A Handbook for Canada's Prairies and Parklands*, Canadian Wildlife Service, Edmonton, 1981.

Proctor, B. R., Thompson, R. W., Bunin, J. E., Fucik, K. W., Tamm, G. R. and Wolf, E. G., Practices for Protecting and Enhancing Fish and Wildlife on Coal Surface-Mined Land in the Powder River-Fort Union Region, FWS/OBS-83/10, USDI, Fish & Wildlife Service, 1983.

Proctor, B. R., Thompson, R. W., Bunin, J. E., Fucik, K. W., Tamm, G. R. and Wolf, E. G., Practices for Protecting and Enhancing Fish and Wildlife on Coal Mined Land in the Uinta-Southestern Utah Region, FWS/OBS-83/12, USDI, Fish & Wildlife Service, 1983.

Reddy, K. R. and Smith, W. H., Eds., *Aquatic Plants for Water Treatment and Resource Recovery*, Magnolia Publishing, Orlando, FL, 1987.

Reed, Jr., P. B., National List of Plant Species That Occur in Wetlands: National Summary, Biol. Rep. 88(24), USDI, Fish and Wildlife Service 1988.

Richardson, C. J., Ed., *Pocosin Wetlands: An Integrated Analysis of Coastal Plain Freshwater Bogs of North Carolina*, Hutchinson Ross Publi. Co., Stroudsburg, PA, 1981.

Riemer, D. N., *Introduction to Freshwater Vegetation*, AVI Publishing Co., Inc., Westport, CT, 1984.

Robbins, E.I., D'Agostino, J.P., Ostwald, J., Fanning, D.S., Carter, V. and Van Hoven, R. L., Manganese nodules and microbial oxidation of manganese in the Huntley Meadows wetlands, Virginia, USA, *In* Skinner, H.C.W. and Fitzpatrick, R. W., Eds., Biomineralization Processes, Iron and Manganese, *Catena Supplement 21*, 179-202, 1992.

Robbins, E.I., Utilization of sedimentological, geochemical and palynological information to determine paleo-functions of ancient coal-forming wetlands, *In* Steadman, D. W. and Mead, J. I., *Late Quaternary Environments and Deep History: A Tribute to Paul S. Martin*, The Mammoth Site of Hot Springs, South Dakota, Inc. Scientific Papers, Vol 3, 1995.

Sanderson, G. C., Ed., *Management of Migratory Shore and Upland Game Birds in North America*, Int'l. Assoc. of Fish and Wildlife Agencies, Washington, DC, 1977.

Sather, J. H. and Ruta Stuber, P. J., Eds., Proceedings of the National Wetland Values Assessment Workshop, FWS/OBS-84/12, USDI, Fish and Wildlife Service, 1984.

Schneller-McDonald, K., Ischinger, L. S. and Huble, G. T., Wetland Creation and Restoration: Description and Summary of Literature, Biol. Rep. 89(3), USDI, Fish and Wildlife Service, 1990.

Sculthorpe, C. D., *The Biology of Aquatic Vascular Plants*, Edward Arnold Ltd., London, 1967.

Sharitz, R. R. and Gibbons, J. W., Eds., Freshwater Wetlands and Wildlife, NTIS, CONF-8603101, USDOE, Savannah River Ecology Laboratory, 1989.

Smith, H. K., An Introduction to Habitat Development on Dredged Material, Tech. Rep. DS-78-19, U.S. Army Engineer Waterways Experiment Station, 1978.

Snyder, B. D. and Snyder, J. L., Feasibility of Using Oil Shale Wastewater for Waterfowl Wetlands, FSW/OBS 84/01, USDI, Fish & Wildlife Service, 1984.

Stephenson, M., Turner, G., Pope, P., Colt, J., Knight, A. and Tchobanoglous, G., The Environmental Requirements of Aquatic Plants, Appendix A of Pub. No. 65, Use and Potential of Aquatic Species for Wastewater Treatment, State Water Res. Control Bd., Sacremento, CA, 1980.

Stuber, P. J., Coord., Proceedings of the National Symposium on Protection of Wetlands from Agricultural Impacts, Biol. Rep. 88(16), USDI, Fish and Wildlife Service, 1988.

Tarver, D. P., Rodgers, J. A., Mahler, M. J. and Lazor, R. L., Aquatic and Wetland Plants of Florida, Florida Dept. of Natural Resources, Tallahassee, FL, 1978.

Terrell, C. R., and Perfetti, P. B., Water Quality Indicators Guide: Surface Waters, SCS-TP-161, USDA, Soil Conservation Service, 1989.

Teskey, R. O. and Hinckley, T. M., Impact of Water Level Changes on Woody Riparian and Wetland Communities, Vol. I-VI, FWS/OBS-77/58 to 78/89, USDI, Fish and Wildlife Service, 1978.

Thunhorst, G. A., Wetland Planting Guide for the Northeastern United States, Environmental Concern, Inc., St. Michaels, MD, 1993.

Tiner, R. W., Jr., Field Guide to Nontidal Wetland Identification, Maryland Department of Natural Resources, Annapolis, MD and U.S. Fish and Wildlife Service, Newton Corner, MA, 1988.

Vymazal, J., *Algae and Element Cycling in Wetlands*, Lewis Publ, Boca Raton, FL, 1995.

Ward, A. and Elliot, B., Eds., *Environmental Hydrology*, Lewis Publ, Boca Raton, FL, 1995.

Weller, M. W., *Freshwater Marshes — Ecology and Wildlife Management, Second Edition*, University of Minnesota Press, Minneapolis, MN, 1987.

Weller, M. W., Ed., *Waterfowl in Winter*, University of Minnesota Press, Minneapolis, MN, 1988.

Wells, G., Landscape Design: Ponds, Landscape Architect Note 2, USDA, Soil Conservation Service, 1988.

Wentz, W. A., Smith, R. L. and Kadlec, J. A., A Selected Annotated Bibliography on Aquatic and Marsh Plants and Their Management, University of Mich., School of Natural Resources, Ann Arbor, MI, 1974.

Wetmacott, R., Landscape and Wildlife Habitat Management in the Countryside, Landscape Architecture Note 3, USDA, Soil Conservation Service, 1987.

Wetzel, R. G., *Limnology*, Saunders College Publishing, Philadelphia, PA, 1985.

Whitlow, T. H. and Harris, R. W., Flood Tolerance in Plants: A State-of-the-Art Review, Tech. Rep. E-79-2, U.S. Army Engineer Waterways Experiment Station, 1979.

Whitman, W. R., Strange, T., Widjeskog, L., Whittemore, R., Kehoe, P. and Roberts, L., Eds., Waterfowl Habitat Restoration, Enhancement and Management in the Atlantic Flyway, Third ed., Delaware Div. Fish and Wildl., Dover, DE, 1995. "a wealth of information on all aspects of wetland restoration, creation and management".

Wilkinson, D. L., Schneller-McDonald, K., Olson, R. W. and Auble, G. T, Synopsis of Wetland Functions and Values: Bottomland Hardwoods with Special Emphasis on Eastern Texas and Oklahoma, Biol. Rep. 87(12), USDI, Fish and Wildlife Service, 1987.

Wolf, R. B., Lee, L. C. and Sharitz, R. R., Wetland Creation and Restoration in the United States from 1970 to 1985: An Annotated Bibliography, Wetlands 6:1, 1986.

Zelanzny, J. and Feierabend, J. S., Proceedings of a Conference: Increasing Our Wetland Resources, National Wildlife Federation, Washington, DC, 1988.

APPENDIX B

COMMON AND SCIENTIFIC NAMES

PLANTS

Alder, Black	*Alnus glutinosa*
Alder, Hazel	*Alnus rugosa*
American Beech	*Fagus grandifolia*
Arrow Arum	*Peltandra cordata*
Arrowhead	*Sagittaria* spp.
Aspen, Bigtooth	*Populus grandidentata*
Aspen, Quaking	*Populus tremuloides*
Barnyard Grass	*Echinochloa crusgalli*
Bay, Loblolly	*Gordonia lasianthus*
Bay, Red	*Persea borbonia*
Bay, Sweet	*Magnolia virginiana*
Basswood	*Tilia americana*
Beggarticks	*Bidens* spp.
Birch, Paper	*Betula papyrifera*
Birch, White	*Betula populifolia*
Birch, Yellow	*Betula alleghaniensis*
Blackberries	*Rubus* spp.
Black Cherry	*Prunus serotina*
Black Walnut	*Juglans nigra*
Bladderwort	*Utricularia* spp.
Bog Laurel	*Kalmia polifolia*
Box Elder	*Acer negundo*
Blueberry	*Vaccinium uliginosum*
Buck Bean	*Menyanthes trifoliata*
Bulrush, River	*Scirpus fluviatilis*
Bulrush, Soft-stemmed	*Scirpus validus (S. actus)*
Burreed	*Sparganium eurycarpum*
Buttonbush	*Cephalanthus occidentalis*
Cattail, Narrow-leaved	*Typha angustifolia*
Cattail, Wide-leaved	*Typha latifolia*
Cattail, Blue	*Typha domingensis*
China berry	*Melia azedarach*
Coontail	*Ceratophyllum demersum*
Cottonwood, Eastern	*Populus deltoides*

Cottonwood, Swamp	*Populus heterophylla*
Cranberry	*Vaccinium macrocarpon*
Cypress, Bald	*Taxodium distichum*
Cypress, Pond	*Taxodium ascendens*
Dahoon Holly	*Ilex cassine*
Dog Fennel	*Eupatorium capilifolium*
Dogwood, Flowering	*Cornus florida*
Dogwood, Red-osier	*Cornus stolonifera*
Duckweed	*Lemna* spp.
Duckweed, Giant	*Spirodela* spp.
Elder	*Sambucus callicarpa*
Elm, American	*Ulmus americana*
Elm, Winged	*Ulmus alata*
Fern, Cinnamon	*Osmunda cinnamomea*
Fern, Marsh	*Thelypteris palustris*
Fern, Royal	*Osmunda regalis*
Fern, Water	*Azolla* spp.
Fetterbush	*Lyonia lucida, ligustrina*
Foxtail	*Alopecurus arundinaceus*
Foxtail Barley	*Hordeum jubatum*
Fragrant White Lily	*Nymphaea odorata*
Frogbit	*Limnobium spongia*
Gallberry	*Ilex glabra*
Giant Reed	*Phragmites australis*
Grass, Cotton	*Eriophorum polystachion*
Grass, Knot	*Paspalum* spp.
Grass, Manna	*Glyceria* spp.
Grass, Marsh	*Scolochloa festucacea*
Grass, Panic	*Panicum agrostoides*
Grass, Prairie Cord	*Spartina pectinata*
Grass, Reed	*Calamogrostis inexpansa*
Grass, Reed Canary	*Phalaris arundinacea*
Grass, Salt	*Distichlis spicata*
Grass, Saw	*Cladium jamaicensis*
Grass, Slough	*Beckmania syzigachne*
Grass, Switch	*Panicum virgatum*
Grass, Wool	*Scirpus cyperinus*
Grape	*Vitis* spp.
Green Ash	*Fraxinus pennsylvanica*
Greenbrier	*Smilax* spp.
Gum, Black	*Nyssa sylvatica*
Gum, Sweet	*Liquidambar styraciflua*
Hackberry	*Celtis occidentalis*
Hardhack	*Spirea douglassii*
Hawthorn	*Crataegus mollis*

Hemlock, Eastern	*Tsuga canadensis*
Hemlock, Western	*Tsuga heterophylla*
Hemp	*Sesbania* spp.
Hempvine	*Mikania* spp.
Hickory, Bitternut	*Carya cordiformis*
Hickory, Mockernut	*Carya tomentosa*
Hickory, Shagbark	*Carya ovata*
Hickory, Shellbark	*Carya lacinosa*
Hickory, Water	*Carya aquatica*
Holly, American	*Ilex opaca*
Holly, Deciduous	*Ilex decidua*
Honeysuckle	*Lonicera* spp.
Iris, Blue	*Iris virginicum*
Iris, Red	*Iris fulva*
Iris, Yellow	*Iris pseudacorus*
Ironwood	*Carpinus caroliniana*
Kentucky Coffee Tree	*Gymnocladus dioica*
Labrador Tea	*Ledum groenlandicum*
Lady's Slipper	*Cypripedium* spp.
Lizardtail	*Saururus cernuus*
Locust, Black	*Robinia pseudocacia*
Locust, Honey	*Gleditsia triacanthos*
Loosestrife	*Lysimachia* spp.
Maidencane	*Panicum hemitomon*
Mallow, Swamp	*Hibiscus moscheutos*
Mallow, Halbeard-leaved	*Hibiscus militaris*
Maple, Red	*Acer rubrum*
Maple, Silver	*Acer saccharinum*
Maple, Sugar	*Acer saccharum*
Marsh Marigold	*Caltha leptosepala*
Melaleuca	*Melaleuca quinquenervia*
Milfoil	*Myriophyllum* spp.
Milkweed, Swamp	*Asclepias incarnata*
Morning-glory	*Convolvulus* spp.
Moss, Sphagnum	*Sphagnum* spp.
Oak, Black	*Quercus velutina*
Oak, Bur	*Quercus macrocarpa*
Oak, Cherrybark	*Quercus falcata* var. *pagodifolia*
Oak, Laurel	*Quercus laurifolia*
Oak, Live	*Quercus virginiana*
Oak, Nuttal's	*Quercus nuttalii*
Oak, Overcup	*Quercus lyrata*
Oak, Pin	*Quercus palustris*
Oak, Red	*Quercus rubra*
Oak, Shingle	*Quercus imbricaria*

Oak, Shumard	*Quercus shumardii*
Oak, Spanish	*Quercus falcata*
Oak, Swamp White	*Quercus bicolor*
Oak, Valley	*Quercus lobata*
Oak, Water	*Quercus nigra*
Oak, White	*Quercus alba*
Oak, Willow	*Quercus phellos*
Orchid, Swamp	*Habenaria* spp.
Pecan	*Carya illinoensis*
Pepperbush, Sweet	*Clethra alnifolia*
Persimmon	*Diospyros virginiana*
Pickerelweed	*Pontederia cordata*
Pine, Pond	*Pinus serotina*
Pine, Slash	*Pinus elliottii*
Pine, Spruce	*Pinus glabra*
Pitcherplant	*Sarracenia* spp.
Poison Ivy	*Rhus radicans*
Pondweed, American	*Potamogeton nodosus*
Pondweed, Sago	*Potamogeton pectinatus*
Popular, Fremont	*Populus fremontii*
Primrose, Willow	*Ludwidgia peruviana*
Quillwort	*Isoetes* spp.
Ragweed	*Ambrosia* spp.
Rice Cutgrass	*Leersia oryzoides*
River Birch	*Betula nigra*
Salt Cedar	*Tamarix tetrandra*
Sassafras	*Sassafras albidum*
Sedge	*Carex* spp.
Sedge, Chufa	*Cyperus* spp.
Sedge, Three-way	*Dulichium arundinaceum*
Skunk Cabbage	*Symplocarpus foetidus*
Skunk Cabbage, Yellow	*Lysichitum americanum*
Smartweed, Pale	*Polygonum lapathifolium*
Smartweed, Pennsylvania	*Polygonum pensylvanicum*
Smartweed, Swamp	*Polygonum coccineum*
Smartweed, Water	*Polygonum amphibium*
Soft Rush	*Juncus effusus*
Spatterdock	*Nuphar luteum*
Spider Lily	*Hymenocallis* spp.
Spikerush	*Eleocharis* spp.
Spruce, Black	*Picea mariana*
Spruce, Red	*Picea rubens*
Spruce, Sitka	*Picea sitchensis*
Sumac, Poison	*Toxicodendron vernix*
Sumac, Smooth	*Rhus glabra*

Sundew	*Drosera* spp.
Swamp Privet	*Forestiera acuminata*
Swamp Rose	*Rosa palustris*
Sweetflag	*Acorus calamus*
Sugarberry	*Celtis laevigata*
Sycamore	*Platanus occidentalis*
Tamarack	*Larix laricina*
Tapegrass	*Vallisneria americana*
Three-square	*Scirpus americanus*
Titi	*Cyrilla raemiflora*
Trumpet Vine	*Campsis radicans*
Tupelo, Swamp	*Nyssa biflora*
Tupelo, Water	*Nyssa aquatica*
Virginia Creeper	*Parthenocissus quinquefolia*
Water Arum	*Calla palustris*
Watercress	*Nasturtium officinale*
Water Buttercup, Yellow	*Ranunculus flabellaris*
Water Buttercup, White	*Ranunculus aquatilis*
Water Elm	*Planera aquatica*
Water Heart	*Nymphoides aquatica*
Water Hyacinth	*Eichhornia crassipes*
Water Locust	*Gleditsia aquatica*
Water Lotus	*Nelumbo lutea*
Watermeal	*Wolffia* spp.
Water Pennywort	*Hydrocotyle umbellata*
Water Plantain	*Alisma* spp.
Water Primrose	*Jussiaea repens*
Water Shield	*Brasenia schreberi*
Water Starwort	*Callitriche* spp.
Water Weed	*Elodea* spp.
Waterwillow	*Justicia americana*
Wax Myrtle	*Myrica cerifera*
White Ash	*Fraxinus americana*
White Cedar, Atlantic	*Chamaecyparis thyoides*
White Cedar, Northern	*Thuja occidentalis*
Widgeongrass	*Ruppia maritima*
Wild Rice	*Zizania aquatica*
Willow, Black	*Salix nigra*
Willow, Dune	*Salix piperi*
Willow, Hooker	*Salix hookeriana*
Willow, Narrow-leaf	*Salix exigua*
Willow, Pacific	*Salix lasiandra*
Wolffiella	*Wolffiella* spp.
Zenobia	*Zenobia pulverulenta*

ANIMALS

Bass	*Micropterus* spp.
Bluegill	*Lepomis macrochirus*
Bowfin	*Amia calva*
Bullhead	*Ictalurus* spp.
Carp	*Cyperinus carpio*
Catfish	*Ictalurus* spp.
Crappie	*Pomoxis* spp.
Fathead Minnow	*Pimephales promelas*
Gar	*Lepisosteus* spp.
Killifish	*Fundulus diaphanus*
Mosquitofish	*Gambusia affinis*
Northern Pike	*Esox lucius*
Pickerel	*Esox americanus*
Shad	*Dorosoma cepedianum*
Shiner	*Notropis* spp.
Sunfish	*Lepomis cyanellus*
Sucker	*Catostomus* spp.
Top Minnow	*Fundulus* spp.
Walleye	*Stizostedion vitreum*
Yellow Perch	*Perca flavescens*
Bullfrog	*Rana catesbeiana*
Green Frog	*Rana clamitans*
Mudpuppy	*Necturus maculosus*
Tiger Salamander	*Ambystoma tigrinum*
Tree Frog	*Hyla* spp.
Water Siren	*Siren* spp.
Alligator	*Alligator mississipiensis*
Caiman	*Caiman* spp.
Snake, Garter	*Thamnophis* spp.
Snake, Mud	*Farancia abacura*
Snake, Queen	*Natrix septemvittata*
Snake, Water	*Natrix* spp.
Turtle, Box	*Terrapene* spp.
Turtle, Cooter	*Pseudemys* spp.
Turtle, Mud	*Kinosternon* spp.
Turtle, Musk	*Sternotherus* spp.
Turtle, Painted	*Chrysemys* spp.
Turtle, Pond	*Clemmys* spp.
Turtle, Slider	*Pseudemys* spp.
Turtle, Snapping	*Chelydra serpentina*
Turtle, Softshell	*Trionyx* spp.
Water Moccassin	*Ancistrodon piscivorus*

American Coot	*Fulica americana*
Bittern, American	*Botaurus lentiginosus*
Blackbird	*Xanthocephalus* spp.
Canada Goose	*Branta canadensis*
Chickadee	*Parus* spp.
Cormorant	*Phalacrocorax* spp.
Crane	Gruidae
Duck, Dabbling	Anatids
Duck, Diving	Aythyatids
Duck, Wood	*Aix sponsa*
Eagle, Bald	*Haliaeetus leucocephalus*
Flycatcher	Tyrannidae
Gadwall	*Anas strepera*
Grebe	*Podiceps* spp.
Gull	*Larus* spp.
Heron	*Ardea* spp.
Loon	*Gavia* spp.
Mallard	*Anas platyrhynchos*
Osprey	*Pandion haliaetus*
Pelican, Brown	*Pelecanus occidentalis*
Pelican, White	*Pelecanus erythrorhynchos*
Peregrine Falcon	*Falco peregrinus*
Pheasant	*Phasianus colchicus*
Rail	Rallidae
Red-headed Woodpecker	*Melanerpes erythrocephalus*
Shorebird	Charadrii
Short-eared Owl	*Asio flammeus*
Snail Kite	*Rostrhamus sociabilis*
Sparrow, Swamp	*Melospiza georgiana*
Swan, Mute	*Cygnus olor*
Swan, Trumpeter	*Olor buccinator*
Warbler	Parulidae
Wigeon	*Mareca americana*
Wren	*Cistothorus* spp.
Bear, Black	*Ursus americanus*
Beaver	*Castor canadensis*
Bobcat	*Lynx rufus*
Coyote	*Canis latrans*
Deer, White-tailed	*Odocoileus virginianus*
Deer, Mule	*Odocoileus hemionus*
Elk	*Cervus canadensis*
Lemming	*Synaptomys* spp.
Marsh Rice Rat	*Oryzomys palustris*
Mink	*Mustela vision*

Muskrat	*Ondatra zibethica*
Moose	*Alces alces*
Nutria	*Myocaster coypus*
Opossum	*Didelphis marsupialis*
Otter	*Lutra canadensis*
Pig, Feral	*Sus scrofa*
Raccoon	*Procyon lotor*
Skunk	*Mephitis* or *Spilogale* spp.
Swamp Rabbit	*Sylvilagus aquaticus*
Water Shrew	*Sorex palustris*
Weasel	*Mustela* spp.
Wolf	*Canis lupus*
Wood Rat	*Neotoma* spp.

APPENDIX C

SOME SOURCES OF SUPPLY

The listing is by no means all inclusive nor does inclusion imply endorsement by the author or publisher.

MAPS, PHOTOGRAPHS, AND SURVEYS

Aerial Photography Field Office
USDA, Agricultural Stabilization and Conservation Service
2222 West 2300 South
P.O. Box 30010
Salt Lake City, Utah 84130-0010
801-524-5856
national repository for ASCS aerial photographs
generally the ASCS office in each county or parish has photos for that county

Earth Science Information Center
USGS
Box 25046 DFC
Denver, Colorado 80225
303-202-5829
for a listing of other centers call 800-USA-MAPS
topographic and other maps

National Climatic Data Center
151 Patton Avenue, Room 120
Asheville, North Carolina 28801-5001
704-271-4800
FAX: 704-271-4876
national repository for climate data

National High Altitude Program
USGS
EROS Data Center
Sioux Falls, South Dakota 57198
605-594-6151
aerial and satellite photos

USGS Map Distribution
Box 25286, Building 810
Denver Federal Center
Denver, Colorado 80225
800-435-7627
FAX: 303-202-4693
topographic and other maps

USGS/ESIC
National Headquarters
507 National Center
Reston, Virginia 22092
800-872-6277 - includes FAX on demand information
703-648-6045
FAX: 703-648-5548
National Wetlands Inventory Digital and hardcopy maps
as well as topographic, hydrologic, geologic maps, and aerial and satellite
photography in digital, hardcopy, and CD-ROM

TVA Map Sales
1101 Market Street
Chattanooga, Tennessee 37402
423-751-6277
FAX: 423-751-6216
topographic, geologic and land use maps, aerial photography, other remote
sensing in the Tennessee Valley region

Wetlands Values Database
USFWS/NWI
9720 Executive Center Drive, Suite 101
St. Petersburg, Florida 33702-2440
813-893-3624
FAX: 813-893-3860
bibliography on wetland values and functions

Wetlands Restoration and Creation Database
Johnson Controls, Inc.
Att: Gerry Horak
4512 McMurry Avenue
Ft. Collins, Colorado 80525
970-226-9413
FAX: 970-226-9230

WETER
U.S.A. Corps of Engineers
Waterways Experiment Station
3909 Halls Ferry Road
Vicksburg, Mississippi 39180
800-522-6937, ext 2393
software for wetland restoration evaluations

WETWorks
P.O. Box 30071
Jackson, WY 83001
800-867-0018
software for wetland evaluations

SUPPLIES AND EQUIPMENT

References

Annual Buyer's Guide
Water Environment & Technology
Water Pollution Control Federation
601 Wythe Street
Alexandria, Virginia 22314
703-684-2400

Pollution Equipment News
8650 Babcock Boulevard
Pittsburgh, Pennsylvania 15237
800-245-3182
FAX: 412-369-9720

Pumps and Systems Magazine
AES Marketing
123 North College Avenue
Suite 260
Fort Collins, Colorado 80524
303-221-2006
FAX: 303-221-2019

Suppliers

Agri-Drain Corporation
1491 340th Street
P.O. Box 458
Adair, Iowa 50002
800-232-4742; 515-742-5211
FAX: 800-282-3353
water control structures (PVC Drop-log), piping, valves, gates

Aquarius Systems
P. O. Box 215
220 North Harrison
North Prairie, Wisconsin 53153
800-328-6555; 414-392-2162
aquatic weed harvesters

Ben Meadows Company
P.O. Box 80549
Atlanta, Georgia 30366
800-241-6401; 404-455-0907
FAX: 800-628-2068
FAX on Demand: 800-765-9698
variety of field, monitoring and drafting/mapping equipment

BonTerra
P.O. Box 9485
Moscow, Idaho 83843
800-882-9489
FAX: 208-285-0201
erosion control materials

Briargreen, Inc.
P.O. Box 6639
Kent, Washington 98064
800-635-8873; 206-630-5024
FAX: 206-630-9124
erosion matting, aerial seeding

Cahoon Farms, Inc.
P. O. Box 243
Arapahoe, North Carolina 28510
919-249-1767
FAX: 919-249-2856
aluminum flashboard risers/stoplog culverts

CID, Inc.
P.O. Box 9008
Moscow, Idaho 83843
208-882-0119
FAX: 208-882-9461
monitoring equipment

Davis Instruments
4701 Mount Hope Drive
Baltimore, Maryland 21215
800-829-7894
FAX on Demand: 800-365-0147
monitoring equipment

Fiberglass Utility Supplies, Inc.
R.R. 1, Box 70
Libertyville, Iowa 52567
515-693-3311
FAX: 515-693-4131
water control structures

Forestry Suppliers, Inc.
Box 8397
205 West Rankin Street
Jackson, Mississippi 39204
601-354-3565
variety of field equipment

GSE Lining Technology, Inc.
19103 Gundle Road
Houston, Texas 77073
800-435-2008; 713-443-8564
FAX: 713-875-6010
liners

Isco, Inc.
P.O. Box 82531
Lincoln, Nebraska 68501
800-228-4373; 402-474-2233
FAX: 402-474-6685
monitoring equipment

Hudgins, Daniel H.
466 Diamond Street
San Francisco, California 94114
DANCAD 3D drawing program

North American Green
14649 Highway 41 N.
Evansville, Indiana 47711
800-772-2040; 812-867-6632
FAX: 812-867-0247
erosion control materials

Polybac Corporation
Courtney Place
3894 Courtney Street
Bethlehem, Pennsylvania 18017-8999
800-523-9385; 215-867-7338
FAX: 215-861-0991
monitoring equipment

Polyfelt Americas
1000 Abernathy Road
Building 400, Suite 825
Atlanta, Georgia 30328
800-458-3567; 404-668-2119
FAX: 404-668-2116
liners

Solar Components Corporation
121 Valley Street
Manchester, New Hampshire 03103
603-668-8186
FAX: 603-627-3110
fiberglas tanks

The Birds' Nest
7 Pattern Road
Bedford, New Hampshire 03102
603-623-6541

USCO
P.O. Box 310
425 Riverside Avenue
Medford, Massachusetts 02155
800-343-7555; 617-395-9023
FAX: 800-232-8726

Verdyol
P.O. Box 605
Pell City, Alabama 35125
205-338-4411
FAX: 205-338-4595
erosion control materials

Weyerhaeuser Company, Inc.
Engineered Fiber Products
Sort CCB5D6
Tacoma, Washington 98477
800-443-9179
FAX: 206-924-7148
erosion control materials

Vegetation: Planting Stock and Seeds

Many also provide consulting and planting services. In addition to the sources
listed below, shrub and tree seedlings can often be inexpensively obtained
from:
State Forestry Division nurseries, U.S. Forest Service nurseries, Champion,
Weyerhaeuser, International Paper, Union Camp, Westvaco, and other timber
and paper company nurseries.

Northeast

Aquatic Images Water Gardens
1221 Crestview Drive
Baraboo, Wisconsin 53913
608-356-9508

Burr Oak Nursery
Route 1, Box 310
Round Lake, Illinois 60073
312-546-4700

Country Wetlands Nursery & Consulting Ltd.
S75 W20755 Field Drive
Muskego, Wisconsin 53150
414-679-1268
FAX: 414-679-1279

Designs on Nature
202 Lincolnway East
Mishawak, Indiana 46544
219-256-2242
FAX: 219-257-1966

Ecoscience
RR # 4, Box 4294
Moscow, Pennsylvania 18444
717-842-7631
FAX: 717-842-9976

Environmental Concern, Inc.
P.O. Box P.
St. Michaels, Maryland 21663
301-745-9620

Ernst Crownvetch Farms
R.D. 5, Box 806
Meadville, Pennsylvania 16335
800-873-3321
FAX: 814-425-2228

Grass Roots Nursery
24765 Bell Road
New Boston, Michigan 48164
654-2405

J & J Transplant Aquatic Nursery, Inc.
P. O. Box 227
Wild Rose, Wisconsin 54984
800-622-5055; 414-622-3552
FAX: 414-622-3660

Kester's Wild Game Food Nurseries, Inc.
P.O. Box 516
Omro, Wisconsin 54963
800-558-8815; 414-685-2929
FAX: 414-685-6727

Lilypons Water Gardens
6800 Lilypons Road
Buckeystown, Maryland 21717
301-874-5133

Maryland Aquatic Nurseries
3427 North Furnace Road
Jarrettsville, Maryland 21084
301-557-2837

Musser Forests, Inc.
Dept. S-95M, P. O. Box 340
Indiana, Pennsylvania 15701-0340
412-465-5685
FAX: 412-465-9893

New England Environmental Wetland Plant Nursery
800 Main Street
Amherst, Massachusetts 01002
413-256-0202
FAX: 413-256-1092

North West Nursery Wholesale, Inc.
9150 Great Plains Boulevard
Chanhassen, Minnesota 55317
612-445-4088

Octoraro Wetland Nurseries
P.O. Box 24
Oxford, Pennsylvania 19363
610-932-3762
FAX: 610-932-2854

Prairie Ridge Nursery
RR2, 9738 Overland Road
Mt. Horeb, WI 53572

Southern Tier Consulting and Nursery, Inc.
P. O. Box 30, 2677 Route 305
West Clarksville, New York 14786
800-848-7614; 716-968-3120
FAX: 716-968-3122

Sylva Native Nursery & Seed Co.
R.D. # 2, Box 1033
New Freedom, Pennsylvania 17349
717-227-0486
FAX: 717-227-0484

Sylvan Nursery, Inc.
1028 Horseneck Road
South Westport, Massacheusetts 02790
508-636-5615

Taylor Creek Restoration Nurseries
P.O. Box 256
Brodhead, Wisconsin 53520
608-897-8641
FAX: 608-897-8486

Vick's Wildgardens, Inc.
Box 115
Gladwyne, Pennsylvania 19035
215-525-6773

Wicklein's Water Gardens
1820 Cromwell Bridge Road
Baltimore, Maryland 21234
410-823-1335

Wildlife Nurseries
P.O. Box 2724
Oshkosh, Wisconsin 54903
414-231-3780
FAX: 414-231-3554

Southeast

A. D. Andrews Nursery
P.O. Box 1126
Chiefland, Florida 32626
904-493-2496

Alabama Forestry Commission
513 Madison Avenue
Montgomery, Alabama 36130
205-240-9345

Apalachee Native Nursery
P.O. Box 204
Lloyd, Florida 32337
904-997-8976

Aqua-Tech Farm
462 Junipter Street
Neeses, South Carolina 29107
803-247-5697 voice and FAX
5128 Mt. Welcome, Christiansted
St. Croix, Virgin Islands
809-773-3246
FAX: 809-773-6829

Beersheba Wildflower Gardens
Beersheba Springs, Tennessee
615-692-3575

Central Florida Lands & Timber, Inc.
Route 1, Box 899
Mayo, Florida 32066
904-294-1211

Central Florida Native Flora, Inc.
P.O. Box 1045
San Antonio, Florida 33576
904-588-3687

Charleston Aquatic Nurseries
674 Ferry Street
Mt. Pleasant, South Carolina 29464
803-881-8843

Creative Native
P.O. Box 713
Perry, Florida 32347
800-628-4831

Florida Aquatics of Avon Park
3846 Menendez Drive
Pensacola, Florida 32503
904-438-2585

Delta View Nursery
Route 1, Box 28
Leland, Mississippi 38756
800-748-9018

Florida Keys Native Nursery, Inc.
102 Mohawk Street
Tavernier, Florida 33070
305-852-2636

Florida Native Plants, Inc.
1170 North Military Trace
West Palm Beach, Florida 33406
305-686-8943

Florida Natives Nursery
255 RAcetrack Road
Odessa, Florida 33556
813-920-5152

Flower City Nurseries
P.O. Box 75
Smartt, Tennessee 37378
615-668-4351
FAX 615-668-3263

Gardens of the Blue Ridge
P.O. Box 10
Pineola, North Carolina 28662
704-733-2417

Hastings
434 Marietta Street, N.W.
P.O Box 4274
Atlanta, Georgia 30302
404-524-8861

Hawkersmith and Sons Nursery, Inc.
Route 4 Box 4155
Tullahoma, Tennessee 37388-9204
800-222-0371; 615-455-5436
FAX: 615-455-3643

Hillis Nursery Co., Inc.
Route 2, Box 142
McMinnville, Tennessee 37110
615-668-4364

Hydro-Perfect Vegetation
P. O. Box 369
Loxley, Alabama 36551
800-367-9331; 205-964-5122
FAX: 205-964-5471

I.T.C. Wholesale Tree Nursery
P.O. Box 6301
Pensacola, Florida 32503
904-476-6193

Jarrell's Aquatic Nursery
Route 2, Box 330
Picayune, Mississippi 39466
601-798-1720

Perry's Water Gardens
191 Leatherman Gap Road
Franklin, North Carolina 28734
704-524-3264

Plants for Tomorrow, Inc.
16361 Norris Road
Loxahatchee, Florida 33470-9430
800-448-2525; 407-790-1422
FAX: 407-790-1916

Riverland Aquafarms
221 Foster Road
Route 1, Box 1220
Orangeburg, South Carolina 29115
800-849-4650; 803-534-4650
FAX: 803-535-4303

Southern Shade Tree Co.
P.O. Box 271
Winona, Mississippi 38967
601-283-5517

Stringfield Farms
Route 1, Box 830
Walterboro, South Carolina 29488
803-893-2001

The Liner Farm
P.O. Box 1369
St. Cloud, Florida 33770
407-892-1484

The Wetlands Co., Inc.
7650S. Tamiami Trace
Sarasota, Florida 34321
813-921-6609

Tommy Savage Nursery
Route 9, Box 57B
McMinnville, Tennessee 37110
615-668-3700

Warren County Nursery, Inc.
Route 2, Box 204
McMinnville, Tennessee 37110
615-668-8941

Wetlands Management, Inc.
2704 Southwest Horseshoe Trace
Palm City, Florida 34990
305-334-1643

Northwest

Bitterroot Native Growers, Inc.
445 Quast Lane
Corvallis, Montana 59828-9406
406-961-4991
FAX: 406-961-4626

Native Seed Foundation
Star Route
Moyie Springs, Idaho 83845
208-267-7938

Pacific Wetland Nursery
7035 crawford Drive
Kingston, Washington 98346
206-297-7575

Plants of the Wild
P.O. Box 866
Tekoa, Washington 99033
509-284-2848

Siskiyou Rare Plant Nursery
2825 Cummings Road
Medford, Oregon 97501
503-772-6846

Sound Native Plants
P.O. Box 10155
Olympia, Washington 98502
206-866-1046
FAX: 206-943-7026

Wetlands Northwest
8414 280th Street East
Graham, Washington 98338
206-846-2774

Wood's Native Plants
5740 Berry Drive
Parkdale, Oregon 97401
503-352-7497

Southwest

Aquatic Plants Nursery, Inc.
P.O. Box 423
Spicewood, Texas 78669
210-693-1225

Environmental Seed Producers, Inc.
P.O. Box 5904
El Monte, California 91734
818-442-3330

Granite Seed
1697 West 2100 North
P.O. Box 177
Lehi, Utah 84043
801-768-4422; 801-531-1245
FAX: 801-768-3967

Hillview Water Gardens
1044 East Hillview
Salt Lake City, Utah 84124
801-261-4912

Melot's Inc.
8900 W. Memorial Road
Oklahoma City, Oklahoma 73142
405-721-4394
FAX: 405-721-4394

Native Plants, Inc.
417 Wakara Way
Salt Lake City, Utah 84124
801-261-4912

Plants of the Southwest
1812 Second Street
Sante Fe, New Mexico 87501
505-983-1548

Progressive Plants
9180 South Wasatch Boulevard
Sandy, Utah 84093
801-942-7333

Seeds of Change
621 Old Santa Fe Trail, #10
Santa Fe, New Mexico 87501
505-983-8956

Sharp Bros. Seed Co.
Healy, Kansas 67850
800-451-3779
316-398-2231
FAX: 316-885-8647
branch offices in Missouri, Colorado, and Texas

Stock Seed Farms, Inc.
28008 Mill Road
Murdock, Nebraska 68407-2350
800-759-1520; 402-867-3771
FAX: 402-867-2442

Van Ness Water Gardens
2460 Euclid Avenue
Upland, California 91786
714-982-2425

Western Seed Company
P. O. Box 1062
Casa Grande, Arizona 85222
602-836-8246

Wildseed, Inc.
1101 Campo Rosa Road
P.O. Box 308
Eagle Lake, Texas 77434
800-848-0078; 409-234-7353
FAX: 409-234-7407

APPENDIX D

SHORT COURSES AND WORKSHOPS ON LEGAL, REGULATORY, RESTORATION, AND MANAGEMENT ASPECTS

Division for Public Service and Extended Education
University of North Carolina at Wilmington
601 South College Road
Wilmington, NC 28403
910-395-3105
wetland identifcation & delineation, soils, hydrology, botany, and ecology

Eagle Hill Field Research Station
Dyer Bay Road, PO Box 9
Steuben, ME 04680
207-546-2821
FAX: 207-546-3042
wetland plants, soils, identification, and delineation

Environmental Concern, Inc.
P.O. Box P
210 West Chew Avenue
St. Michaels, Maryland 21663
410-745-9620
FAX: 410-745-5517
wetland design, evaluation, mitigation, identification and delineation, soils, vegetation, horticultural techniques

Environmental Education Enterprises, Inc.
2764 Sawbury Boulevard
Columbus, Ohio 43235-4580
800-792-0005;
FAX: 614-792-0006
wetland design, hydrology, soils, wetlands for wastewater treatment

Environmental Services
1942 Oak Ridge Turnpike
Oak Ridge, Tennessee 37830
423-482-4337
FAX: 423-482-4514
assessment and management

Environmental Technology Center
8413 Laurel Fair Circle
Ste. 200
Tampa, FL 33610
800-348-8848
wetland identification & delineation

Georgia Center for Continuing Education
The University of Georgia
Athens, Georgia 30602-3603
800-884-1381; 706-542-2134
FAX: 800-884-1419; 800-884-1381
wetland identification & delineation, wetland functional assessment

Institute for Wetland & Environmental Education & Research
P. O. Box 331
Sherborn, Massachusetts 01770-0331
508-653-1877
wetland identification & delineation, restoration & creation, soils and
hydrology

International Erosion Control Association
P.O. Box 774904
Steamboat Springs, Colorado 80477-4904
800-455-4322; 303-879-3010
FAX: 303-879-8563
bioengineering, bank stabilization

Learning Center of Applied Environmental Technology
R. E. Wright Associates, Inc.
3240 Schoolhouse Road
Middletown, Pennsylvania 17057-9956
800-944-5735; 717-944-5501
FAX: 717-944-5642
wetland identification & delineation

Michael Baker Jr., Inc.
1800 Water Place, Suite 170
Atlanta, Georgia 30339
404-850-8770
wetland identification & delineation

National Biological Service, U.S. Geological Survey and University of
Southwestern Louisiana
National Biological Service
700 Cajundome Boulevard
Lafayette, LA 70606
318-266-8699
FAX: 318-266-8513
wetlands classification, identification, GPS, GIS and photo interpretation

National Wetland Science Training Cooperative
L. C. Lee & Associates
221 First Avenue West, Suite 415
Seattle, Washington 98119
800-810-9052
FAX: 206-283-0627
wetland design & construction, jurisdictional delineation, assessment of
wetland functions, plant identification

PRC Environmental Management, Inc.
330 South Executive Drive
Suite 203
Milwaukee, WI 53005
wetland delineation

PRTL Environmental Services
1942 Oak Ridge Turnpike
Oak Ridge, Tennessee 37830
423-346-7406
FAX: 423-482-4514
wetland assessment and management

Rutgers Office of Continuing Professional Education
Cook College
P.O. Box 231
New Brunswick, New Jersey 08903-0231
908-932-9271
FAX: 908-932-8726
plant identification, soils techniques, wetland identification & delineation,
wetland creation/restoration, bioengineering, wetland hydrology,
environmental law

SCWA Wetland Wildlife Center
P. O. Box 450
Pinewood, South Carolina 29125
803-551-4616
wetland identification & delineation

Texas Engineering Extension Service
The Texas A &M University System
College Station, Texas 77843-8000
800-252-2420, ext 140
FAX: 409-845-3419
wetland identification & delineation, wetland soils & hydrology

U.S. Army Engineer Waterways Experiment Station
WRTC
3909 Halls Ferry Road
Vicksburg, Mississippi 39180
601-634-2935
wetland design & construction, jurisdictional delineation, assessment of
wetland functions, plant identification

Wetlands Biogeochemistry Institute
Louisiana State University
Baton Rouge, Louisiana 70803-7511
wetland identification & delineation

Wetland Resources
P.O. Box 694
Daphne, Alabama 36526
334-626-3408
wetland identification & delineation

Wetland Training Institute, Inc.
P. O. Box 1022
Poolesville, Maryland 20837-1022
301-972-8112
FAX: 301-349-2154
wetland identification & delineation, plant identification, soils & hydrology,
wetland construction & mitigation, ecology, regulatory

APPENDIX E

NEWSLETTERS AND JOURNALS

NEWSLETTERS

Aguaphyte
V. Ramey and K. Brown, Eds,
Center for Aquatic Plants
University of Florida
7922 N. W. 71st Street
Gainesville, FL 32653
904-392-1799
FAX: 904-392-3462

Wetlands
The Wetlands Conservancy
PO Box 1195
Tualatin, OR 97062
503-691-1394

Center for Forested Wetlands Research Newsletter
USDA, Forest Service
Southeastern Forest Experiment Station
2730 Savannah Highway
Charleston, SC 29414
803-727-4271

Waterfront
Center for Wetlands & Water Resources
Phelps Lab
PO Box 116350
University of Florida
Gainesville, FL 32611
904-392-2424
FAX: 904-392-3624

Fish and Wildlife Reference Service Newsletter
FWRS
Suite 110, 5430 Grosvenor Lane
Bethesda, MD 20814
800-582-3421

Restoration and Management Notes
Society for Ecological Restoration
University of Wisconsin-Madison Arboretum
1207 Seminole Highway
Madison, WI 53711
608-262-9547
FAX: 608-262-7560

The Wetlands Research Program Bulletin
Wetlands Research Program,
CEWES-EP-D
U.S. Army Engineer Waterways Experiment Station
3909 Halls Ferry Road
Vicksburg, MS 39180
601-634-2569
601-634-3664

Wetland Journal: Research, Restoration, Education
E. W. Garbisch, Jr. Ed.
Environmental Concern, Inc.
PO Box P
St. Michaels, MD 21663
410-745-9620
FAX: 410-745-3517

Land and Water
K. M. Rasch
P.O. Box 1197
Fort Dodge, IA 50501
515-576-3191
FAX: 515-576-2606.

Erosion Control
J. Trotti
Forester Communications
5638 Hollister # 301
Santa Barbara, CA
805-681-1300
FAX: 805-681-1312

National Wetlands Newsletter
Environmental Law Institute
1616 P St., NW
Suite 200
Washington, D.C. 20077
800-433-5120; 202-328-5150
FAX: 202-328-5002

Watershed Events
Office of Wetlands, Oceans and Watersheds (4501F)
401 M Street, SW
Washington, D. C. 20460
202-260-8076
FAX: 202-260-2529

JOURNALS

Wetlands
Society of Wetland Scientists
PO Box 1897
Lawrence, Kansas 66044
913-843-1221
FAX: 913-843-1274

Ecological Engineering
W. J. Mitsch, W. J. Editor-in-Chief
School of Natural Resources
The Ohio State University
2021 Coffey Road
Columbus, OH 43210
614-292-9774
FAX: 614-488-0136

Wetlands Ecology and Management
SPB Academic Publishing
P O Box 11188
1001 GD Amsterdam
The Netherlands

The Journal of Wildlife Management
The Wildlife Society
5410 Grosvenor Lane
Bethesda, Maryland 20814
301-897-9770
FAX: 301-530-2471

Ecology
The Ecological Society of America
2010 Massachusetts Avenue, NW
Suite 400
Washington, D. C. 20036
202-833-8773
FAX: 202-833-8775

Restoration Ecology
Society for Ecological Restoration
University of Wisconsin-Madison Arboretum
1207 Seminole Highway
Madison, WI 53711
608-262-9547
FAX: 608-262-7560

INDEX

.

T - #0337 - 071024 - C456 - 234/156/20 - PB - 9780367401177 - Gloss Lamination